X-ray Microscopy

Instrumentation and Biological Applications

Edited by
Ping-chin Cheng and Gwo-jen Jan

With 180 Figures and 16 Plates

Springer-Verlag
Berlin Heidelberg New York
London Paris Tokyo

Professor Dr. Ping-chin Cheng
Department of Electrical and Computer Engineering
State University of New York at Buffalo
Buffalo, NY 14260, USA

Professor Dr. Gwo-jen Jan
Department of Electrical Engineering
School of Engineering
National Taiwan University
Taipei, Taiwan, 10764, Republic of China

ISBN 3-540-18148-2 Springer-Verlag Berlin Heidelberg New York
ISBN 0-387-18148-2 Springer-Verlag New York Berlin Heidelberg

Library of Congress Cataloging-in-Publication Data.
X-ray microscopy: instrumentation and biological applications: proceedings of the X-ray microscopy 86, Taipei, Taiwan, Republic of China, August 13–15, 1986 / edited by Ping-chin Cheng and Gwo-jen Jan. Bibliography: p. Includes index. ISBN 0-387-18148-2 (U.S.). 1. X-ray microscopy–Congresses. I. Cheng, Ping-chin 1952-. II. Jan, Gwo-jen, 1946-. [DNLM: 1. Microscopy–methods–congresses. 2. Radiation, Ionizing–congresses. QH 212.X2 X12 1986]. QH 212.X2X23. 1987. 87-28443.

Printed in Germany

Offsetprinting: Druckhaus Beltz, Hemsbach/Bergstr.
Bookbinding: J. Schäffer GmbH & Co. KG., Grünstadt
2131/3130-543210

Preface

In 1979, a conference on x-ray microscopy was organized by the New York Academy of Sciences, and in 1983, the Second International Symposium on X-ray Imaging was organized by the Akademie der Wissenschaften in Göttingen, Federal Republic of Germany. This volume contains the contributions to the symposium "X-ray Microscopy '86", held in Taipei, Taiwan, the Republic of China in August 1986. This is the first volume which intends to provide up-to-date information on x-ray imaging to biologists, therefore, emphasis was given to specimen preparation techniques and image interpretation. Specimen preparation represents a major part of every microscopy work, therefore, it should be strongly emphasized in this emerging field of x-ray microscopy. Theoretically, x-ray microscopy offers the potential for the study of unfixed, hydrated biological materials. Since very few biological system can be directly observed without specimen preparation, we would like to emphasize that new information on biological specimens can only be obtained if the specimen is properly prepared. In the past decade, many of the published x-ray images were obtained from poorly prepared biological specimens, mainly air-dried materials. Therefore, one of the goals of this conference is to bring the importance of specimen preparation to the attention of x-ray microscopy community.

X-ray microscopy can be subdivided into several major areas. They are the classic x-ray projection microscope, x-ray contact imaging (microradiography) and the more recent x-ray scanning microscope, x-ray photoelectron microscope and x-ray imaging microscope. The classic x-ray projection microscopy and x-ray microradiography used much shorter wavelengths than that used by the x-ray microscopy today. The advantage of shorter wavelengths is in the higher penetration abilities of the photons, permitting the observation of bulk specimens. With modern x-ray detectors and computer image proccessors, it should be possible to revitalize the x-ray projection microscope as a realtime instrument for the study of bulk specimens. In this proceedings, a paper by S. Newberry gives a review on the history of x-ray microscopy. It is important to point out that the majority of the recent x-ray microscopy activities focus on the wavelength region of 20-50Å, in contrast to the early works which primarily concentrated on the wavelength region of fractions of an angstrom.

In the field of x-ray contact imaging (microradiography), the lack of knowledge on the properties of x-ray resist (photochemical detector) represents a serious problem for the further advancement of the technology. Magnifying the contact image by electron microscopes could seriously damage the x-ray resist. In this volume, D. M. Shinozaki gives a detailed review on the examination of topographic images in resist surfaces by electron microscopy. The so-claimed high resolution (50Å on PMMA resist) of contact imaging should be reevaluated. Since the 1983 meeting in Göttingen, Federal Republic of Germany, tremendous progress has been made both on the scanning and imaging x-ray microscopes. The resolution of the Göttingen imaging microscope has been significantly improved. It should become a useful instrument for cell biology. Furthermore, progress on the fabrication of zone plates by holographic, electron beam lithography and electron beam contamination writing promise better zone plates in the near future. A paper by G. Schmahl *et al.* in this proceedings proposes the possibility of a phase contrast x-ray microscope which could be used with x-rays of shorter wavelengths.

The advances in x-ray sources have been a major factor behind the activity in the x-ray microscopy field. Besides the traditional electron impact sources, such as the stationary and rotational target sources, both synchrotron radiation storage rings and plasma sources offer high intensity x-rays. Just a few days before the conference, the construction of a 1.3GeV synchrotron storage ring began in Hsing-chu, Taiwan, the Republic of China. This synchrotron storage ring, when completed, will provide the needed photons for the Chinese x-ray microscopy community. The Synchrotron Radiation Research Center (SRRC) will become the first national laboratory of the Republic of China. A paper by E. Yen reports the present status of the synchrotron construction.

An extensive bibliography is compiled to provide a useful source of reference for the biologists who are interested in working with x-ray microscopy, . The bibliography not only contains the cited references of all the papers in this proceedings, but also contains many uncited references which are relevant to x-ray microscopy. Although, it is by no mean complete, the bibliography should provide a useful source of references for biologists who intend to use the technologies in their research. A mini atlas also has been complied at the end of this volume. It is intended to provide sample images obtained from various imaging techniques.

This preface could only be written after the rest of the proceedings were completed, and at this point our appreciation crystallized. We would like to thank many organizations in both public and private sectors for their contributions to the symposium. Special thanks to the following organizations for their financial support.

Ministry of Education, Government of the Republic of China
National Science Council, Republic of China
Synchrotron Radiation Research Center, Republic of China
National Taiwan University, College of Engineering
Physics Society of the Republic of China

It could not have been possible to have a successful conference without the assistance from various commercial groups and government agencies. We are grateful to the following organizations for their kind assistance before, during and after the conference.

Coordination Council for North American Affair, NY Office
Science Division, CCNAA, Washington D.C. Office

Taiwan Tourism Bureau, New York Office
Bureau of Information, Taiwan Provincial Government, ROC
International Conference Hall, National Taiwan University
China Air Lines (Taipei head office and Chicago branch office)
Taipei United Hotel

There were many individuals who helped during the planning and execution of the conference. This conference would not have been possible without their support. Mr. C. S. Huang of China Air Lines, Chicago Branch Office, has spent many hours arranging travel plans for many of the invited speakers. Prof. Dr. B. K. Tseng of the Department of Physics, National Taiwan University, and Prof. Dr. D. M. Shinozaki of the Faculty of Engineering, University of Western Ontario, gave us much needed encouragement and endless help. We would like to acknowledge many of the graduate students who helped the preparation of coffee and donuts, as well as maintaining audio-visual equipment. The symposium was held at the newly constructed International Conference Hall, National Taiwan University. We are grateful for their hospitality and assistance during the symposium. Thanks are due to the authors, reviewers and to Springer-Verlag for their combined efforts. In order to achieve uniformity in format and type faces, all of the manuscripts were retyped by the editors, and the final camera ready copies were produced on the IBM Document Composition Facility at the T. J. Watson Research Center.

Taipei, Taiwan, Rep. of China
September 1987

Ping-chin Cheng
Gwo-jen Jan

Contents

Contributors

Sayre

永懷師恩
Shinozaki
Tan

Buckley Kirz

Jan
Cheng

DuKe
Feder

Twu
Jan
Chen

Schmahl
Wu
Yuan
Kirz

X光顯微術 86

中華民國台北市

X-ray Microscopy 86

Taipei, Taiwan, Republic of China

X光顯微術 86

中華民國台北市

X-ray Microscopy 86

Taipei, Taiwan, Republic of China

國際X光顯微術研討
X-RAY MICROSCOPY '86
X-ray Microscopy '86

List of Participants

The number which follows each name corresponds to the number in the group photo

Akio, Saito	
Aoki, Sadao	6
Baldwin, Kenneth	64
Buckley, Christopher	68
Caldorolo, Marilyn	63
Chang, Shin-lin (張石麟)	32
Chang, Tsu-liang	
Chen, Bih-yueh	
Chen, Chang-po (陳章波)	17
Chen, chao-ying	
Chen, Jia-chyuan (陳家全)	48
Chen, Jian-guang	
Chen, Liang-feng (陳良豐)	19
Chen, Shyang-hwang (陳翔鳳)	10
Chen, Yang-fang (陳永芳)	42
Cheng, An-nien	
Cheng, Chu-nan (張秋男)	55
Cheng, Li-lin (陳麗玲)	24
Cheng, Ping-chin (鄭炳今)	28
Cheng, Yu-kuei (成游貴)	11
Chew, Chen-hung (邱鎮宏)	65
Chiou, Henry (邱瑞隆)	47
Chou Chan-shin (周傳心)	31
Duke, Philip	61
Fan, Ining	
Feder, Josephine	
Feder, Ralph	57
Fujisaki, Hisao	
Heish, Chung-tien	
Hiroak, Aritom	
Hiroshi, Kihara	39
Hisao, Fujisaki	
Hong, Chen-tang	
Hou, Chung-jen (侯崇仁)	1
Hsieh Ju-chi (謝如姬)	21
Hsu, Kuo-tung	
Hsue, Yian-jen (徐衍珍)	16
Huang, Dah-yeou	23
Huang, Ken-ning	
Huang, Li-hung	
Huang, Tung-wuu (黃銅武)	40
Huang, Ying-sheng (黃鸞謦)	30
Hung, Duan-yuo (黃大有)	
Hung, Hsueh-hsing (洪 雪行)	2
Hwang, Hwei-tein	
Jan, Gwo-jen (詹國禎)	26
Kirz, Janos	60
Kenney, John	56
King, Chang-jong (金昌中)	43
Kong, Hsien-ping (孔憲苹)	5
Kuo, L. T. (郭深潭)	7
Kuo, Mei-jung (郭美蓉)	27
Law, Shein-fong (盧順芳)	38
Lee, Kuo-ping	
Lin, Hsi-che (林希哲)	49
Lin, Inho (林銀河)	44
Lin, Jenn-an	
Lin, Kuang-chieh (林廣傑)	52
Lin, Ming-yen	
Liu, Ling-kang	
Liu, Yuen-chung (劉遠中)	51
Lo, Tung-bin	
Nagai, Yoshinori	
Ou-yang, Liang (歐陽亮)	46
Quo, Wen-yen	
Ro, Chin-sheng (樂錦盛)	36
Ruei, Eric	
Sayre, Ann	

1. Introduction to X-ray Microscopy

Ping-chin Cheng[1] and Gwo-jen Jan[2]

[1]IBM T. J. Watson Research Center, Yorktown Heights, NY 10598 USA

[2]Department of Electrical Engineering, National Taiwan University, Taipei, Taiwan, 10764, Republic of China

"If there is magic on this planet,
it is contained in water" (Loren Eiseley)

It has long been the dream of the biologists to observe living cells with a high resolution, much beyond the limits of the optical microscope. However, due to the inherent difficulties of electron microscopy (EM) with hydrated specimens, our knowledge of cellular ultrastructure is primary based upon fixed and dehydrated materials.

Some prior studies have been done on the imaging of wet specimens by electron microscopy. Parsons *et al.* (1974) demonstrated that drying caused an obvious change in shape and mass thickness profile in red blood cell compared to wet cell images taken by electron microscopy. Fernandez-Moran (1965) imaged wet T4 bacteriophages claiming that this the helical nature of the tail was more detailed than dry, negatively stained images. The resolution of most of these images of hydrated samples was poor. Theoretical calculations indicate that unstained wet biological materials will give very little image contrast when imaged by electrons, due to the similarity in electron scattering properties of water and proteins (Kirz and Sayre, 1980). However, recent advances have shown excellent promise using cryo-electron microscopy (Dobochet *et al.*, 1985; Lepault, 1985). Lepault and

X-ray Microscopy
Ed. by P.C. Cheng and G.J. Jan

Leonard (1985) demonstrated the three-dimentional structure of unstained, frozen hydrated extended tails of bacteriophage.

For many decades, alternative microscopy methods including x-ray microscopy have been considered by physicists for the imaging of hydrated biological specimens. Due to the short wavelength of x-rays, a better resolution than light microscopy can theoretically be achieved (Cosslett and Nixon, 1960). The short wavelength of x-rays and the low attenuation of x-rays by water (just below the oxygen edge) make x-ray microscopy a potential tool to study wet specimen (Sayre and Feder, 1980; Schmahl *et al.*, 1980). The absorption coefficient of water, calculated from the mass fraction of oxygen and hydrogen (Henke and Ebisu, 1974; Henke *et al.*, 1982; McCorkle, 1980) indicates that water possesses an x-ray transparent window in the soft x-ray spectrum (between oxygen K edge and carbon K edge). Maximum transmittance is located just below the oxygen K edge (~ 2.4nm) at which a 1μm thick water film possesses an approximately 80% transmittance. The difference of absorption coefficient between water and protein at this 2.4nm wavelength (McCorkle, 1980; Kirz and Sayre, 1980) should provide sufficient image contrast. Therefore, it should be possible to image wet specimens. In addition, between 31 and 43.6Å wavelength the path-lengths of photons in air are 10-20 times larger than those of even 10 MeV electrons. This provides that, with x-rays, the observation of thicker specimens is possible. At present, five imaging techniques, projection x-ray microscopy (Newberry, 1954; One, 1967), imaging x-ray microscopy (Rudolph *et al.*, 1984), scanning x-ray microscopy (Horowitz and Howell, 1972; Rarback *et al.*, 1984; Niemann, 1984; Burge *et al.*, 1985; Kirz, 1985; Spiller, 1984), photoelectron x-ray microscopy (Polack and Lowenthal, 1984) and contact imaging, are available (Goby, 1913; Sayre, 1980; Cheng *et al.*, 1986).

Various types of x-ray microscopy have been developed by various investigators (Engström, 1966 and 1980; Schmahl *et al.*, 1980;

Goby, 1913; Horowitz and Howell, 1972; Polack, 1984), and potential applications of x-ray microscopy have been proposed by Pfannkuch *et al.* (1984) and Sarafis (1984). Although the resolution of x-ray microscopy will be less than that of electron microscopy, the potential application to wet cell imaging (Robinson, 1982) will provide a means to eliminate artifacts attributed to specimen fixation, staining and dehydration in electron microscopy.

The intention of this article is to introduce various types of x-ray microscopy to biologists. Schematics of the instruments and examples of resulting images are given for each type of imaging technique.

(1) X-ray contact imaging (x-ray microradiography): Contact imaging involves the use of a photochemical detector which is placed in contact with the specimen. A contact print is produced and a magnified image is obtained by subsequently magnifying the contact images. Traditionally, high resolution silver halide photographic emulsion was used as the photochemical detector, and the contact image is enlarged to the required size by normal photographic means.

High resolution polymer photoresist as the photochemical detector is used by modern contact imaging, which is magnified by means of a scanning or transmission electron microscope. Figure 1.1 shows the basic set-up of x-ray contact imaging. Figure 1.2 is an example of modern x-ray contact image obtained on poly-methyl methacrylate (PMMA) photoresist, and magnified by a transmission electron microscope.

A commercial sealed-off x-ray tube system for x-ray contact imaging was developed by Philips in the 50's. This system utilizes much shorter wavelengths of x-rays (8Å or shorter) than modern x-ray contact imaging (20Å or longer). For a given x-ray source size, there is less geometric blurring than projection microradiography.

(2) Projection x-ray microscope: This technology was developed in the 50s, a commercial model was built by General Electric, and a production run of approximately 100 units was made (S. P. Newberry,

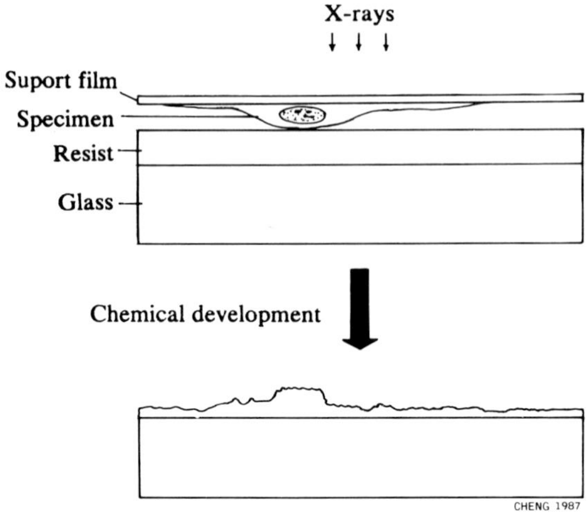

Figure 1.1. X-ray contact imaging set-up.

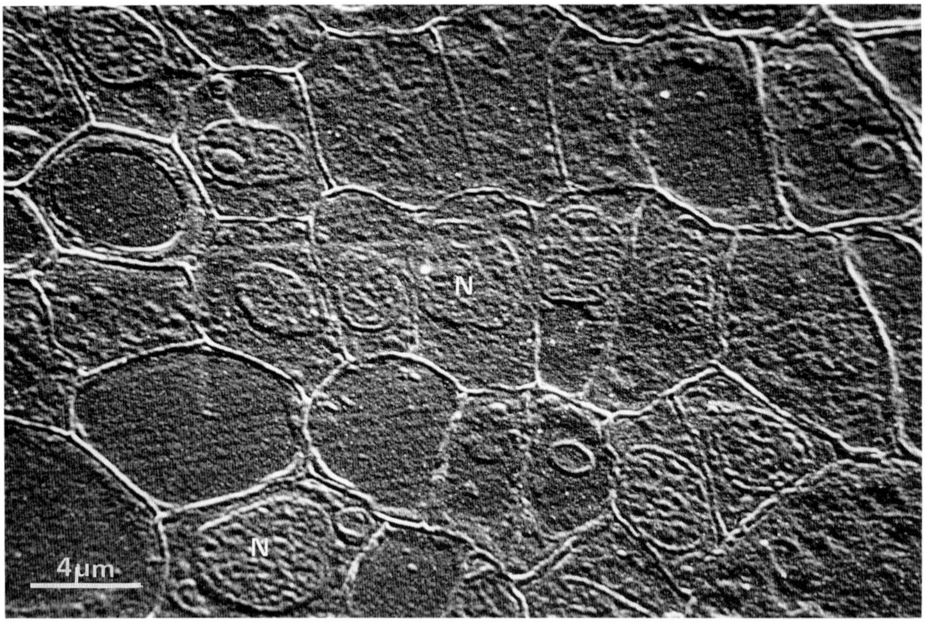

Figure 1.2. X-ray contact image of the anther tissue of *Caltha palustris* L.: The specimen was fixed in Cheng's fixative (Cheng *et al.*, 1979), post-fixed in OsO_4, dehydrated in ethanol series, embedded in Spurr's medium. X-ray contact image was obtained by using PMMA as photochemical detector, and the contact image was magnified by using a scanning electron microscope.

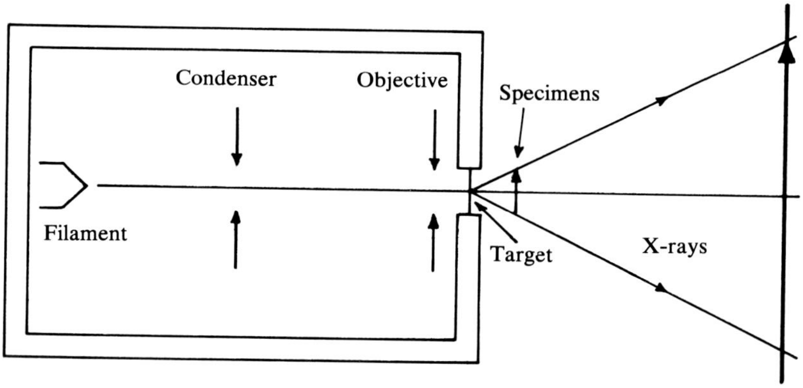

Figure 1.3. Schematic of x-ray projection microscope.

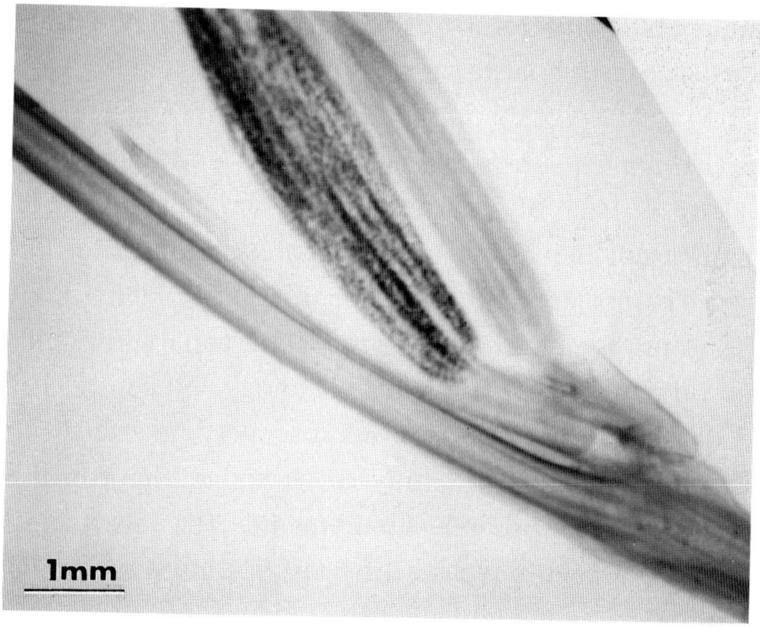

Figure 1.4. X-ray image of the tassel spikelet of Nothern teosinte (*Zea mays* subsp. *mexicana*, race Norbogame): The tissue was fixed in Cheng's fixative (1979), dehydrated in acetone series, and critical-point dried in CO_2. The acceleration voltage of the electron gun was 20kV, and a tungsten target was used. Note the developing pollen grains. (S. P. Newberry and P. C. Cheng)

personal communication). The microscope consists of an electron gun and a set of electrostatic lenses which focus the electron beam to a tight spot. The electron beam is focused on a thin film target, hence, there is considerable intensity of x-ray emission from the back side of the target. Because of the tight electron focus, the x-ray emitting spot is very small. Biological specimens are placed near the target material, and a photographic plate is placed behind the specimen (Figure 1.3). A projection image of the specimen is formed. The resolution depends on the x-ray spot size. In order to obtain sufficient intensity of x-rays passing through the target materials, a typical acceleration voltage of 20kV - 40kV is used. The x-ray projection microscopes use much shorter wavelengths of x-rays than any other modern x-ray microscope. The advantage of using high energy x-ray photons is the high penetration ability, therefore, it is possible to study thick biological specimens such as whole tissues. For instance, Figure 1.4 shows an image of a male spikelet of the terminal inflorescence of Northern teosinte (*Zea mays* subsp. *maxicana*, race Norbogame). Note the higher x-ray density of meristematic tissue. Because of the short wavelength used, the image density represent the total mass of the specimen. The projection microscope, if equipped with modern image intensifiers and computer image processing capability, could become a very useful real-time instrument for biology. Recently, Takahashi *et al.* (1983) used x-ray projection microscope converted from a scanning electron microscope to study Golgi-stained neurons. Similar x-ray projection microscope attachments for scanning electron microscope have been developed by various investigators (Horn and Waltinger, 1978; Yada and Ishikawa, 1980)

(3) The imaging microscope is an instrument using x-ray focusing elements, zone plates, as both condenser and objective "lenses" to magnify images of miniature specimens. X-ray microscopy with zone plates as x-ray The Göttingen group at DESY were the first to use zone plate as x-ray imaging elements for x-ray microscopy in 1976.

The microscope consists of a grazing incidence grating monochromator, a condenser zone plate and a micro-zone plate. The monochromator provides monochromatic x-ray radiation from a synchrotron storage ring. A typical optical arrangement of an imaging x-ray microscope is shown in Figure 1-5. Figure 1-6 shows an x-ray image of cultured human fibroblast taken with 35Å x-rays. The advantage of this technology is that the time required to obtain an image is relatively short (~ 0.5-3 sec typical). Therefore, it is suitable for the study of living biological specimens. Real-time images can be obtained if a suitable two dimensional array detector (*e.g.*, microchannel plate) is used.

(4) Scanning x-ray microscopes (SXM) use a small x-ray beam as a probe, and scan the specimen by either scanning the x-ray beam or by moving the specimen against the x-ray beam. Transmitted (or emission of secondaries) signals are detected, and the image is formed on a CRT. The method using mechanical scanning stage was used by the SUNY group (State University of Nwe York, Dr. Kirz). Scanning of the specimen was achieved by using two piezo-electric benders.

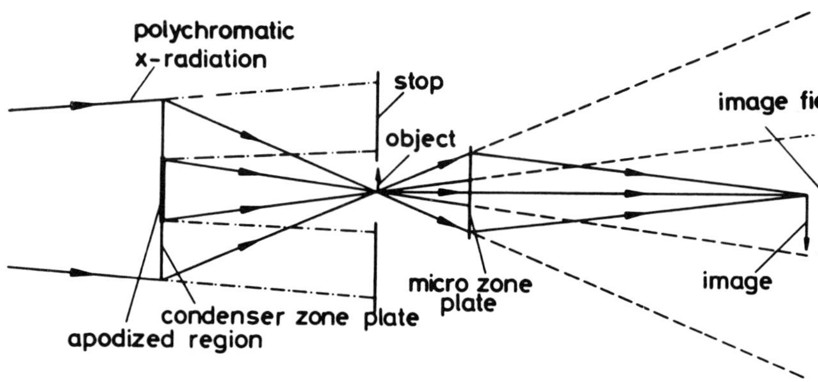

Figure 1.5. Imaging x-ray microscope. (Courtesy G. Schmahl)

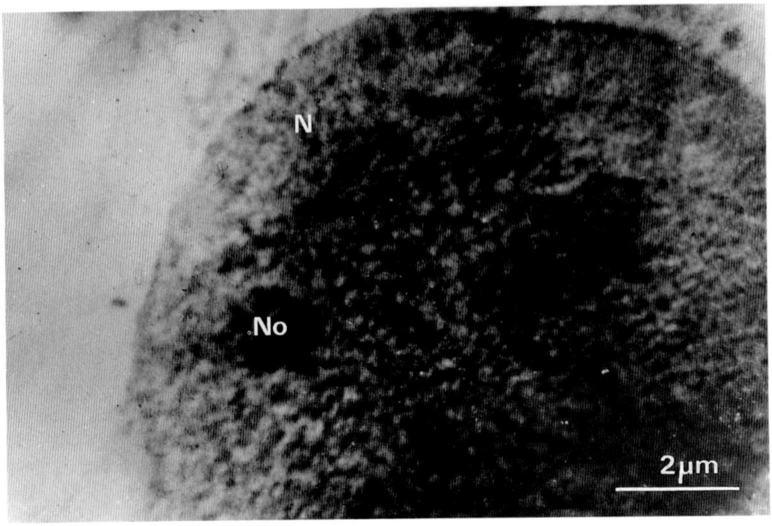

Figure 1.6. X-ray image of cultured human fibroblast.: The cell was cultured on 1000Å thick Formvar film in Dulbecco's modified Eagles's medium supplemented, with 10% fetal calf serum and antibiotics. The cells were fixed in 1% glutaraldehyde, dehydrated in ethanol series, critical-point dried in CO_2. The image was taken by the Göttingen x-ray microscope at BESSY with 35Å soft x-rays. (W. Meyer-Ilse, G. Nyakatura, P. Guttmann, B. Niemann, D. Rudolph, G. Schmahl and P. C. Cheng)

The scanning x-ray microscope developed at the University of Göttingen used a different scan mechanism, for a detailed description, reader should refer to the article by Niemann in this volume. A small x-ray probe can be obtained either by using a pinhole (Horowitz and Howell, 1972) or more recently, by using a zone plate (Kirz, 1986; Kenney *et al.*, 1987; Aoki *et al*, 1987) and normal incidence multilayer optics (Spiller, 1984). Figure 1.7 shows a typical setup of a scanning x-ray microscope. A typical image from the scanning x-ray microscope is shown in Figure 1.8.

(5) Photoelectron x-ray microscopy is similar to the contact microscopy method but replaces the photochemical detector (film or resist) by a photocathode. Photoelectrons emitting from the

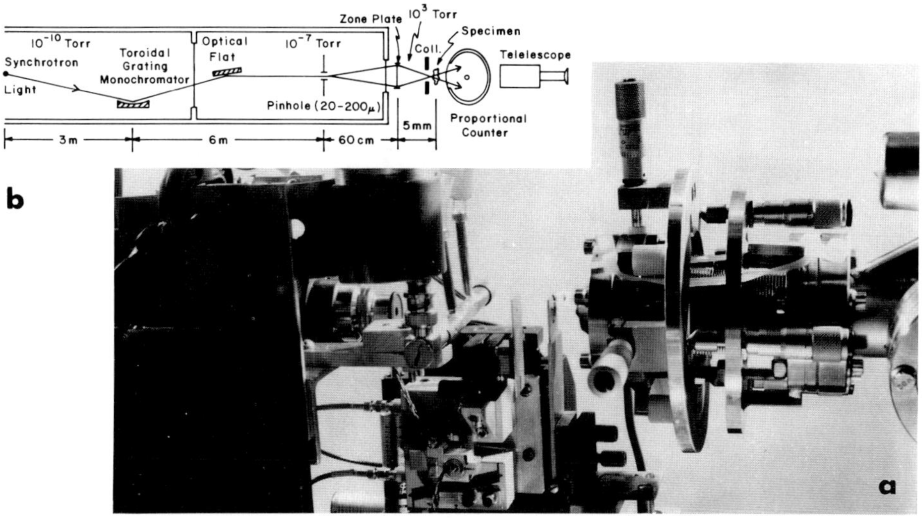

Figure 1.7. Scanning x-ray microscope.: Diagrammatic representation of the Stony Brook scanning microscope (a). and a close-up view of the microscope (b).

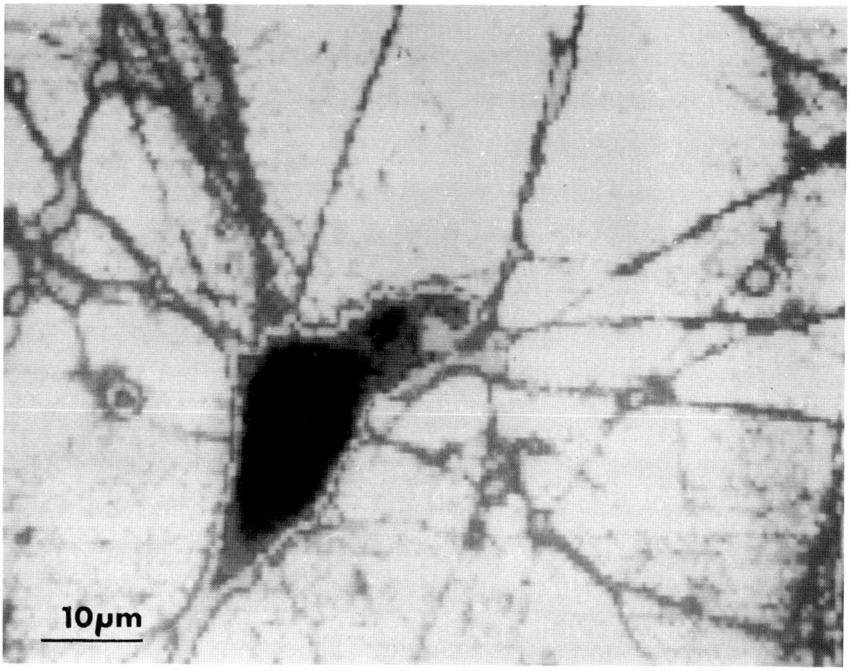

Figure 1.8. X-ray image of neuron outgrowth obtained by using the Stony Brook scanning x-ray microscope. (Courtesy J. Kirz.)

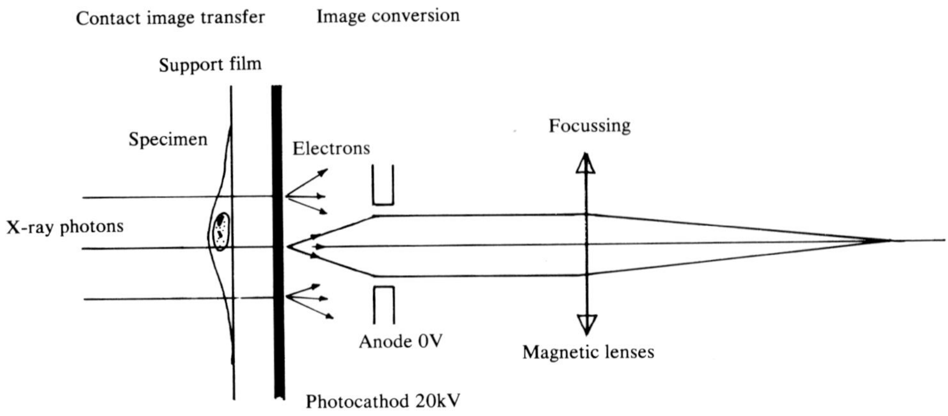

Figure 1.9. Schematic of a photoelectron x-ray microscope

photocathode are focused by a series electromagnetic lenses, filtered by a magnetic prism. the magnified image is formed on a phosphor screen or film. The photoelectron microscope is still being developed. Figure 1.9 shows a schematic of the photoelectron x-ray microscope developed by Polack (1984).

Radiation damage is an important topic for all forms of microscopy employing ionizing radiation as a light source. The energy deposition on a specimen is dependent upon the radiation used and the elemental composition of the specimen. Very little is known about how biological specimens behave under x-radiation. After absorption of the incident photon, various events take place within the absorbing atom. These events include the excitation of electrons to higher orbital levels and the ejection of electrons. The production of photons and Auger electrons is associated with the return to ground state of the excited atom. For soft x-rays, the formation of the image and its contrast are primarily dependent on the inner shell ionization of the elements which make up the specimen. The inner shell ionization invariably causes the loss of several electrons from the outer shell of the atom resulting in breakage of the molecular bond. The study of $C_4H_{12}Pb$ (Carlson and White, 1968) indicates that subsequent L-shell ionization of lead causes the molecule to break into its atomic constituents. Ultrasoft x-rays produce, within an irradiated cell, only electrons of very low energy and very short range. Ultrasoft x-rays are more effective than hard x-rays in inducing inactivation and mutation in Chinese hamster cells, human diploid fibroblasts, and to a lesser extent, in inducing chromosome aberrations in human lymphocytes (Goodhead and Thacker, 1977; Goodhead, 1982). In other words, the use of C-K radiation, at energy levels too low to cause inner shell ionization of carbon, causes more damage than hard x-rays. Therefore, radiation damage to the sample by ultrasoft x-rays should be seriously considered.

Results obtained from absorption spectroscopy of organic materials such as gelatin, agar, Formvar, poly vinyl chloride (PVC), poly methyl methacrylate (PMMA) and poly (butene-1) sulfone (PBS), have shown significant mass loss upon soft x-ray irradiation (Cheng, 1985; Cheng *et al.*, 1985). Furthermore, the x-ray dose required to obtain a contact image on PMMA resist at 35Å monochromatic light is enough to cause the browning of thick (40μm) plant tissue (P. C. Cheng and D. M. Shinozaki, unpublished results). As Kirz and Sayre (1984) have pointed out, a realistic assessment of the radiation damage problem in x-ray imaging is a sizable challenge for all of us.

Conclusions

The present status of soft x-ray microscopy is reminiscent of electron microscopy during the 1940's when most of the related techniques were under development (Kirz and Sayre, 1984; Feder *et al.*, 1984). Considerable refinement is needed before this technique can become a useful tool in biological research.

2. Imaging Properties of the Soft X-ray Photon

David Sayre

IBM T. J. Watson Research Center Yorktown Heights, NY 10598, USA

Introduction

X-ray microscopy uses photons in the approximate wavelength range 1 to 10nm (soft x-ray photons) as imaging particles. In this article we describe first the basic properties of these particles which are relevant to this use; next, the characteristics of the microscopy which result; third, the several different forms which the microscopy can take; and finally some considerations concerning soft x-ray sources for x-ray microscopy. The article is intended as a brief introduction to x-ray microscopy and to the subsequent articles in this book.

An excellent general reference to the older work in x-ray microscopy is that of Cosslett and Nixon (1960); the reader will also find of interest the historical article by Newberry in this volume. In recent years there has been a large increase of activity in the subject; for a guide to the newer work see Parsons (1980), Schmahl and Rudolph (1984), Kirz and Rarback (1985), and Howells *et al.* (1985).

Basic Properties of the Soft X-Ray Photon

We note generally that photons are chargeless, massless particles which have spin 1 and are thereby bosons; they are the bosons which mediate the electromagnetic force. Although without mass, they have energy and wavelength. The properties of the soft x-ray photon which are most important in connection with its use as an imaging particle are summarized in Table 2.1.

X-ray Microscopy
Ed. by P.C. Cheng and G.J. Jan

Table 2.1. Some Basic Properties of the Soft X-Ray Photon

Wavelength	1-10nm	
Reactions	Photoelectric absorption of photon by atom	
	Near-field: Modulation of intensity	Strong
	Far-field: Reduction of direct beam	Strong
	Coherent scattering (imaginary)	Medium
	Production of secondary particles	
	Bond-rearrangement in absorber	
	Elastic scattering of photon by atom	
	Coherent scattering (real)	Medium
	Inelastic scattering	Weak
Particle statistics	Consistent with high source brilliance	

Wavelength

The approximate range of wavelengths used in x-ray microscopy, 1 to 10nm, has been previously noted. Occasionally wavelengths somewhat outside this range are also used. The range 2 to 5nm is particularly favored for biological specimens because of its relationship to the absorption edges of carbon, nitrogen, and oxygen (see section 2).

Reactions with Matter

The most frequent reaction of soft x-ray photons with matter is photoelectric absorption, in which the photon disappears in an atom. The observable effects are of several kinds. First, the beam as it exits from any partially transmitting object shows spatial modulation of the intensity, due to the discrete and usually also nonuniform distribution of the absorbing sites (*i.e.* atoms) in the absorber. This near-field effect is strong and is readily observed by placing a high spatial-resolution detector close to the exit surface of a specimen (Figure 2.1). If on the other hand the field is observed at a large distance from the object, the effects are different. There is a strong effect in the form of a reduction in the intensity of the direct beam (Figure 2.2) and a weaker effect in the form of photons scattered out of the direct beam (Figure 2.3). The scattered photons are scattered coherently, and are a diffraction by-product of the near-field intensity modulation arising from the absorption; because of this relation to absorption the scattering coefficient for this scattering is imaginary. For specimens in which there is only the discreteness of the atom distribution to cause the near-field modulation, the scattering is about 3 orders of magnitude weaker than the other effects, but it becomes stronger when there is nonuniformity of the atom distribution as well.

The coherent scattering effects may also be observed at intermediate distances (Figure 2.4). This case is discussed more fully in the subsequent paper by the author in this volume.

There remain two other significant effects of the photoelectric absorption reaction. The atom which absorbs the photon is raised to an excited state and returns to the ground state through the emission of secondary particles, namely a photoelectron and frequently Auger electrons and/or fluorescence photons. In addition, permanent effects may remain in the absorber in the form of chemical bond rearrangements at or near the sites where photons were absorbed.

A second reaction of soft x-ray photons with matter is elastic scattering, in which a photon recoils elastically from an atom. These photons too appear as coherently scattered photons. The scattering coefficient in this case is real, and for soft x-rays is generally of the same order of magnitude as that arising from absorption. The exact ratio varies considerably with the wavelength of the photons, however, especially near an absorption edge of the atom.

Lastly, soft x-ray photons also scatter inelastically from the electrons in matter, but the reaction is approximately 3 orders of magnitude weaker than the coherent scattering reactions and is not currently significant in x-ray microscopy.

We may summarize these remarks by saying that an object placed in a beam of soft x-rays produces three distinct though interrelated effects (or signals) in the x-ray field surrounding it: the intensity-modulation signal in the near field, the direct-beam reduction-of-

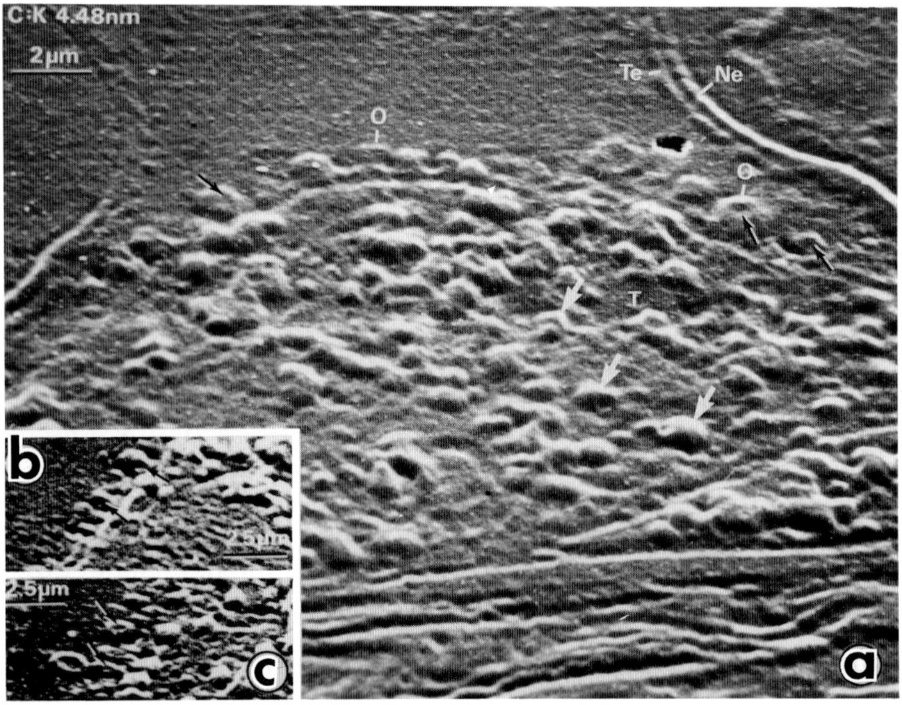

Figure 2.1. Intensity modulation of soft x-rays exiting from a specimen. The effect is observed by placing a high-spatial-resolution soft x-ray detector close to the exit surface of the specimen and subsequently examining the detector under high magnification in an electron microscope. The method constitutes one form of x-ray microscopy (contact microscopy). The images here show anther tissue of *Caltha palustris* L. at several magnifications. The x-ray wavelength was 4.48nm and the detector was an x-ray sensitive film of polymethylmethacrylate (PMMA). (Courtesy P.C. Cheng)

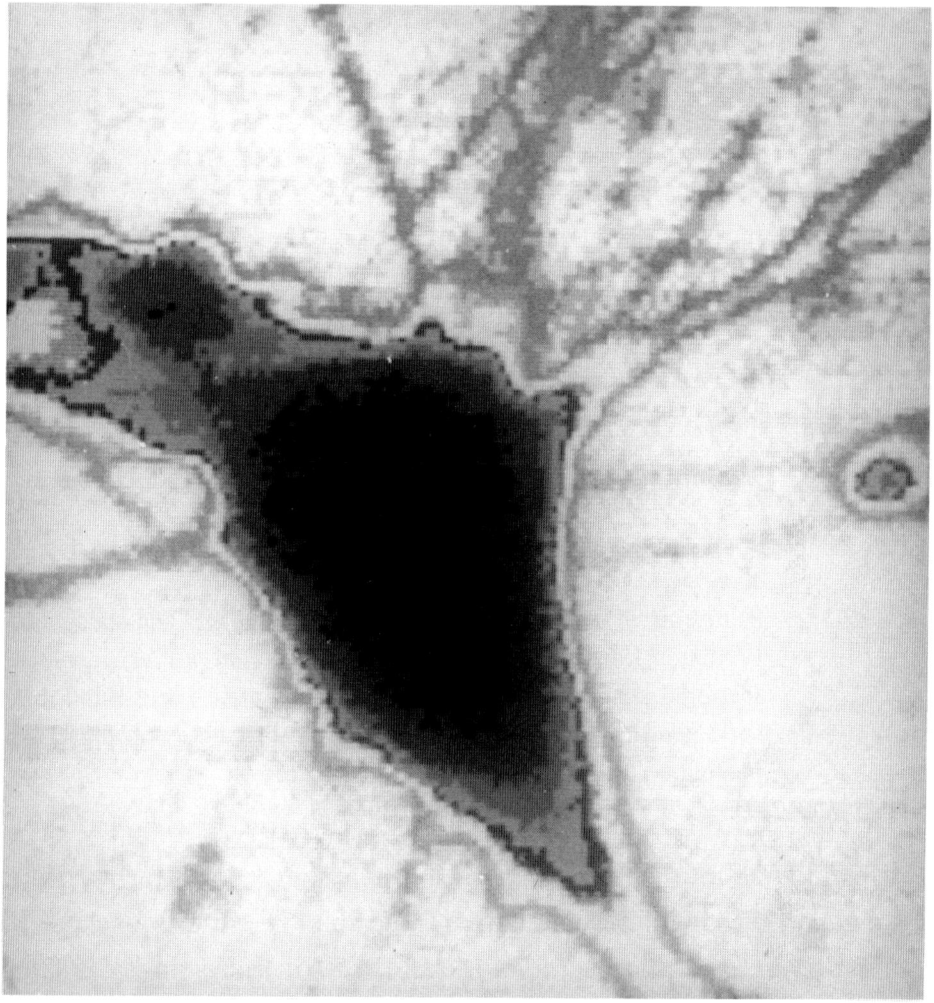

Figure 2.2. Reduction of the intensity of the direct beam after passing through a specimen. The image is a display of the output of a proportional counter placed several centimeters behind the specimen as the specimen was scanned over a narrow incident beam. The method constitutes a second form of x-ray microscopy (scanning microscopy). The image shown here is of a nerve cell. (Courtesy J. Kirz.)

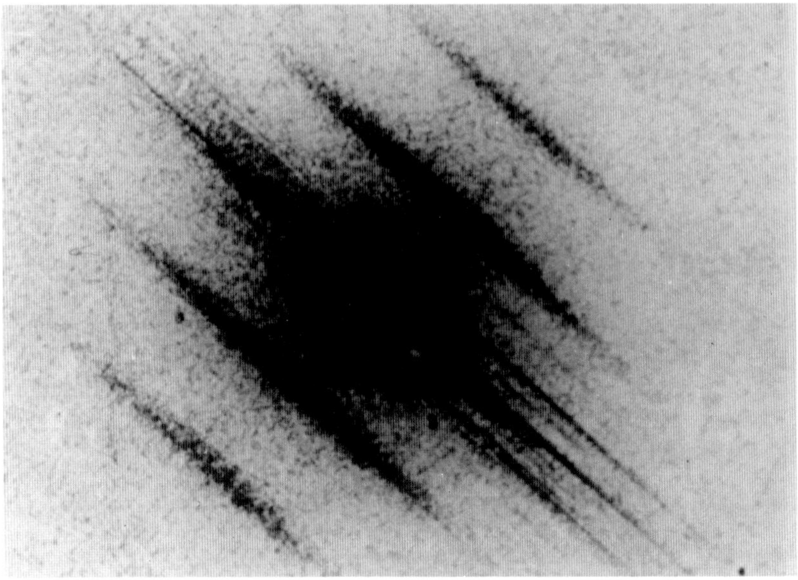

Figure 2.3. Scattering of soft x-rays outside the direct beam. The specimen (a diatom) was placed in a beam of 3.2nm x-rays and the scattered x-rays recorded on a silver halide plate 2cm behind the specimen. Exposure was approximately 30 minutes on the U15 beamline at the National Synchrotron Light Source (Brookhaven). (From Yun *et al.*, 1987)

intensity signal in the far field, and the coherent-scattering signal (often called the diffraction signal). As we shall see, methods exist today for doing x-ray microscopy with all three of these signals. There is also a fourth signal, consisting of the secondary particles produced in the absorption reaction. There are plans for a microscopy based on this signal as well.

Particle Statistics

As bosons, photons obey boson particle statistics, which permit a given volume of space to contain an unlimited number of particles of identical wavelength and momentum. Translated into other terms, there is no fundamental limit on source brilliance for photons.

Characteristics of X-Ray Microscopy

Based on the above, we can now list the principal characteristics of x-ray microscopy (Table 2.2).

Type

In its current forms, x-ray microscopy is a transmission or volume microscopy (*i.e.* one sees the interior of the specimen); this has the effect of giving its images the general appearance of those of transmission electron microscopy.

This characteristic arises basically from the volume property of absorption as the principal imaging reaction. In a partially transmitting specimen, every atom has the opportunity of being an absorption or

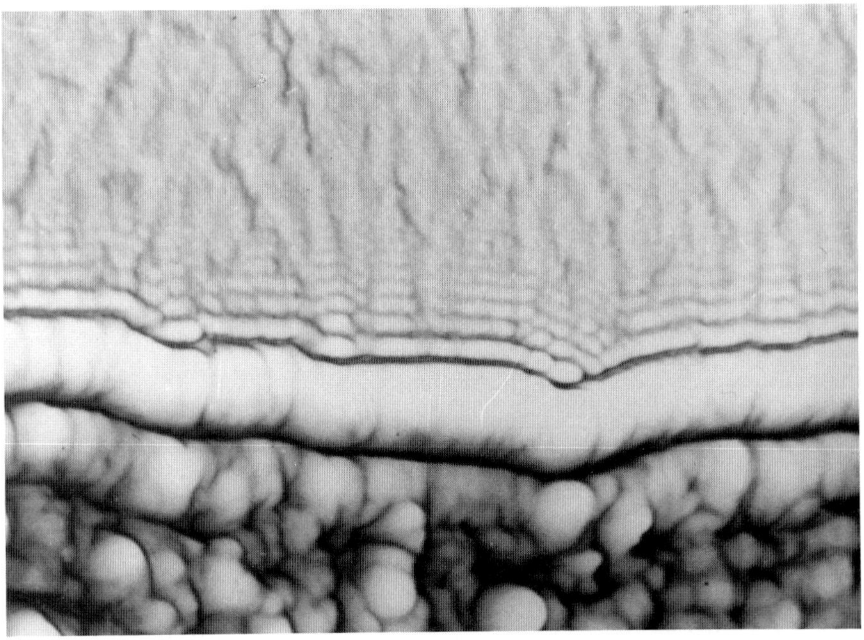

Figure 2.4. Scattering of soft x-rays in the intermediate field. The arrangement was similar to that of Figure 2.1, except that the distance from specimen to detector was slightly increased. The specimen was a section of rabbit psoas muscle. (Courtesy R. Feder.)

Table 2.2. Characteristics of x-ray microscopy

Type	Transmission microscopy; broadly resembles transmission electron microscopy Future potential for surface microscopy	
	From: Properties of imaging reactions	
Resolution	Maximum theoretical resolution	1nm
	Best present resolution 5 - 50nm	
	From: Wavelengths used	
Specimens	Increase in range of specimens	
	No requirement on:	specimen thickness (up to 2-3μm) vacuum compatibility immobilization high-Z content for contrast
	Intact imaging of:	microdevices, *etc.* small biological cells
	Cells have been imaged while in living state Specimen design for information and convenience	
	From: Variation of absorption cross-section Boson nature permitting high intensity sources	
Elemental mapping	Potential for high sensitivity Increased specimen range as above	

scattering site and hence of contributing to the imaging signal, and thus being seen.

We note, however, that in the case of the secondary-particle signal, the low-energy secondaries have only a short path length in the specimen, and hence will not be detected unless they originate near the specimen surface. By making use of this property, a surface x-ray microscopy may also be possible in the future.

Resolution

The maximum theoretical resolution of x-ray microscopy is approximately 1nm, much higher than that for light microscopy, but about an order of magnitude poorer than that of electron microscopy.

This characteristic arises from the wavelengths of the photons used in x-ray microscopy and the fundamental (diffraction) limit on the resolution of imaging imposed by them. It follows that x-ray microscopy will never quite equal electron microscopy in resolution.

The best resolutions actually achieved to date in x-ray microscopy range from approximately 5 to 50nm, depending on the form of microscopy used. (See Figure 2.5 for an example of imaging at the high-resolution end of this range.) Thus, even to reach the diffraction limit of resolution, there is still much work to be done. The key advantage of x-ray microscopy comes in the next topic.

Specimens

Figure 2.6 plots the cross-section of the photoelectric absorption reaction for a typical soft x-ray photon wavelength against atomic species Z. From the general placement of the curve, it is seen that x-ray microscopy is suitable for imaging specimens which are an order of magnitude thicker than those of electron microscopy, namely up to 2-3 μm specimen thickness for x-ray microscopy, compared with 0.2-0.3 μm for electron microscopy. In addition, from the shape of the curve, with its abrupt changes arising from the absorption edges of the atoms, it is seen that specimen contrast can exist without the need for high-Z atoms; *e.g.*, for the particular wavelength shown, 2.4nm, carbon, nitrogen, and calcium (Z = 6, 7, and 20) are strongly visible against a background of water ($Z = 1, 8$). By changing the wavelength, other contrasts can be elicited. Finally, as noted in the previous section, very high imaging intensities can be obtained with soft x-ray sources. The result is that such specimens as small whole cells in biology, complex composites in materials science, and microdevices in technology, can be imaged intact, without need for sectioning, dehydration, staining, immobilization, or other modification of their original state. This is illustrated in Figures 2.7 and 2.8. In Figure 2.8 the concept is pushed to its limit in the imaging of a living cell.

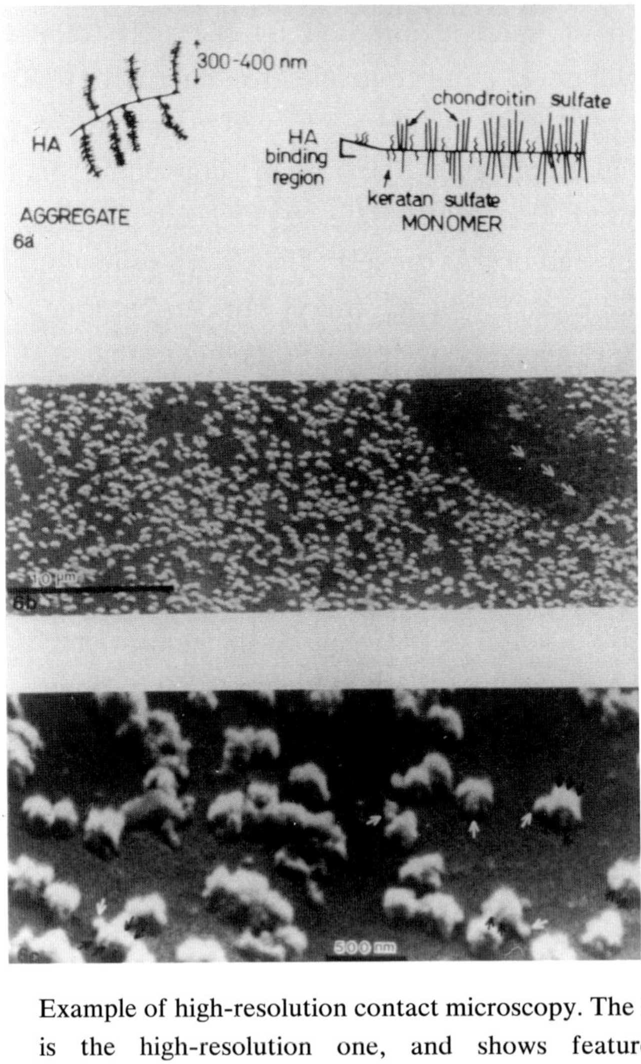

Figure 2.5. Example of high-resolution contact microscopy. The lower image is the high-resolution one, and shows features of size approximately 5nm; some of these are indicated by arrows. The individual specimens in the field are molecular monomers (upper right) of the macromolecular assembly proteoglycan (upper left). (From Panessa *et al.*, 1981)

This property of x-ray microscopy, that it is a minimal-demand microscopy, not demanding the preparation of specimens in such a way as to satisfy requirements imposed by the microscopy itself, is the

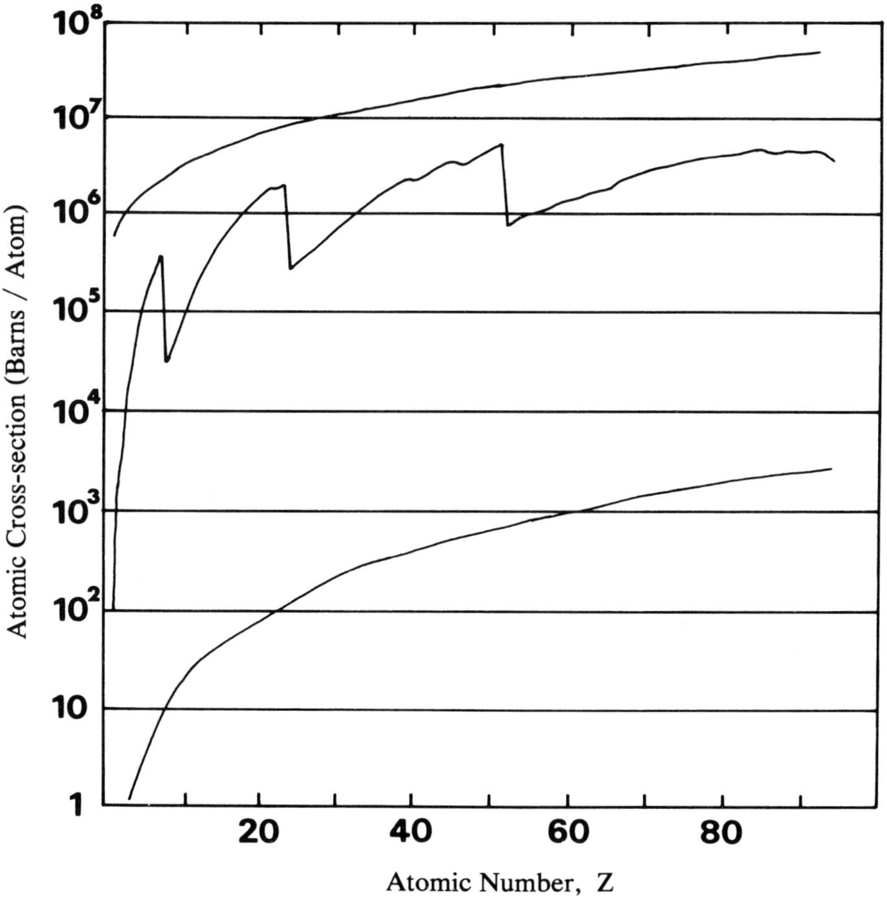

Figure 2.6. Cross-sections of important imaging reactions, versus atomic species *Z*. The upper, middle, and bottom curves relate respectively to electron microscopy, x-ray microscopy, and x-ray structure analysis. (Upper curve: total scattering of 100keV electrons. Middle curve: photoelectric absorption of 2.4nm photons. Lower curve: coherent scattering of 0.15nm photons.) The favorable consequences of the position and shape of the middle curve are discussed in the text.

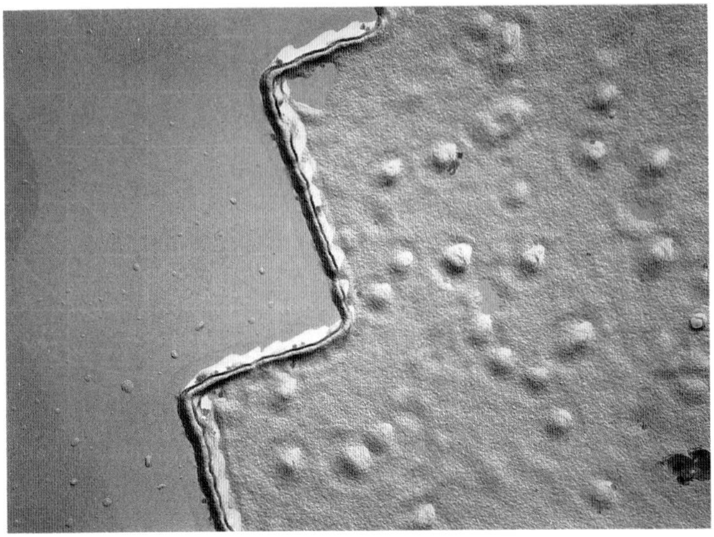

Figure 2.7. Contact soft x-ray microscopy image of a simple microdevice (connection pad). The device can be imaged in its intact state. The image was taken as part of a study of electromigration. (Courtesy D. Shinozaki.)

special characteristic of x-ray microscopy. It is not shared by the other nanometer-resolution microscopies, electron microscopy and x-ray structure analysis. It is shared by light microscopy, which however operates at much lower resolution. Its implications, by no means fully explored as yet, appear to be far-reaching: the ability to adopt a new and freer approach to specimen design, with emphasis on realism, informativeness, and process rather than static structure.

Elemental Mapping

X-ray microscopy also has the potential for producing high-resolution images of the distribution of individual elements in a specimen. The potential sensitivity can be high, and the advantages of extended specimen range just noted also hold. For more detail see

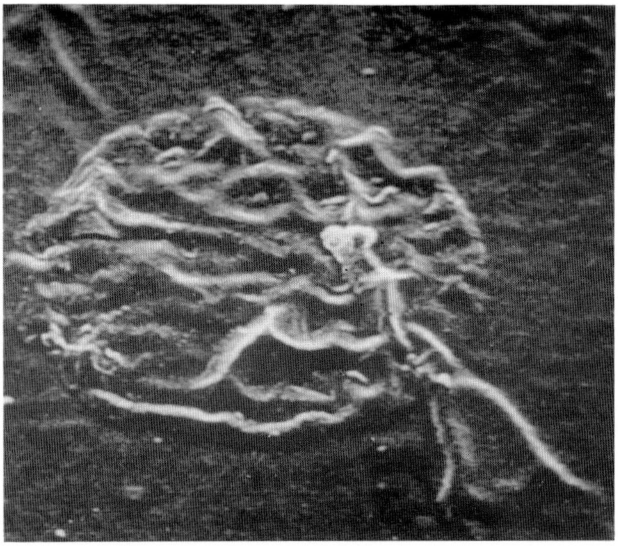

Figure 2.8. Contact soft x-ray microscopy image of a living blood platelet. The image was taken with a single 100ns flash of soft x-rays. (From Feder *et al.*, 1985)

Beese *et al.*, (1986) and the articles by Cheng, Nagai *et al.* and Kirz *et al.* in this volume.

Forms of X-Ray Microscopy

We are now in a position to summarize briefly the rather large number of different forms which x-ray microscopy can take (Table 2.3). Our discussion here will be brief, serving merely to introduce the techniques for more detailed discussion in the subsequent papers.

In this classification, the four major forms of microscopy are as follows. In scanning microscopy, the intensity-reduction signal of the direct beam in the far field is used. A small incident beam of photons is moved raster-wise over the specimen, and a detector located at some distance behind the specimen simultaneously measures the intensity of the transmitted direct beam. Displaying the intensity on a similar raster produces the image of the specimen (Figure 2.2). For further details see the article by Kenney *et al.*, and Niemann in this volume. The same technique, with a suitable change of detector, is potentially capable of imaging from the secondary-particle signal.

The remaining microscopies use a quite different imaging philosophy: where scanning microscopy so to speak dissects the specimen via the small size of the incident beam, they illuminate the entire specimen and rely on the spatial complexity of the resultant signals to supply the structural details of the specimen. In contact microscopy, the signal is the intensity modulation in the near field (Figures 2.1, 2.5, 2.7, and 2.8). Imaging microscopy uses the coherent scattering, or diffraction, signal with direct optical reconstruction of the image (Figure 2.9),

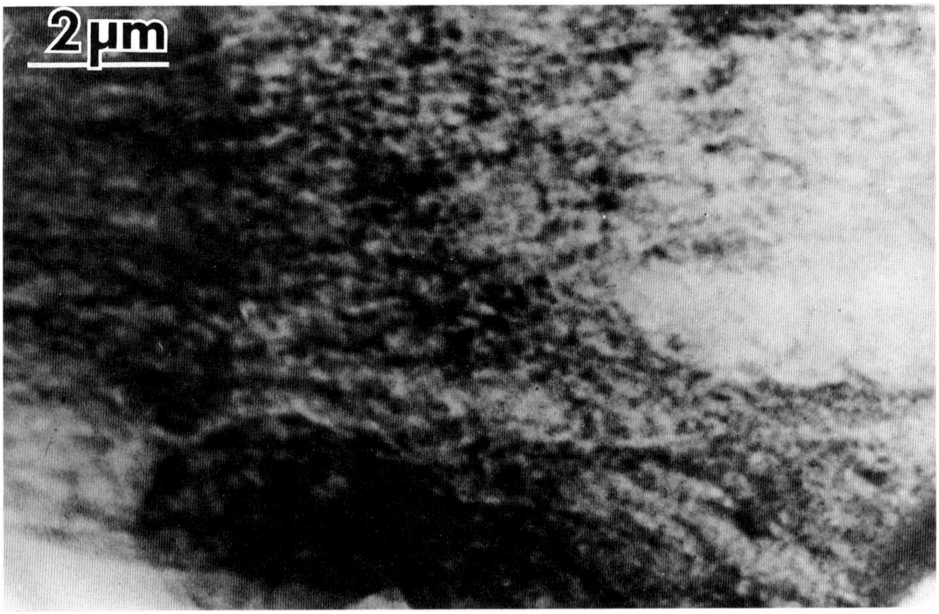

Firure 2.9. Imaging soft x-ray microscopy image of part of a critical point dried human fibroblast. The x-ray wavelength was 45Å The image was made with the Fresnel zoneplate imaging microscope at the BESSY storage ring in Berlin. (Courtesy W. Meyer-Ilse, G. Nyakatura, P. Guttmann, B. Niemann, D. Rudolph, G. Schmahl and P. C. Cheng)

Table 2.3. Forms of x-ray microscopy

	Scanning	Contact	Imaging	Diffraction-Imaging
Signal	A2,(PS)	A1	Sc	Sc
Beam	S	L	L	L
Optics	O	De	O	De
Type	T,(Su)	T	T	T
Status	Y,(N)	Y	Y	N
Resolution	50nm	5*	50	-
Limit	10?	5*	10?	1-3?
Cause	Te	Te,D	Te	D

*For thin specimens. Resolution worse for thick specimens.
A1=near-field intensity modulation PS=production of secondaries
A2=far-field reduction of intensity of direct beam Sc=scattering
S=small beam L=large beam
De=detector O=optics plus detector
Te=transmission imaging Su=surface imaging
Available now? Y=yes N=no
T=technological limit D=diffraction limit

while diffraction-imaging microscopy uses the diffraction signal but records it for subsequent offline reconstruction of the image. The latter is still only a potentiality at the present time, useful images not as yet having been produced by it in the soft x-ray regime. Further articles on these microscopies are found in this volume as follows: contact (Cheng *et al.*, and Jan *et al.*), imaging (Meyer-Ilse *et al.*), and diffraction-imaging (Sayre).

Two of the microscopies (scanning and imaging) depend upon focussing optics, in the first case to form the small scanning spot, and in the latter to do the image reconstruction from the diffraction signal. Given perfect optics, both these microscopies could reach resolutions comparable to the wavelengths used. In fact the optical devices available today fall very short of this performance (Table 2.3), and indeed in the foreseeable future it does not seem likely that technological limits in the optics will allow resolutions better than about 10nm to be realized for these microscopies. However, as the convenience, versatility, *etc.* of these microscopies would make them highly desirable even at that resolution, efforts and progress toward better optics continue to be made. Several of the articles in this volume (Buckley *et al.*, Nagai *et al.*, Aoki *et al.* and Schmahl and Rodulph) report on this subject.

The two remaining microscopies (contact and diffraction-imaging) avoid optics, with the aim of achieving higher resolution, although at the expense of more inconvenience in operation. The simpler of the two is contact microscopy, which records the near-field intensity modulation signal with a high spatial-resolution detector. With very thin specimens, as in Figure 2.5, the limit on resolution is set by the detector resolution, which is approximately 5nm. With thicker specimens, the detector is in the intermediate-field region, at least for the parts of the specimen furthest from the detector, and the diffracted photons begin to confuse the signal, as in Figure 2.4. The result is an apparent loss in resolution (to about 50nm at a specimen thickness of 1μm). This loss of resolution is at least theoretically avoided in diffraction-imaging microscopy, where the diffraction signal is made the basis of the microscopy.

Soft X-Ray Sources for Microscopy

Finally, in this section we examine briefly the radiation requirements of the several forms of x-ray microscopy discussed above. The major points are summarized in Table 2.4.

The existence of fairly broad time-ranges in the table reflects the fact that there are significant variations in detail in the radiation requirements of the different forms of the microscopy. These are due in part to fundamental, and in part to technical, factors. For example, the requirements of contact microscopy in terms of photon energy and

Table 2.4. Properties of soft x-ray sources for microscopy

Brilliance*	Most of the imaging methods can only use radiation in a small volume of photon energy and phase space; i.e. they require high source brilliance. Typical imaging times:	
	Storage-ring bending magnets	secs-ksecs
	Storage-ring undulators	msecs- secs
	Soft x-ray lasers (when developed)	μsecs-msecs
	Soft x-ray sources are now beginning to exceed electron sources in brilliance.	
	The method which currently uses the largest volume of energy-phase space is contact microscopy. Here	
	Soft x-ray plasma sources	nsecs-μsecs
Tunability	Also important.	
Summary	Best general source today is the storage-ring undulator. But need for laboratory-scale sources.	

*Brilliance is the number of photons emitted per unit bandwidth, emitting area, solid angle, and time.

phase space are generally not very stringent. It has available to it at present, however, only rather inefficient high spatial-resolution detectors, resulting generally in medium imaging times. Imaging microscopy uses efficient detectors but has a technical limitation on photon energy acceptance, arising from the fact that it is normally based today on optics (Fresnel zoneplates) having large chromatic aberration. Even so it tends to be the fastest of the microscopies on a typical storage-ring beamline. Scanning microscopy is similar, but in addition has an acceptance limitation on phase space volume, arising from the need to concentrate photons into as small a beam as possible. The situation with diffraction-imaging microscopy is roughly similar, and is discussed in more detail in the later paper by the author in this volume. The latter two thus tend to be the slowest of the microscopies, and also those which most demand high source brilliance.

With storage-ring undulators, imaging speeds in x-ray microscopy generally are equal to or greater than those of electron microscopy. Speeds will increase further if soft x-ray laser sources are developed. Heated-plasma sources, which emit intense radiation into an extended volume of energy and phase space, are currently giving nanosecond imaging speeds with contact microscopy. For further detail see the article by Baldwin in this volume.

In certain applications (*e.g.*, mapping of the distribution of individual elements) the need for a narrow energy bandwidth arises from the application itself, and not simply from the requirements of the microscopy. In these cases tunability of the energy is usually also a requirement. Either a tunable source is needed, or a tunable beamline with broadband source.

Finally, although most of the technical needs of x-ray microscopy are satisfied by storage-ring undulator sources, these sources are not at all widely available. The development of a laboratory-scale x-ray source with similar characteristics would be of extreme value in x-ray microscopy.

Summary and Conclusions

The effects of inserting a specimen in a beam of soft x-rays have been described and are found to include at least four effects suitable for realizing microscopies: modulation of the field intensity near the exit surface of the specimen (contact microscopy), reduction of the intensity of the transmitted direct beam (scanning microscopy), coherent scattering of the incident radiation (imaging microscopy, diffraction-imaging microscopy), and the emission of secondary particles from the specimen (another form of scanning microscopy). The magnitudes of these effects and their variation with atomic species and x-ray energy, combined with the high x-ray intensities available from current and future photon sources, enable the microscopies to image important classes of specimens in their natural state. At the same time, the microscopies are in a relatively early stage of development, with imaging resolutions still quite far from those theoretically attainable, and with suitable photon sources accessible only on a limited basis.

3. Status of X-ray Microscopy Experiments at the BESSY Laboratory

W. Meyer – Ilse[1], G. Nyakatura[1], P. Guttmann[1], B. Niemann[1], D. Rudolph[1], G. Schmahl[1] and P. C. Cheng[2]

[1]Forschungsgruppe Röntgenmikroskopie, Universität Göttingen, Geismarlandstr. 11, D-3400 Göttingen, Fed. Rep. Germany

[2]IBM T. J. Watson Research Center, Yorktown Heights, NY 10598, USA

Introduction

Soft x-ray microscopy provides higher resolution than visible light microscopy combined with the advantage of an inherent contrast mechanism. Unlike electron microscopy, x-ray microscopy can be used to investigate comparably thick objects of several μm in thickness. The reason of the natural contrast is, that the atomic scattering factors strongly depend on the different atomic species contained in the object under examination. Between 2.4nm and 4.4nm there is about an order of magnitude difference between the absorption coefficients for water and organic components. The natural contrast mechanism and the relative low radiation damage (Niemann *et al.* 1986) makes the x-ray microscope an useful tool in biology. The absorption is only one effect described by the complex atomic scattering factor, the second effect (the phase shift) is an even more valuable contrast mechanism (Schmahl and Rudolph, 1987).

Zone Plate X-ray Optics

Soft x-ray microscopes are constructed using zone plate diffraction optics. There are two x-ray microscopes of the Göttingen

X-ray Microscopy
Ed. by P.C. Cheng and G.J. Jan

x-ray microscopy group in operation at the BESSY Synchrotron Laboratory in Berlin. The present status of the scanning x-ray microscope (SXM) is reported by Niemann (1987) in this volume. This paper reports the status of the x-ray microscope (XM). The optical arrangement of this microscope is depicted in Figure 3.1. It makes use of a condenser zone plate and a micro zone plate. The currently applied condenser zone plate is the so called KZP6 as described by Schmahl *et al.* (1983). This condenser with diameter of 9mm consists of 38,000 zones. The width of the outermost zone of KZP6 is 60nm. The focal length at 2.4nm and 4.5nm wavelength are 222.4mm and 118.6mm respectively. The sophisticated fabrication process of this zone plate is also described in detail by Schmahl *et al.* (1983).

The condenser zone plate focuses the incident synchrotron beam into a hole of about 20μm diameter in the object plane. This arrangement - a linear monochromator provides the desired object illumination. The micro zone plate generates the magnified image of the object. Currently the so called micro zone plate MZP3 is used. It is described by Schmahl *et al.* (1983) and Guttmann (1983).

The Imaging X-ray Microscope

The x-ray microscope is installed at BESSY in Berlin. The arrangement of the beam line is described by Rudolph *et al.* (1983). One of the aims of improvements of this system is to minimize the radiation dose applied to the specimen under examination. Possible ways to attain this goal at a desired x-ray wavelength are: maximizing the efficiency of zone plates and detector, and leave out all the radiation which doesn't contribute to the final image. A reasonable amount of radiation dose is applied to the object if one searches for the proper focus of the image using x-rays. In order to certify the correct focus with x-rays one has to take the image, and in order to find the correct focus distance, one has to take image series. This results in an appli-

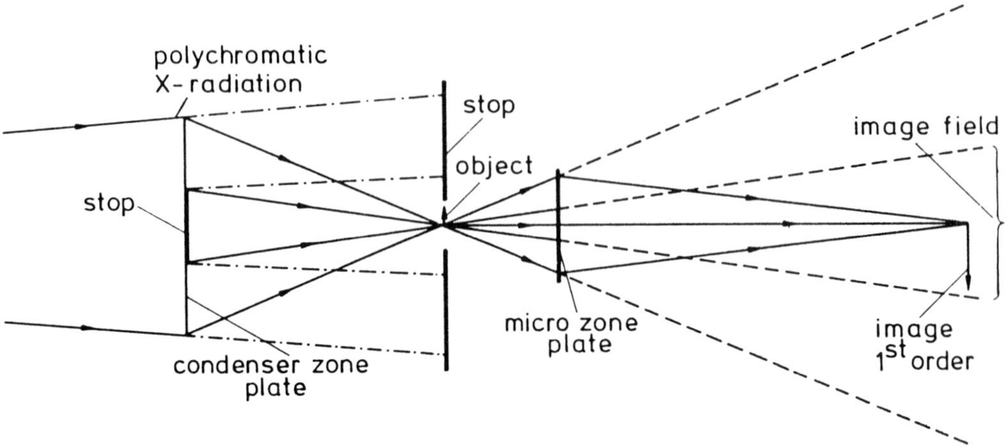

Figure 3.1. Optical arrangement of the x-ray microscope (XM). The synchrotron radiation meets the condenser zone plate (KZP6) and is focused onto the object, and a micro zone plate creates the magnified image of the object.

cation of an total radiation dose, which is a multiple of the dose necessary to take the final image. To avoid this problem, an optical prefocus is included in the imaging x-ray microscope.

Optical Prefocus

The basic idea of the optical prefocus is to incorporate an optical microscope into the imaging x-ray microscope and to link the focusing components of these microscopes. An arrangement, which meets these requirements, is incorporated in the x-ray microscope at BESSY in Berlin. This system is sketched in principle in Figure 3.2. The main portion of the optical part of the microscope is a microscope objective with a central hole of about 2mm in each lens. It provides a high magnification (40 times) with a long working distance (3mm). These features are necessary to get a good focusing accuracy of less than one

μm and to leave enough space to mechanically fix the micro zone plate in front of the objective.

The object under examination is supported by a thin foil (0.3μm polyimide) in back of the 20μm diameter monochromator hole. The x-rays are focussed onto the object by the condenser zone plate. The stop with the monochromator hole is made of glass, to allow an illumination of the object with visible light. The object environment can be separated from the vacuum of the microscope with a second glass pinhole of about 0.2mm diameter, which is also covered with a thin foil, in a distance of about 0.16mm behind the object. This environment chamber is followed by the micro zone plate and the optical objective. The micro zone plate and the objective are linked on one support carriage, providing a joint focussing of the instrument.

Experiments with Wet Objects at 2.4nm

Getting microscopic images with a high resolution and valuable contrast is the principal advantage of x-ray microscopy. In biological applications one demands a contrast of water against protein. Using an absorption contrast mechanism, suitable differences of the absorption coefficients of water and protein are expected at the oxygen K-edge (2.4nm). Present experiments in x-ray microscopy make use of this absorption mechanism.

Up to now there are only few x-ray microscopic images of wet objects published by our research group (Schmahl *et al.* , 1980; Schmahl *et al.*, 1986), and these images were taken at 4.5nm. Present experiments at 2.4nm are performed with hydrated *in vitro* cultured KB cells, a cell line derived from a human mouth epidermal carcinoma. The cells were grown on 0.3μm thick polyimide film mounted on a quartz ring. The film was irradiated with UV light (about 2 hr) to facilitate wetting and heat sterilized (2 hr, 180°C) before being introduced into a culture vessel containing 2ml cell culture medium (Eagle's

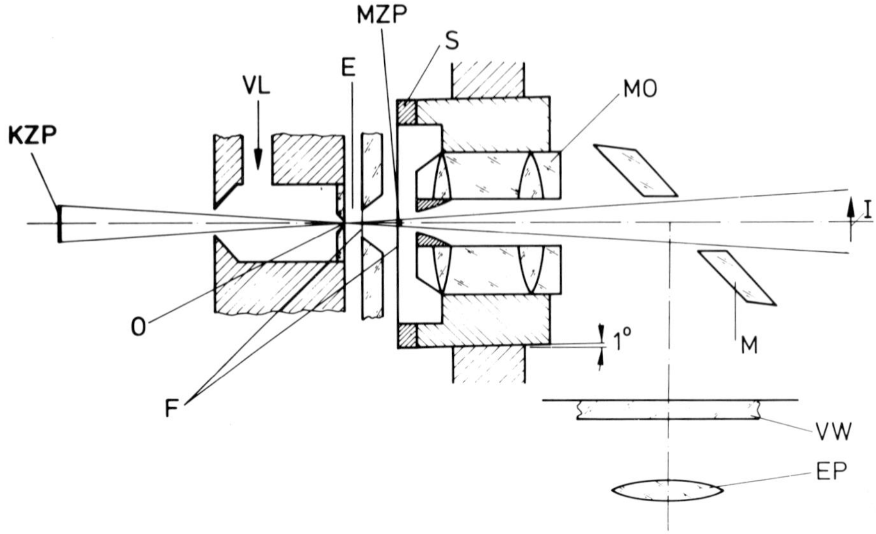

Figure 3.2. Schematic of the optical prefocus in the XM: KZP: condenser zone plate, MO: microscope objective, MZP: micro zone plate, M: mirror, VL: visible light, VW: vacuum window, O: object on thin foil, EP: eyepiece, E: object environment, I: X-ray image of object, F: thin foils, S: support of micro zone plate.

minimum essential medium supplemented with 10% fetal calf serum and antibiotics). After introducing the glass rings with the film into the vessel, the medium was inoculated with about 100 thousand cells to give a final concentration of about 30 - 40 thousand cells per ml and incubated at 37°C. After three days an almost confluent cell lawn had been formed. The quartz rings, which are the specimen holders for the microscope, were carefully withdrawn from the culture vessel, the excess medium allowed to drain away into tissue paper leaving a thin residual water film. The hydrated cells were put into an airtight environment chamber and then transferred into the microscope. The im-

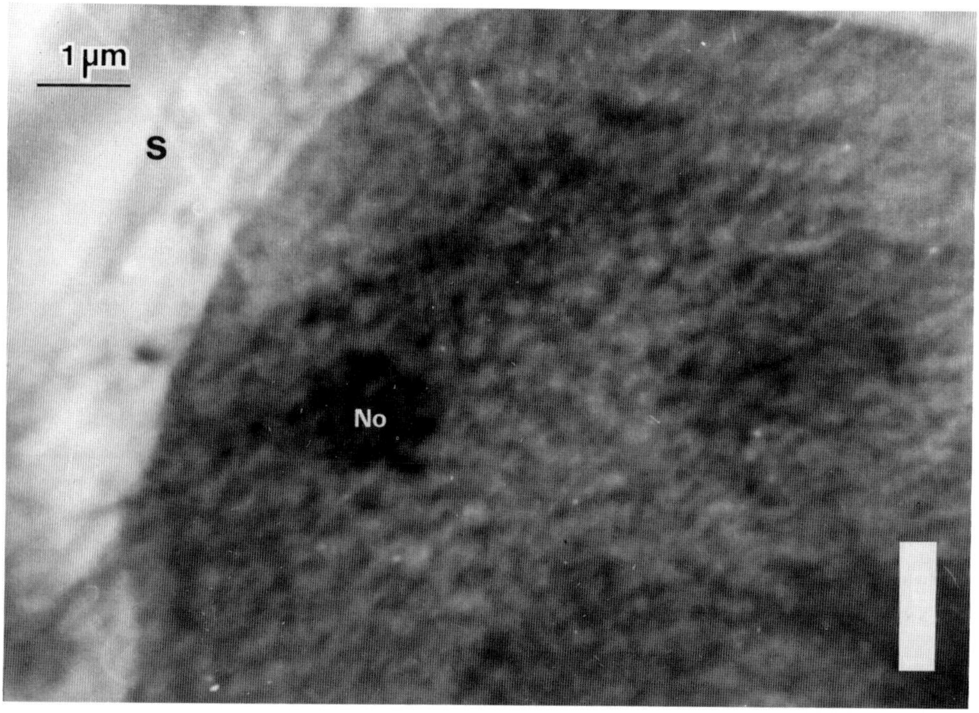

Figure 3.3. Nuclear region of a cultured human fibroblast imaged at 4.5nm: Note the granular appearance of the nucleoplasm and X-ray dense nucleolus (No). Stress fiber bundle (S) can be seen in the cytoplasm.

ages show the expected high contrast. The method, however, is up to now not reproducible and requires further effort.

X-ray Images of Human Fibroblasts Imaged at 4.5nm

Unstained dehydrated human fibroblasts were imaged at 4.5nm wavelength. These cells have been grown in Dulbecco's modified Eagles's medium supplemented with 10% fetal calf serum and antibiotics on a thin (100nm) carbon-coated Formvar film. They were then washed with phosphate buffer saline, and fixed with 1% glutaraldehyde in phosphate buffer (pH7.2), dehydrated in ethanol and critical-point dried with CO_2. The micrograph (Figure 3.3) exhibits enough contrast to allow easy recognition of some of the cellular components. The nucleus with a well defined nuclear envelope, the nucleoplasm with it's granular appearance and x-ray dense nucleoli are clearly discernible. Stress fiber bundles and parts of the endoplasmic reticulum can be seen in the cytoplasm region.

Acknowledgements

This work has been funded by the German Federal Minister for Research and Technology (BMFT) under contract number 05320DAB. A part of the developments of zone plates and x-ray microscopes have been supported by the Stiftung Volkswagenwerk.

4. Current Status of the Göttingen Scanning X-ray Microscope - Experiments at the BESSY Storage Ring

B. Niemann

Forschungsgruppe Röntgenmikroskopie, Universität Göttingen Geismarlandstr. 11, D-3400, Göttingen, Fed. Rep. Germany

Introduction

The period of development and installment of a scanning x-ray microscope at the BESSY storage ring is finished and first tests were performed. The concept of this scanning x-ray microscope has already been published earlier (Niemann *et al.*, 1981; Niemann, 1984).

The instruments hardware and software can also be used, when higher spatial resolution and higher count rates must be handled. This will be essential in future, if the instrument would be run at an undulator, which delivers radiation of higher spectral brilliance than the bending magnet used now at BESSY. In the following text a brief description of the scanning x-ray microscope is given, and first results are reported.

The assembly of the scanning x-ray microscope contains two separated optical arrangements of different purpose: a monochromator and a scanning stage. Figure 4.1 shows a schematic of the microscope and Figure 4.2 shows the scanning x-ray microscope at BESSY.

The Monochromator

A zone plate linear monochromator is used. It simply consists of three elements: an entrance pinhole near the source, which is comparable to the entrance slit of a grating monochromator, a condenser zone plate,

X-ray Microscopy
Ed. by P.C. Cheng and G.J. Jan

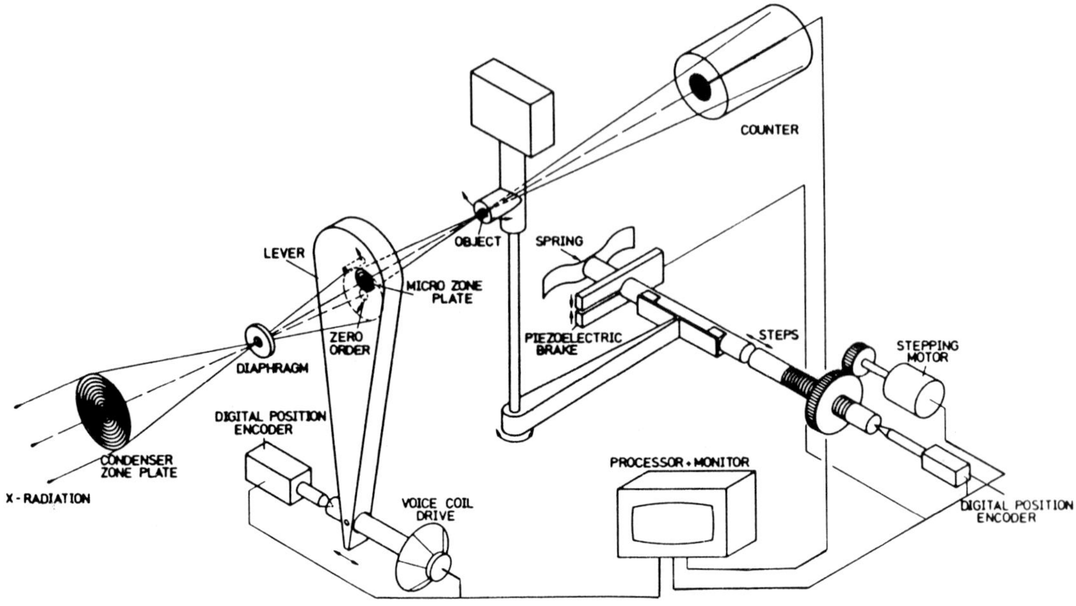

Figure 4.1. Construction of the scanning x-ray microscope.

which is the dispersing element, and a monochromator pinhole, which corresponds to the exit slit of a grating monochromator. The source is formed by the electron beam in the bending magnet of the BESSY electron storage ring. It delivers polychromatic synchrotron radiation through the entrance pinhole to the condenser zone plate in the microscope, which is the located at 12 meter distance. The condenser zone plate KZP4 has D = 2.8 millimeter diameter and 554 millimeter focal length at 4.5 nanometer wavelength. At 2.4 nanometer wavelength condenser zone plate KZP5 with 556 millimeter focal length and 2.5 millimeter diameter can be used (Thieme, 1984). The condenser zone plate images the source to a spot, 20 times demagni-

Figure 4.2. Photograph of the scanning x-ray microscope at BESSY. The left part of the vacuum chamber contains the zone plate linear monochromator. At the right end the housing of the mechanical scanner is to be seen.

fied in diameter. In the image plane of the condenser a monochromator pinhole of the same diameter d is placed. The radiation passing the pinhole gets a monochromaticity of approximately D/2d.

Behind the monochromator pinhole there is a small area, which also contains zero order radiation of the condenser zone plate. This area has got the same diameter as the monochromator pinhole and is located concentric to the optical axis through the condenser and monochromator pinhole, see Figure 4.1.

Estimation of the Coherent Flux

In the first experiments the monochromator was used without entrance pinhole. The source diameter of BESSY at METRO optic configuration - about 0.3 millimeter x 0.6 millimeter at some hundred milliamperes beam current - was imaged to a diameter of about 15 micrometer x 30 micrometer. Therefore the monochromator delivered a monochromaticity of roughly $\lambda / \Delta \lambda = 70$. The monochromator pinhole has got 18 micrometer diameter, delivering a spatial coherent flux of 1.7×10^8 photons per second. This calculation is based on the loss factors mentioned below.

Future Improvements of the Monochromator

It is planned in further experiments to use the zone plate monochromator with the entrance pinhole of 100 micrometer in diameter , which is already installed 0.5 meter away from the source point in the bending magnet exit chamber. It can be moved remote controlled. Using this entrance pinhole at the source, the monochromator pinhole diameter - for best monochromaticity - can be reduced to 5 micrometer, delivering a monochromaticity of roughly 200 or coherence length of 200 λ, as was shown in detailed calculations (Thieme, 1984). This is sufficient, because the following scanning stage contains micro zone plate MZP3 with 250 zones, requiring a minimum coherence length of roughly $250\lambda/2$.

Using the entrance pinhole the total flux, and thus the heat load applied to the condenser zone plate, will be reduced to about 1/10 or less, increasing its stability. The total flux through the monochromator pinhole is reduced to the same amount. The coherent flux emerging from the monochromator per unit bandwidth will not be changed, but it will be spread in a larger solid angle.

Only coherent flux from the monochromator must be collected by the micro zone plate in the connected scanning stage, if it shall be run at its highest, diffraction limited resolution.

The Scanning Stage

The scanning stage consists of two independent mechanical motions in perpendicular directions, each with a maximum amplitude of 10 micrometer. The scanning stage only contains a single optical element: a micro zone plate, namely MZP3, which can generate a diffraction limited spot with a diameter in the order of 60 nanometer (Schmahl *et al.*, 1984). By varying the distance between the monochromator pinhole and the scanning stage, the degree of coherence of the collected photon flux can be varied.

The micro zone plate is located in the upper part of the radiation cone emerging from the monochromator pinhole, above the small area containing zero order radiation from the condenser. Therefore only 1st order radiation from the condenser zone plate meets the micro zone plate.

The micro zone plate is moved continuously up and down at a rapid speed, whereas the object, located in its focal plane, is moved in horizontal direction only when the zone plate ends a line scan and changes direction.

Both motions are done with levers. The levers are moved with amplitudes of some millimeters at their free ends and the reduced amplitudes near their turning points are approximately a thousand times smaller and are used for scanning. The lever mechanisms are constructed to be free of backlash and creeping. Two different types of lever mechanisms do the motions and will be tested to find out which one works best.

Mechanism for a Line Scan

The motion of the micro zone plate is done with a single lever mechanism in vertical direction. The motion is circular, but as the angle of rotation is 0.5 degree at the maximum, the deviation from a straight line is negligible. The lever bearing consist of a torsion joint. The free end of the lever is moved with a voice coil of a loudspeaker, and its amplitude can be measured with a Heidenhain digital position encoder. The encoder has a positional resolution of 0.1 micrometer and a minimum electrical respond time of 50 nanoseconds. The lever is one meter long. The micro zone plate is fixed at one millimeter distance from the turning point. Therefore the zone plate motion is a thousand times smaller than the value measured with the digital position encoder. Thus a movement of one increment of the digital position encoder in principle corresponds to a 1 Angstrom displacement of the micro zone plate and produces one electronic pulse at the output of the position encoder. Surely the mechanical stability does not reach 1 Angstrom - the acoustic noise will still be higher - but the electronic hardware can handle the output pulses errorless and without missing one. For example, a movement of +100 Angstrom of the scan spot produces a net of +100 forward pulses at the output of the digital position encoder. They are counted in a self-made free programmable CAMAC up-down counter with asynchronous reset within 50 nanoseconds. Vibrations may be superimposed to the linear movement and will produce an additional number of forward and an equal number of backward pulses - so they cancel in the counter and do not alter the resulting measured position.

Presetting "100" to the counter, it produces interrupts in the CAMAC - system every 10 nanometer during the scan. The CAMAC system subsequently counts the number of x-ray photons detected between two interrupts and measures the belonging collecting time. It produces a pixel of corresponding intensity and position on the monitor. The correct position of the lines on the monitor is determined by a reference point signal, which is generated by the digital position encoder during each line scan.

Mechanism for Scanning Perpendicular to a Line Scan

The mechanism for moving the object includes a long steel rod, which is fixed with one end to the case of the microscope. At the other end of the rod a lever arm is connected, which is used to twist the rod to and fro. The degree of twisting decreases linearly to the fixed end. At a distance of only a few centimeters from the fixed end, the object holder is clamped to the rod. With this mechanism the object will be moved a thousand times less in amplitude than the free end of the lever.

Scanning an Image

Each time when the zone plate movement ends a line scan and the zone plate movement reverses its direction, a piezoelectric brake, which is attached to the lever of the mechanism moving the object, is loosened for only a few milliseconds. Hereby, a spring load pushes the lever arm to the new position of a stepping motor driven spindle, and thus the object moves to the next line to be scanned. Next the brake clamps again and the object is held tight in the new position. Now the stepping motor starts to align the position of the next line scan. This alignment is done within 200 milliseconds, the minimum time for a line scan.

For a picture of 200 x 200 image points the minimum scan time is 40 seconds, corresponding to a scan time for one image point of one milisecond. Therefore the time scale for the mechanical motion and the data processing is matched to one millisecond for one image point. though this could be increased by simply using advanced models of the used microprocessors. The computer comprises a CAMAC crate with

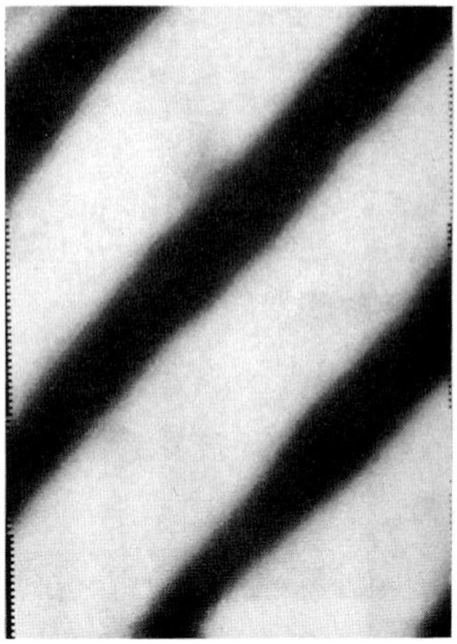

Figure 4.3. Picture taken at photon rate millisecond: Therefore the time scale for the of 10^6 per second showing a part of mechanical motion and the data processing is a grating. Grating constant: about 2.5 matched to one millisecond for one image micrometer. Scan time: about 3 minutes.

dedicated controller and a memory, an additional host computer with a LSI 11/23 microprocessor and a DE ANZA ID 2000 black and white monitor system. The system can store 160 images of 200 x 200 pixels with 8 bit gray levels. A Honeywell Color Graphic Recorder is used to produce black and white or false color prints after processing.

Test of the Scanning Mechanism

Figure 4.3 shows an image of some transparent holes which were scanned at a low speed for testing the stability and reproducibility of the scanning mechanisms. The size of one pixel is 500 Angstrom in vertical direction and 300 Angstrom in horizontal direction.

Each line scan of the micro zone plate is represented by one vertical line in the image. The horizontal white line indicates the reference point position in each line scan; this line is to be seen only in this image for demonstration. The intensities of the gray levels on the image are I, I/2, I/4, and I/8. A white spot represents the intensity range I to I/2, a dark spot represents the intensity I/8 or less. When the measured intensity drops to half the maximum intensity or below, the image point gets the gray level I/2 or less. I/2 still has got a comparably small photon noise. The boundary between two image areas with intensities I and I/2 should be a continuous and smooth line, if in the ideal case the scanning mechanism works reproducible and no photon noise or fluctuations of the beams spectral brilliance occur.

A smooth line, which is not lying parallel to one of the axis of the scanning mechanisms will be shown as a regular staircase profile on the monitor. As can be seen on the picture the limit between I and I/2 has got a continuous, regular shape, whereas the boundaries between I/4 and I/8 mostly show an irregular profile because of an increased photon noise. The 22 lines to the left of the picture show identical vertical line scans with no horizontal movement. In the ideal case only straight horizontal lines should occur. In fact the limit between I and I/2 again is nearly perfect and straight, whereas the limits of I/4 and I/8 are not separated sharply because of an increased photon noise.

It can be concluded from this test that the scanning mechanisms are suitable for at least 500 Angstrom steps.

Figure 4.4b shows the same image as Figure 4.4a, but it was median - filtered with a computer to reduce the photon noise. Hereby also the reference points were filtered out. Figure 4.3 and 4.5 were also processed with the same filter.

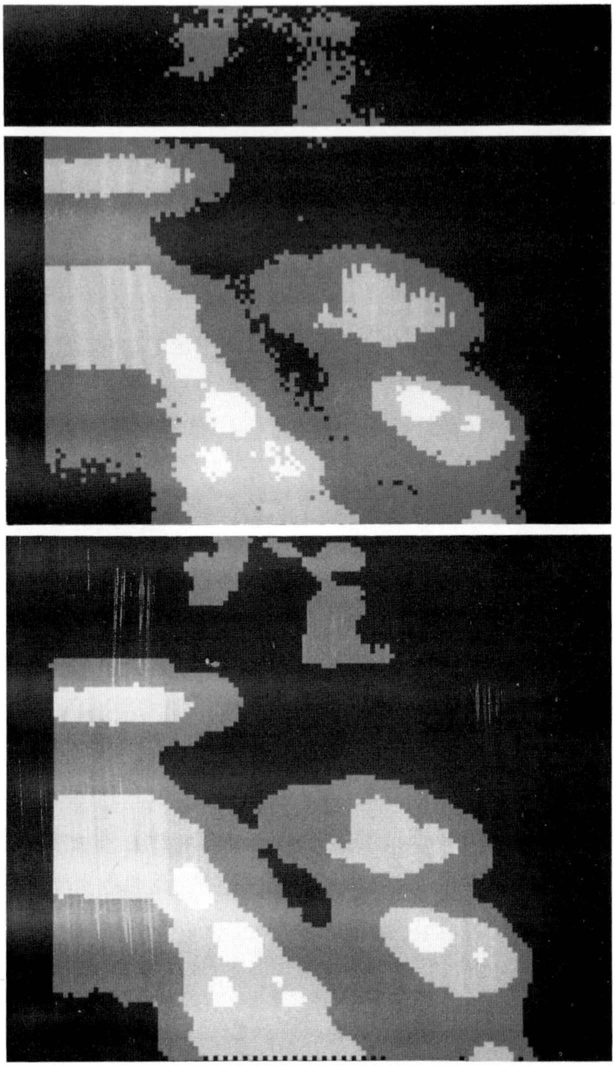

Figure 4.4. a:(left) and b: (right): Scanned image showing only 4 gray levels and the position of the reference points.: The size of one pixel is 500 Angstrom in vertical direction and 300 Angstrom in horizontal direction. The right part of the picture represents a two dimensional scan, whereas the 22 lines on the left show 22 different scans of the same line. (a): the same as (b), but median filtered.

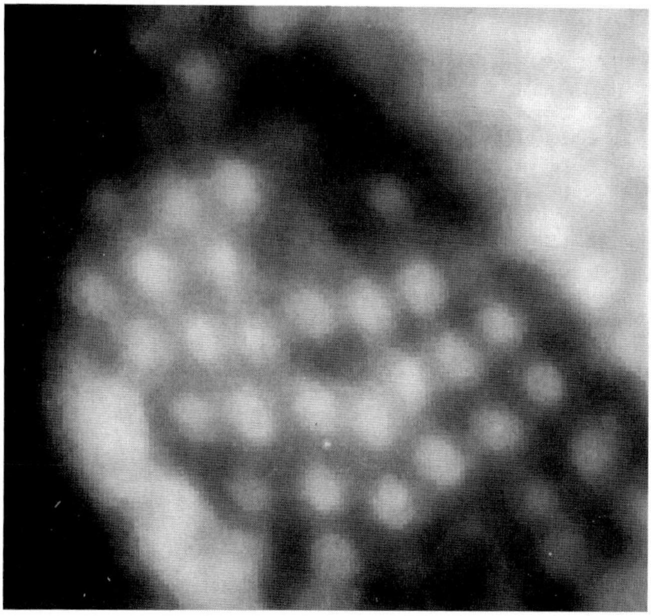

Figure 4.5. Diatoms imaged at 4.5 nanometer wavelength: The smallest holes are 0.3 micrometer in diameter. Fine focussing was done by simply changing the monochromator wavelength a few tenth of an Angstrom. The image data were stored and processed in the following manner: As there was some 1st order background radiation superimposed to all the image, a constant offset was subtracted from all data. Thus the minimum displayed intensity level was reduced to zero. The maximum intensity of the data obtained was only 1/3 of the maximum possible intensity. Therefore all the data were multiplied by a constant factor to use the full range of the video look up table of the image processor. Then the data were median filtered (size 3x3) and photographed. The size of one pixel is about 500 Angstrom by 500 Angstrom.

The Proportional Counter

A proportional counter was developed, which was investigated in detail by Guttmann (1987). This special design uses a 100 millimeter long counter wire, which is parallel to the optical axis of the microscope and has an entrance window at its front. The center of the window contains a central stop. Therefore no zero order radiation of the micro zone plate enters the counter. The counter is run with methane gas at 90 millibar pressure. As the absorption of photons is a statistic process, it can take place anywhere in the counter and the following gas amplification process produces a space charge in a small region. Here no other gas amplification process can take place during the next microsecond until the electrical field has been restored. As the space charge region is small compared to the dimension of the counter, most of the counter volume remains unaffected and is ready for the next gas amplification process. Guttmann (1987) showed that 6 x 10^6 photons per second can be measured with this counter.

Measured Photon Flux

Images with a measured photon rate of slightly more than 10^6 photons per second were obtained under the following conditions: The microscope was run with a resolution of 0.1 micrometer. The monochromator pinhole was not used, so the monochromaticity was only about $\lambda/\Delta\lambda = 70$ at 4.5 nanometer wavelength. The BESSY source was about 0.3 millimeter x 0.6 millimeter in diameter at 400 milliamperes electron beam current. Condenser zone plate KZP4 accepted about 0.27 milliradians of the radiation and illuminated a monochromator pinhole 18 micrometer in diameter. The scanning system was evacuated. The object was surrounded by a small pinhole, which eliminated most of the -1. order radiation of the micro zone plate.

The number of photons per second delivered from BESSY at 400 milliamperes electron beam current, 1% bandwidth and 0.63 milliradians x 0.63 milliradians beam divergence at 4.5 nanometer wavelength is about 3.2×10^{13}. Flux losses occur at the following elements:

	loss factor
Entrance window, chromium on Capton, in front of the condenser zone plate 50% transmittance	2
solid angle of BESSY radiation / solid angle accepted by the condenser zone plate KZP4	5.4
8% efficiency of KZP4, germanium on Capton	12.5
image area / pinhole area	2
Capton foil on pinhole, 70 % transmittance	1.4
1.4 % monochromator bandwidth	0.7
2 - 3% efficiency of MZP3, estimated	35
solid angle of monochromator output / solid angle accepted by MZP3 = $\{(20 \times 0.27\text{mrad})/(0.0556\text{mm}/122\text{mm})\}^2$	140
80% transmittance of foil carrying the object	1.25
40% transmittance of entrance window including a central stop to suppress zero order radiation of MZP3 98 % absorption efficiency of counter gas	2.5
total loss factor:	4×10^6
maximum predicted count rate:	8×10^6/sec

The maximum count rate measured is a factor of 7 less than this estimation. Dead time effects of the charge sensitive preamplifier, which was not run with its minimum dead time during these first tests, may have contributed to the discrepancy. In fact during these tests the dead time of the counter V, the charge sensitive preamplifier V_1 and the timing-filter amplifier summed up to 600 nanoseconds, and limited the maximum measurable count rate to 1.7 x 10^6 photons per second (Guttmann, 1986). The measured maximum count rate of 1 - 1.1 x 10^6 corresponds to a true count rate of 2.5 - 3.2 x 10^6 photons absorbed in the counter per second. This will be verified in the future with an electronic system of a shorter dead time. Also the effective source size of BESSY was not accurately determined. It was derived from the FWHM of the measured counting rate, when the condenser zone plate was shifted perpendicular to the optical axis.

As soon as the microscope will run with its designed monochromaticity of 200 and its ultimate resolution of 60 nanometers, the number of detectable photons will be reduced by a factor of 10, resulting in 7 x 10^5 photons per second detectable in the counter at 4.5 nanometer wavelength.

This flux can be increased mainly by improving the efficiency of MZP3 using reactive ion etching techniques, which were already used for processing the condenser zone plates. Also the spectral brilliance of the BESSY electron beam can be approximately 1.5 - 2 times higher, as during this test its brilliance was reduced in favor of an increased life time.

Acknowledgement

This work has been presented at the SPIE conference at Berlin, december 1986. I thank my colleagues of the Göttingen x- ray microscopy group very much for numerous discussions during the time of building the microscope. Especially I thank D. Rudolph and G. Schmahl, who initialized this project. I thank J. Thieme for calculating the condenser zone plates, and R. Hilkenbach for developing highly efficient condenser phase zone plates (Hilkenbach, 1987). I thank G. Schmahl, D. Rudolph , P. Guttmann and O. Christ, who developed the micro zone plate, and P. Guttmann for his work on the flow proportional counter and for programming the interlock system of the beam line. I thank W. Meyer-Ilse for programming the image processor, and the members of the mechanical and the electronic workshops for their constant work. This work was supported by the Stiftung Volkswagenwerk and has been funded by the German Federal Minister for Research and Technology (BMFT) under contract number 05 320 DA B 7.

5. The Beginning of Scanning X-ray Microscopy at Daresbury

J. M. Kenney, M. T. Browne, C. J. Buckley, R. E. Burge, R. C. Cave, P. Charalambous, P. J. Duke, A. G. Michette, G. R. Morrison, K. Ogawa, and A. Rogoyski

Physics Department, King's College, London, U.K.

Introduction

At the Synchrotron Radiation Source (SRS) of the Science and Engineering Research Council's (SERC) Daresbury Laboratory, England, the U5 undulator beamline has been built as a tunable source of soft (200-2000eV) x-rays appropriate for a scanning transmission x-ray microscope (STXM). Here we discuss the motivation, concept, design, and experimental program for the STXM with particular attention given to those features which are peculiar to this instrument. Finally, we take a look at the future of scanning x-ray microscopy.

There are many references (Kirz and Sayre, 1980; Spiller, 1983; Howells *et al.*, 1985) which discuss the motivation for the development of the STXM. Some experiments using an STXM with high resolution zone plate optics have already been performed in the USA (Cinotti *et al.*, this volume), and scanning x-ray microscopy is being developed in Germany (Niemann, this volume) and Japan (Aoki *et al.*, this volume). The STXM is well suited to the interactive study of specimens (particularly biological ones) in their natural state with high spatial resolution while delivering the minimum radiation dose.

The fact that in all materials the soft x-ray scattering cross section is much lower than the absorption cross section has two important

X-ray Microscopy
Ed. by P.C. Cheng and G.J. Jan

consequences. Firstly, the insignificant amount of soft x-ray scattering means that the probe size is not affected by the specimen; hence, specimens up to several micrometers in thickness can be examined in transmission with high resolution. Secondly, the ability to tune the x-ray energy to different elemental absorption edges (especially in the low Z elements) reduces the need to stain the specimen to provide contrast. In particular, over the x-ray energy range known as the water window, between the oxygen K absorption edge (532eV) and the carbon K absorption edge (284eV), the absorption by organic materials is much higher than that of water, allowing wet specimens to be studied.

A scanning system has a number of advantages over imaging (Meyer-Ilse *et al.*, this volume) and contact (Cheng *et al.*, this volume) methods. Firstly, locating the monochromating and focusing elements before the specimen and using an efficient detector ensures that most of the radiation that is transmitted by the specimen is used to form an image - the percentage lost is determined by the detector efficiency and the absorption in the atmosphere between the specimen and the detector. Thus, the radiation dose delivered to the specimen is minimized. Secondly, the image can be formed directly in a digital form, which allows on-line and off-line quantitative analysis of the data and is especially useful for determining chemical composition on a microscopic scale (Kirz, 1980). Finally, real-time data acquisition gives the experimenter interactive control over the image forming process.

The theoretical resolution limit for soft x-rays (2-4nm in wavelength) is about two orders of magnitude smaller than that of visible light (400-600nm in wavelength). At present, the spatial resolution (>100nm) is limited by the optical elements. As the quality of the optics is improved, it is expected that, for biological specimens, the resolution will be limited by radiation damage, as is the case in electron microscopy. However, this limitation is believed to be less severe for x-rays than it is for electrons, especially for unfixed specimens and specimens primarily composed of the light elements (Sayre *et al.*, 1977).

The Concept

The concept of the STXM is simple. A soft x-ray source is focused to a small spot. The specimen is scanned in a two dimensional raster across the optical axis at this focused spot, and the transmitted x-ray flux is recorded. An image is formed which displays the transmissivity of the specimen as a function of position.

Over ten years ago, Horowitz and Howell (1972) reported on experiments using a scanning x-ray microscope. Their instrument did not have any focusing optics; instead, they used pinhole optics which limited spatial resolution to worse than one micrometer. In addition, they used synchrotron radiation with the x-ray energies in the range outside the water window, which reduced the effectiveness of their microscope for examining biological specimens in transmission. This was the state of scanning soft x-ray microscopy until a few years ago when bright sources of soft x-rays and optics capable of focusing soft x-rays with reasonable efficiency and high spatial resolution became available.

Synchrotron radiation is a suitable source of soft x-rays for the STXM. A limited number of bending-magnet radiation beamlines at various electron storage rings have been used as tunable sources of soft x-rays. Now undulators (insertion devices installed on the straight sections of electron storage rings) are being developed that promise to provide soft x-ray sources which are orders of magnitude brighter.

Unconventional optical elements are needed to focus soft x-rays. The refractive index of materials for x-rays is very nearly unity; no refractive optics can be used and conventional reflective optics can be used only at grazing incidence so that very high figure and surface tolerances are required. So far, the result has been expensive reflective optical elements of only moderate resolution. However, high resolution, normal incidence, diffractive optical elements have been developed. These are Fresnel zone plates (Hecht and Zajac, 1976). The Fresnel zone plate, as in the case of a linear diffraction grating, throws radiation into various positive (real) and negative (virtual) orders as well as passing a large amount of unfocused, zero order radiation. For a perfect Fresnel zone plate the ratio of zero order to positive first order radiation is 2.5:1 and the most efficiently focused radiation is the positive first order. Another type of near normal incidence soft x-ray optical element, the multilayer mirror, is being developed by a variety of groups (E. Spiller at IBM, NY; J. Underwood at Center for X-ray Optics, Berkeley, CA; T. Barbee at Lawerence Livermore Nat. Lab., also see Spiller, 1984), and may be used in future x-ray microscopes.

The Design

The STXM requires that we overcome some peculiar design challenges: to bring soft x-rays from a UHV environment to atmosphere; to focus the soft x-rays; to align accurately several small objects over relatively large distances; to control the instrument; to acquire the image data; and to display the image in real-time. The instrument can be considered in separate sections: the source of x-rays, the vacuum system, the alignment system, the optics, and the control system.

The source of soft x-rays will be the U5 beamline at the SRS (Duke, this volume). At U5 the undulator radiation is tuned by adjusting the gap between the pole pieces with further monochromation provided by an SX700 design monochromator (Petersen, 1982). Together, the undulator and the monochromator are designed to provide 200-2000eV x-rays with a bandwidth of 0.1-0.2%. Adjusted independently in height and width from about 10μm to 5mm, the exit slit of the monochromator acts as the x-ray source for our focusing system. At present, this instrumentation is being characterized.

The mechanical configuration of the STXM is shown in Figure 5.1. The ultra high vacuum (UHV) environment of the SRS and our monochromator is preserved by a differential pumping system, designed so that a vacuum of only 10-5 to 10-6 torr is required in the chamber downstream (the direction in which the x-rays travel) of the monochromator exit slit. This moderate vacuum requirement should allow rapid evacuation and the use of non-UHV compatible materials (such as epoxy).

We intend to align the chamber with respect to the x-ray beam emerging from the exit slit with the aid of x-ray sensitive phosphor screens and a transmission x-ray photodiode. This photodiode will be used to monitor the flux emitted from the exit slit so that the detected intensity transmitted by the specimen can be normalized to remove the effects of beam intensity fluctuations from the final image.

The x-rays leave the vacuum chamber through a thin silicon nitride window, approximately 100nm thick and 300μm square. This window separates the high vacuum environment of our chamber from the atmospheric environment of the specimen, and transmits about 70% of 400eV x-rays.

The zone plate is supported on a similar silicon nitride window which is rigidly fixed with respect to the exit window about 200μm downstream (see Figure 5.2). Master zone plates have been fabricated in carbon at King's College using a specially modified scanning transmission electron microscope (Buckley *et al.*, 1985). These carbon zone plates are used as x-ray lithographic masks to make gold zone-plate replicas which focus with much higher efficiency (Buckley *et al.*, 1986). With the present replication scheme we have only been able to produce gold zone plates with outer zone-widths down to 75nm (Buckley *et al.*, this volume), even though carbon zone plates with much finer zone-widths have been made.

The first order focus of the zone plate will be used as the x-ray probe across which the specimen is to be scanned. The other orders

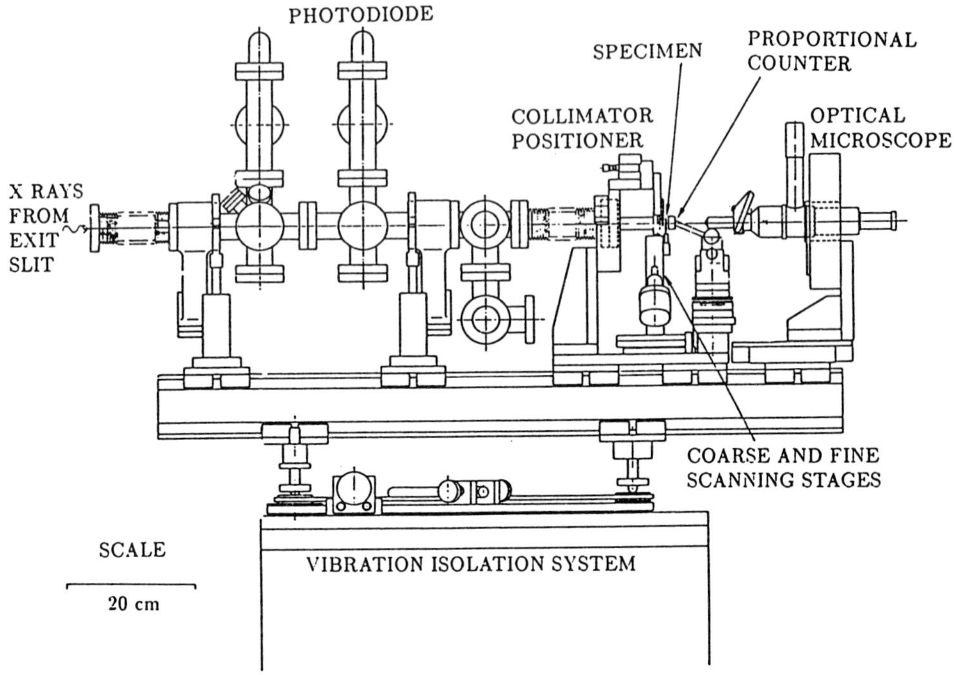

Figure 5.1. Scale drawing of the STXM showing the vacuum system between the exit slit of the monochromator and the zone plate, the mechanical support and alignment system, and the components of the STXM.

do not come to a focus at the first order focus point and, as the detector has no position resolving capability, these other orders, especially the zero order, need to be removed before they reach the specimen. In this way, the signal-to-noise ratio (S/N) can be improved and the x-ray dose delivered to the specimen is minimized. A central stop (blocking about 20% of the collection area) is applied to the zone plate and in the shadow of this stop a pinhole collimator is positioned downstream of the zone plate and upstream of the first order focus, reducing the unwanted background radiation and providing several

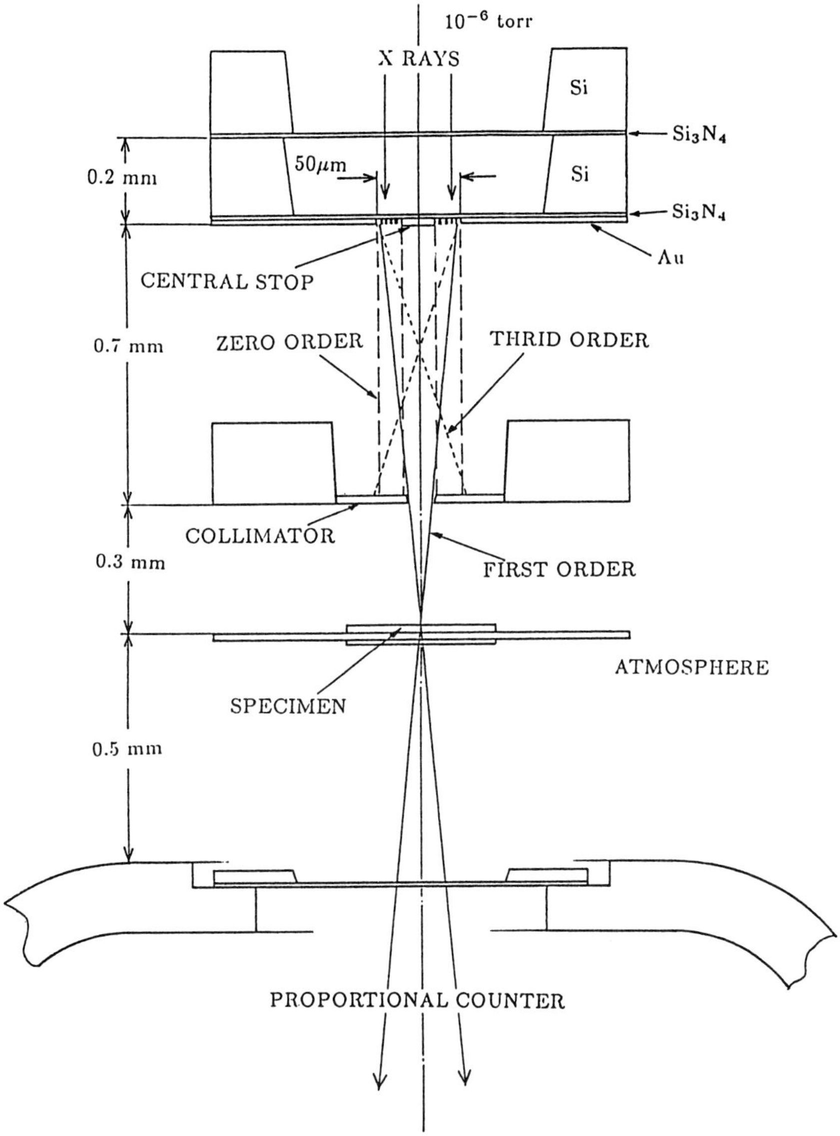

Figure 5.2. Diagram of the relative positions of the Si_3N_4 vacuum barrier, zone plate, collimator, specimen, and proportional counter. Note how the central stop of the zone plate and the collimator are arranged so as to pass the focused first order radiation to the specimen, while blocking the zero order radiation and most of the higher order radiation.

hundred microns of working distance. The zone plate will be placed about one meter downstream of the monochromator exit slit so that it would form a probe which is a 1:1000 demagnified image of the exit slit.

The specimen is mounted on a fine stage which scans in the plane perpendicular to the optical axis using a pair of piezo-electric positioners and a pair of inductive position transducers coupled together through an analogue feedback circuit. The fine stage is mounted on a stepper motor driven coarse stage which is used to bring the specimen into focus and to locate those areas of interest in the specimen which are subsequently to be scanned by the fine stage.

There are two methods for specimen mounting: dry and wet. For dry mounting, the specimens can be either sections supported on open mesh grids, as with electron microscopy, or whole specimens supported by a thin foil, for example collodion. For wet mounting the specimens are held in an aqueous environment by a cell which restricts water loss, but allows the x-rays to pass through the specimen and into the detector. This cell is made by by using two thin foils, such as boron nitride, to sandwich the specimen.

Downstream of the specimen is the gas flow proportional counter used to detect the soft x-rays which are transmitted by the specimen. This detector and its associated electronics provide separate voltage pulses for each photon which converts in the body of the counter. The volume between the exit window and the detector can be flushed with helium, thus improving the transmission of 400eV x-rays by about a factor of two.

Stationed downstream of the detector, a long working distance optical microscope will be used for roughly aligning the zone plate, collimator, specimen and detector. It can also be used, in conjunction with an interferometer attached to the objective, to bring the specimen to the x-ray focus. This reduces the time spent in focusing the specimen; and, because visible light is used to focus the specimen instead of x-rays, the total radiation dose delivered to the specimen is significantly reduced

The STXM is controlled by a BBC micro-computer coupled to an IEEE interface system which communicates with the scanning stages, the detectors, and the image display. Figure 5.3 shows a block diagram of the control system. This system uses a menu-driven program to interact with the experimenter. The experimenter can control the positioning of the stages, select the scan conditions, and start and interrupt the scan through a video terminal. The system controls the operation of the microscope in real time to generate, store and display a two dimensional image of the x-ray transmission of the specimen. These stored images may be re-displayed, processed and analyzed off-line.

Even though we have not been able to use (as of August 1st, 1986) the U5 beamline as a source of x-rays, we have tested various components of the STXM system. The photodiode and the proportional counter have been used to detect aluminium K line radiation (1.5keV). The optical microscope alignment and focusing system has been checked for accuracy and reliability down to 1um. The scanning stages have been checked and calibrated using independent measuring systems. The coarse stage has a range of 3mm and a position resolution $<1\mu$m. The fine stage has a range of about 40μm, an expected positioning accuracy of 10-20nm and a settling time of 1-2ms. These characteristics are more than adequate given that we expect 10-100ms dwell per picture element and a resolution limit $>$ 75nm.

The Proposed Experimental Programme

The immediate experimental programme has two major objectives: to characterize the performance of the instrument and to demonstrate its capabilities using well known, non-biological and biological

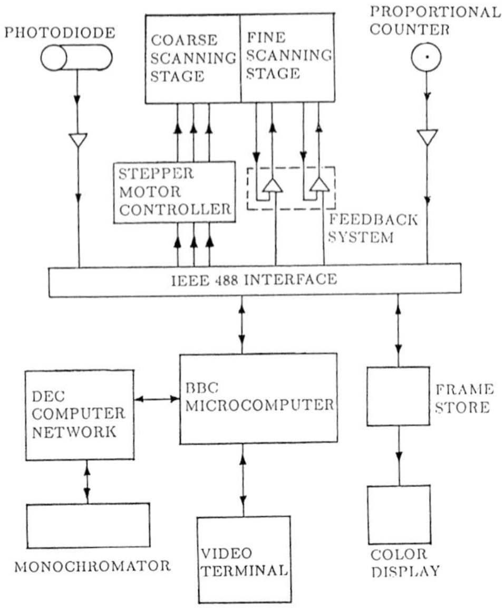

Figure 5.3. Block diagram of the STXM control system.

specimens with the aim of determining how the STXM should be used as an instrument for research in biology and materials science.

The performance characteristics of the STXM which we intend to measure are throughput, resolving power and S/N. The throughput depends on x-ray source characteristics, the monochromator performance, the zone plate focusing efficiency, the composition of the specimen, and the desired level of detail to be observed in the specimen. To have a single estimate of the throughput of the instrument we will neglect the effect of the specimen and instead we simply measure the number of photons per second in the diffraction-limited focus. This is observed by adjusting the size of the source (exit slit), such that its demagnified image is equal to the diffraction limit of the zone plate, then measuring the count rate with the specimen removed. Under these conditions, with the present highest resolution gold zone plates, we expect 10^4 to 10^6 photons per second in the focus, depending on the undulator and monochromator performance. We estimate that we should be able to examine several specimens per day.

The resolving power of the STXM depends on the scanning stage, the temporal coherence of the monochromator, the size of the demagnified image of the source, and the zone plate. Of these, the zone plate is expected to be the limiting factor. Using the Rayleigh criterion the first order, diffraction limited, spot size (σ) for a perfect Fresnel zone plate is given by:

$$\sigma = 1.22 \times \text{the outermost zone width} \tag{5.1}$$

For the zone plates we expect to use the resolution limit should be 80-100nm.

Resolution is not a single, precise number, and is determined not only by the optical system, but also by the method of characterization and the choice of specimen. A simple measure of the line spread function of our probe is made by scanning a knife edge across the focus spot so that we measure the count rate as a function of position. We also plan to image biological specimens containing repeat structures. Fourier analysis of these images will contain information on the probe size, giving a measure of the resolution in an image of a biological specimen.

The S/N is a function of the integrated flux detected, the inherent contrast of the specimen, and the amount of background radiation (those x-rays which pass through the collimator, but do not come to the first order focus). The background radiation intensity distribution can be measured by scanning a small (1μm diameter) pinhole in the focal plane. However, the total S/N will need to be determined for each image, and can be improved by having longer scan times and by proper preparation of the specimen to maximize the contrast.

We wish to exploit the ability of the microscope to image wet specimens. We will study specimens with simple, robust structures, which are of interest at sub-optical resolution, such as hydrated cement grains. We hope also to detect the presence of calcium in spores (both hydrated and dry) by taking advantage of the tunability of the x-ray energy and sharp spectral features at the calcium $L_{II,III}$ edges (near 350eV). (eds.' note: please refer to Cinotti *et al.* in this volume)

These first experiments will help to determine the direction of our future research in x-ray microscopy. We will have one month (September 1986) of beam time at the U5 beam line, after which the SRS will shut down for at least six months for installation of a high brightness lattice. During this shutdown we will analyze our results and plan for improved x-ray microscopy experiments in the Summer of 1987.

Acknowledgements

This research is supported by the SERC. We would like to acknowledge Rigaku Corp. for their generous financial and technical assistance. We would also like to thank: the director and staff of the SRS for implementation of the U5 beamline; the Physics technical and machine shop staff of King's College, especially T. Kuriyama, A. Freake, and W. Luckhurst for the engineering, designing and machining we required for the STXM; J. Kirz and his collaborators at Stony Brook Univ. and Brookhaven National Lab. for co-operation in earlier exploratory work (in particular zone plate replication) on the U15 beamline at the NSLS; and R. Feder and his colleagues at IBM, Yorktown Heights (especially for providing silicon nitride membranes).

6. Recent Advances in Contact Imaging of Biological Materials

P. C. Cheng[1], D. M. Shinozaki[2] and K. H. Tan[3]

[1]IBM T. J. Watson Research Center, Yorktown Heights, NY 10598 USA

[2]Faculty of Engineering, University of Western Ontario, London, Ontario, Canada N6A 5B9

[3]Canadian Synchrotron Radiation Facility, University of Wisconsin, Stoughton, WI 53589 USA

Introduction

X-ray contact imaging (microradiography) was pioneered by Goby (1913a; 1913b; 1913c; 1925) in the study of the structures of plant tissue, insects and foraminifera. He also developed a method of obtaining three dimensional views from stereo-pairs of thick specimens (Goby, 1925). Later, Dauvillier (1927; 1930) and Lamarque (1936) continued such work with other biological specimens. Using high resolution emulsions (Lippmann type, with silver grain size less than 1μm), Dauvillier (1927; 1930) obtained contact micrographs with magnifications as high as 600X on some histological sections. He also observed that the contrast improved as the x-ray wavelength increased. Engström (1946) and his associates used silver halide photographic emulsions as the recording material to study a variety biological samples. However, the resolution was limited by the size of the silver grains, and the study of subcellular structure was therefore not possible.

Different materials to record the image have been used in the search for higher resolution. For example, x-irradiation of alkali

X-ray Microscopy
Ed. by P.C. Cheng and G.J. Jan

halides such as NaCl or CsBr produces color centers which change the local electronic structure in the crystal. This radiation damage manifests itself as a change in the optical absorption spectrum of the crystal. Pattee (1960) has used this phenomenon to record x-ray images (with radiation from a Cr target stationary source) with a resolution of 1500 bars / inch. Other recording media include photoreactive dye derivatives (for example, pararosanaline leuconitrile) (Pattee, 1960) and polymer films.

The use of Saran film (vinylidene chloride) as an image storing material (Pattee, 1960) preceded the perceived need for such grainless "emulsions" in the microelectronics industry. The irradiation induced molecular changes (crosslinking or chain scission) produce local solubility changes in the polymer. The differential dissolution of the Saran in NH_4OH produced the image. More recent developments in the microelectronics industry have resulted in the use of polymethylmethacrylate (PMMA) as a material with particularly high resolution. However the the sensitivity of such photoresists is much less than that of the silver bromide film ($X10^{-4}$), and the net gains in resolution are balanced somewhat by the need for high intensities of soft x-rays. The concurrent developments in high intensity sources (dedicated synchrotron radiation sources in the optimum energy range; pulsed plasma sources, both laser generated and constrained gas plasma sources) have resulted in a renewed interest in the application of microradiography using soft x-rays for the study of biological materials.

The fundamental advantages of soft x-rays over electrons in the examination of microstructures have been described in detail (Spiller *et al.*, 1976; Sayre *et al.*, 1977; Feder *et al.*, 1977; Spiller and Feder, 1978; McGowan *et al.*, 1979; Manuelidis *et al.*, 1980). The methods of contact imaging, and some of the recent results demonstrating its efficacy in a variety of applications have been well documented

(McGowan *et al.,* 1979; Cheng *et al.,*, 1984; 1985; 1985; Shinozaki *et al.,* 1986).

There are two essential stages to the introduction of a novel method of microscopy: the initial simple demonstration of the experimental feasibility of the method and the subsequent demonstration that some questions in biology can be addressed uniquely with the novel method. The first stage has been accomplished, and the second stage now requires a larger number of diverse examples which begin to illustrate applications of images made with soft x-rays. The purpose of the present paper is to provide such further examples of the study of microstructure in biological materials using soft x-ray contact imaging, and to emphasize some peculiar advantages in the use of soft x-rays which have already been discussed in the literature.

Selected Experimental Problems in Contact Imaging

The modern techniques of contact imaging using soft x-rays are derived from microradiography described by Cosslett and Nixon (1960). The specimen is placed in close contact with an x-ray sensitive polymer resist and the sandwich irradiated by soft x- rays (Figure 6.1). After exposure, the resist is developed in a appropriate solvent. For a positive resist, the more heavily irradiated areas of the resist have a lower molecular weight and dissolve more readily, and thus the resultant image is map of the relative x-ray opacity of the different regions of the specimen. The topography of the developed resist surface is examined in a light microscope (LM), a scanning electron microscope (SEM), or a transmission electron microscope (TEM), and it is at this stage that the image is magnified. The information in the image is essentially contained in the topography of the exposed and developed resist surface and the resolution of the method depends on the resist, the irradiation conditions and the method of examination of the resist.

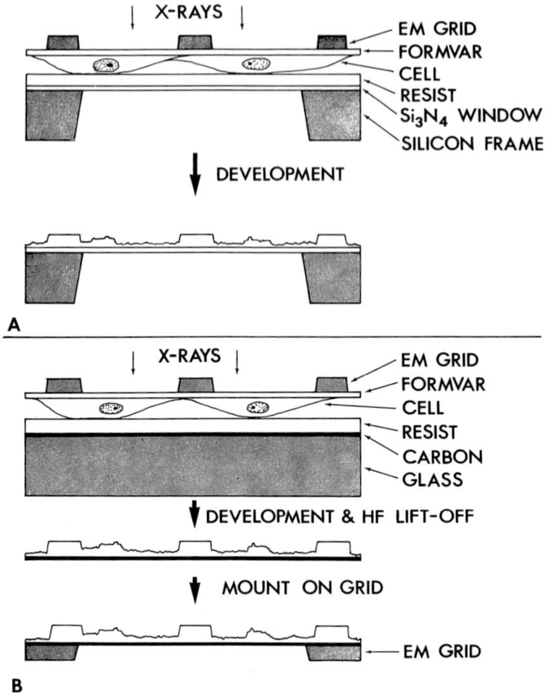

Figure 6.1. Typical experimental arrangements for contact imaging. A. Standard exposure set-up. The cell-bearing EM grid is placed on top of a resist-coated window; after exposure, the resist is developed and viewed in a TEM.

B. Fabrication of a self-supporting PMMA film for TEM observation. The x-ray resist was coated on a carbon-coated glass wafer. After exposure, the resist was developed and lift-off from the supporting glass wafer by using a 2% HF solution.

The techniques of contact imaging have been described in a number of other papers, and in the following sections some important details which facilitate the experiment will be discussed. The experimental examples used here have involved primarily the stationary source (a 5 to 10 keV electron beam incident on a carbon, aluminum, titanium or other solid surface), and the synchrotron radiation sources at the University of Wisconsin (Canadian Synchrotron Radiation Facility on the 240 MeV ring and lately on the 800 MeV ring) and at Brookhaven (U15, courtesy of Kirz and co-workers, on the 740 MeV ring). Preliminary experiments have also been run on pulsed plasma sources (both laser generated and pinched gas plasmas) at the Rutherford laboratory (U.K.), the University of Rochester, Laboratory for Laser Energetics, and at the T.J. Watson Research Center of IBM. The observations and conclusions are therefore based on a wide variety of experiments.

Preparation of the X-ray Resist

Polymethylmethacrylate (PMMA) and polymethylmethacrylate-methacrylic acid co-polymer (CoP(MMA-MAA)) have been widely used in the present experiments. Both resists are dissolved in chlorobenzene (*e.g.*, 6% solution). A drop of the solution is then placed on the center of a Si_3N_4 window wafer, a silicon wafer or a glass wafer (Figure 6.2) and spun at high speed (typically at 5000 RPM) for 60 seconds. A standard resist spinner used for microelectronics fabrication (Figure 6.3a) is needed to form the uniform thin film of resist on the surface of the window/wafer. The thickness of the resist can be controlled by the changing the viscosity of the resist solution (by adjusting the concentration of polymer in the solvent), and by varying the speed of the spinner. A typical thickness-speed- concentration curve is shown in Figure 6.3b. Such calibration curves are best determined empirically for each batch. After spinning, the resist coated window/wafer is placed in a 150°C oven for 1 hour. The PMMA resist used for contact imaging typically has a thickness in the range from $0.4\mu m$ to $1.2\mu m$.

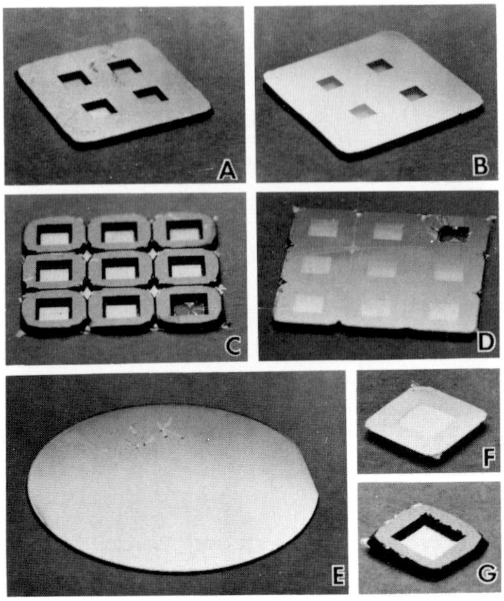

Figure 6.2. Various window material and resist-supporting substrata.(A) and (B): Large Si_3N_4 windows, this window is used for specimen supports. (A): Wafer side, (B): Window side. (C) and (D): Si_3N_4 windows used for TEM observation. note: one of the windows has been broken. (C): wafer side, (D): resist side. The window-bank can be broken into individual pieces as shown in (E) and (F). (G): resist-coated silicon wafer.

Specimen Holders for Laboratory Stationary Sources and for Synchrotron Radiation Sources

Specially made specimen holders are required for holding the x-ray resist-bearing window/wafer in intimate contact with the specimen without breaking the window or damaging the specimen. The contact is necessary to reduce the blurring due to source size and exposure geometry (penumbral effects), and to minimize the extent of the diffraction around the opaque edges of the microstructure.

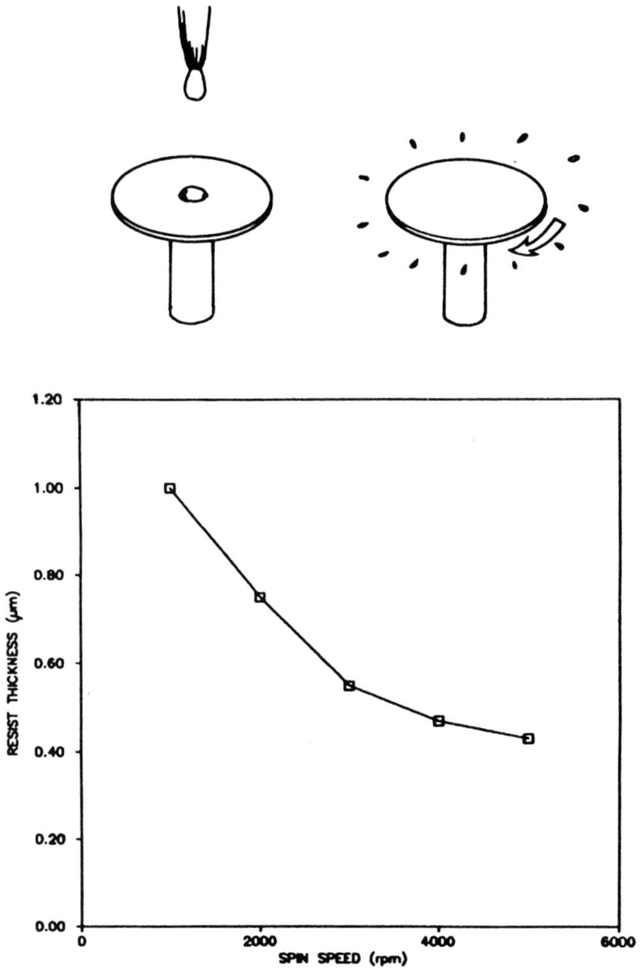

Figure 6.3. (a): Methods for preparing thin PMMA film on silicon wafers or on windows. A drop of PMMA solution (in chlorobenzene) is dropped on the center of a wafer or banks of windows, and the wafer is spun at high speed for a period of time. A thin uniform layer of resist is thus formed. The thickness of the resist coating depends on the concentration of the solution and the speed of the spinner.

(b): Typical resist thickness curve. The curve is used to serve as a reference guide during resist coating process. The actual thickness of the coated resist may be measured with an optical interferometer attached to a microscope.

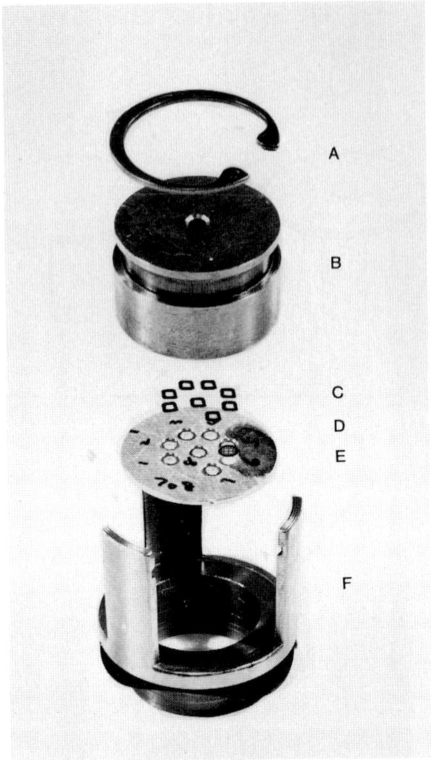

Figure 6.4. (a): Window holding device used for stationary target x- ray source. The holder consists of a base piece (F), the specimen holding plate (E) with eight holes to hold electron microscopy grids (D) and resist-coated windows (C), the spring- loaded pressure plate (B), and a locking spring (A).

The use of very thin brittle Si_3N_4 windows to support the resist or the specimen makes handling a very delicate chore. Although quite large uniform elastic loads can be carried by such membranes, point loads which can easily be generated by very small dust particles or in-homogeneous specimens can cause cracks to propagate through the window. The presence of such cracks in the resist substrate makes uniform developing by the solvent impossible. These observations suggest that the mounting of the specimen and resist should be done in a relatively clean dust- free environment and the specimen itself should be as flat and uniform as possible. Improper handling of the specimen (usually mounted on a TEM grid) results in a rumpled specimen which cannot be held in intimate contact with the resist.

The specimen holders also have to be compatible with the specimen chamber of the x-ray source used. A typical holder for the stationary target source was shown in Figure 6.4. The holder is constructed with a base piece (F), the specimen holding plate (E) containing eight holes to hold electron microscopy grids (D) and resist-coated windows (C), the spring-loaded pressure plate (B), and a locking spring (A). By means of the spring pressure, this device provides excellent specimen-resist contact.

The specimen holder for the synchrotron radiation source was quite different. It was necessary to design a high vacuum compatible holder which can be handled easily with gloves and which has no blind holes or other design features which increase time to pump the chamber with specimen inserted to the required pressure. Typical times to reach pressures of the order of 10^{-5} torr or lower with the holder in Figure 6.5 were in the range of tens of minutes. This simple holder fits onto a 2-inch linear feedthrough on a standard 2 3/4" ultra-high vacuum (UHV) flange.

Areas of P-31 phosphor (Figure 6.5) can be painted on the specimen holder to visually align the x-ray beam with the samples on the holder. For more precise positioning of the sample in the incident beam, a small alignment hole can be machined into a measured position next to the specimen itself, and a detector placed behind the aperture used to measure the position of the center of the beam. Micrometer drives on a the specimen manipulator can be used to then move the specimen itself into the exact position (relative to the center of the beam) desired. This is necessary for quantitative imaging since the incident beam from the monochromator in the beamline has a non-uniform intensity in cross-section. Indeed, in the beamlines used

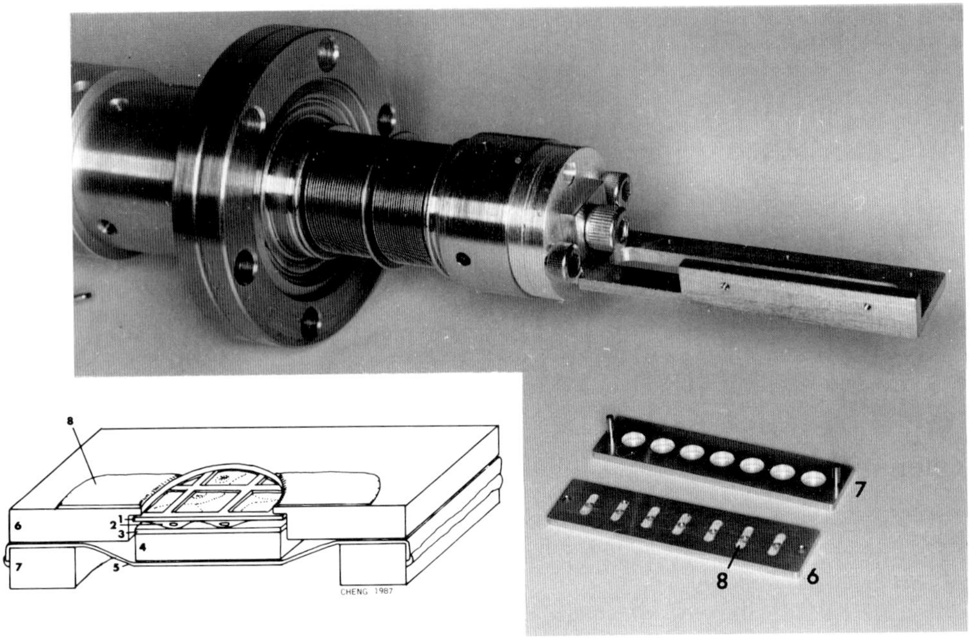

Figure 6.5. Window holding device used for synchrotron radiation at the Canadian Synchrotron Radiation Facility. The specimen (2)(on a EM grid (1)) is kept in intimate contact with the resist-coated wafer (4) by a tight plastic film (5). The whole assembly is kept together by the upper (6) and lower (7) plates. The tight plastic film is glued onto the lower plate on the edges of the plate. Strips of P-31 phosphor (8) are painted on the upper plate for proper alignment of the beam to the specimen. (3): photo resist.

in these cases, the overall vertical extent of the beam is significantly less than the 3mm diameter of the specimen TEM grid.

The ease of handling the specimen mounting depends on the vacuum requirements. At U-15 beamline of Brookhaven, the usual relatively grease-free surfaces are necessary but for a windowless beamline such as CSRF at Wisconsin, where pressures of the order of 10^{-9} torr are necessary, more care in handling is necessary to prevent contamination of the beamline optical elements. In such a case, the total polymer and specimen mass should be kept as low as possible to allow the pressure to reach the required pressure range rapidly.

Qualitative Examination of Resist Development

Following the x-ray irradiation, the resist is developed in an appropriate solvent and dried with dry N_2 gas. For PMMA, methyl isobutylketone (MIBK), or solutions of MIBK with isopropanol (IPA) are used. MIBK is a relatively good solvent while IPA is a poor solvent for PMMA. A mixture of the two is used, the proportions of the two determining the dissolution rate (*e.g.*, a 1:1 v/v mixture). For the development of PMMA-MAA co-polymer resist, ethyl cellosolve acetate (ECA) and MIBK can be used. Since the rate of development also depends on the local molecular weight of the irradiated PMMA, for a given exposure system empirical development curves of development rate *vs.* dose can be measured and used to estimate the depth of development for a given exposure using a given developer solution. Figure 6.6 shows an apparatus designed to achieve reproducible development conditions.

With some experience, the observation of the optical microscope image of the developed resist surface can be used to evaluate the quality of the image. The low contrast optical images are best examined using phase contrast microscopy. The progressive development of the image can then be monitored by starting the dissolution using a light development. It is necessary to combine these optical observations with later electron microscopy of the resist surface to obtain an empirical understanding of the relationship between the low resolution image with the high resolution image finally desired.

For those resists which had cells grown directly on their surface, the cells should be first removed by means of a 1% Chlorox solution (sodium hypochlorite 5.25% w/w) before development. It should be emphasized that thin biological specimens are very often difficult to distinguish from the developed topographic image in the resist surface. This obviously has significant implications for the subsequent electron microscopic examination of the resist. In the absence of other experience, control experiments to ensure the complete removal of the original biological specimen, or to allow detection of the remnants on the

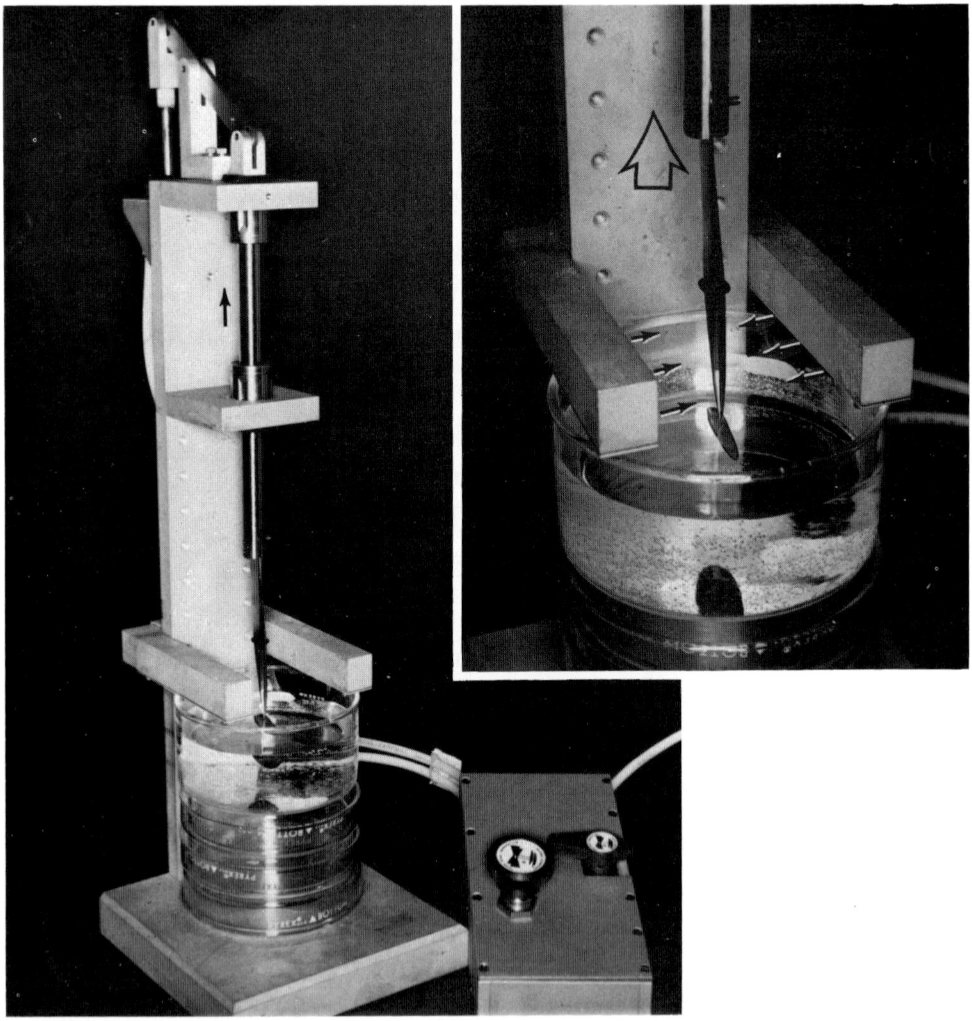

Figure 6.6. Apparatus for developing contact image on photo resist. Drying of the resist surface is achieved by the two flat nitrogen nozzles on both side of the vertical rod.

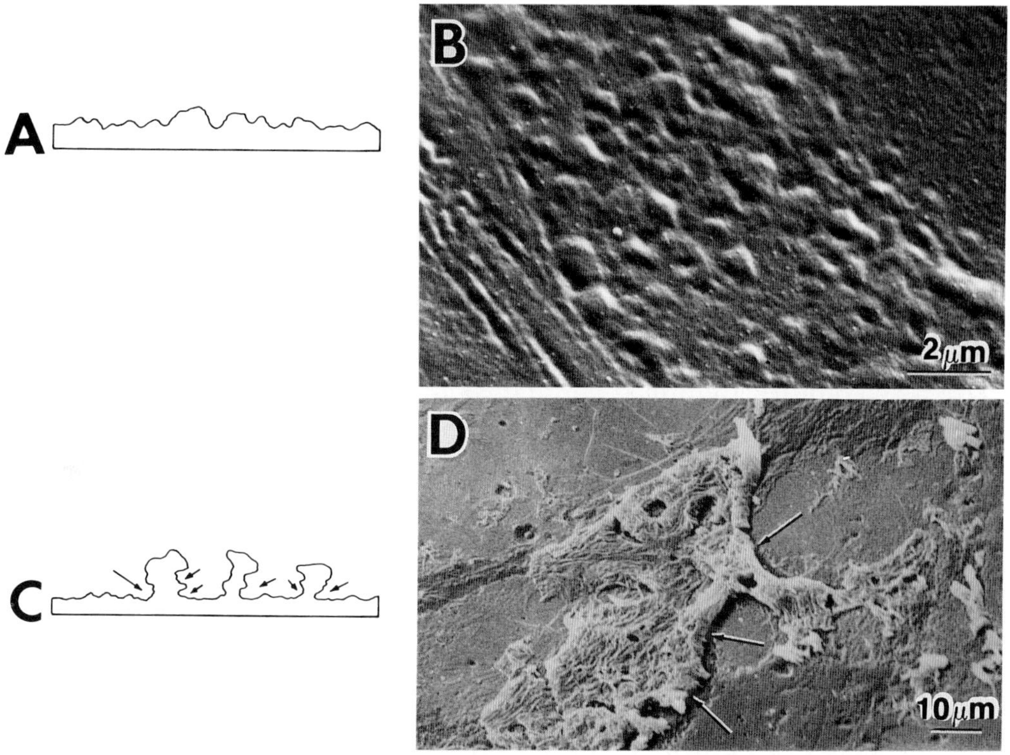

Figure 6.7. One of the methods to identify real contact image from stuck biological samples. (A): Diagrammatic representation of the cross-section view of a real contact image. (B): SEM magnified contact image. (C): Diagrammatic representation of a resist surface with stuck samples. Note: frequently, the biological specimen stuck on the resist shows complex 3D structures. (D): Typical example shows cell remnants on the resist surface.

resist surface must be included. This is partially important when attempt on imaging macromolecules, because it is very difficult to separate real image and stuck specimens. Figure 6.7 shows one of the methods to identify real contact image from stuck biological samples. The stuck specimens often show complex 3D structures (Figure 6.7, arrows), in contrast, the contact image on the resist surface should not have any undercut structure.

The problem of biological specimens adhering strongly to the polymer surface is not surprising in the light of observations that polymers can be made to adhere strongly to other materials under irradiation of the interface. The radiation damage of the hydrocarbon polymer can result in chemical modifications which enhance interface bonding. In these cases, it is often impossible to separate the biological materials from the resist by mechanical means.

High Resolution Examination of the Resist Image

Since the magnification of the soft x-ray image is accomplished in the final examination stage of the process, the absolute resolution of the contact method is strongly dependent on the electron microscopic methods used. A detailed discussion of the this is found in another paper in this volume (Shinozaki and Robertson). The present paper will describe selected practical aspects of the problem.

The study of polymers in the electron microscope (SEM and TEM) is always limited by: (1) the radiation damage induced distortion of the surface, and (2) the electrical charging of the inherently non-conductive material. The manifestation of the latter is seen particularly at high magnifications as image instability and complete loss of contrast of the observed image. This results from the distortion in the electric field near the non-conducting material as a result of a build-up of charge. The field strongly affects the incident and scattered electron beams used to image the resist. At high magnifications, this becomes even more serious since the electron dose required to observe smaller structures is higher.

The surface of the resist is therefore coated with a conducting metal layer, which is kept thin (of the order of 20nm) to minimize the obscuring of the fine surface relief. In general, both TEM and SEM specimens should be coated to prevent charging artifacts. The resist surface must be coated uniformly, and for this reason, sputter coating

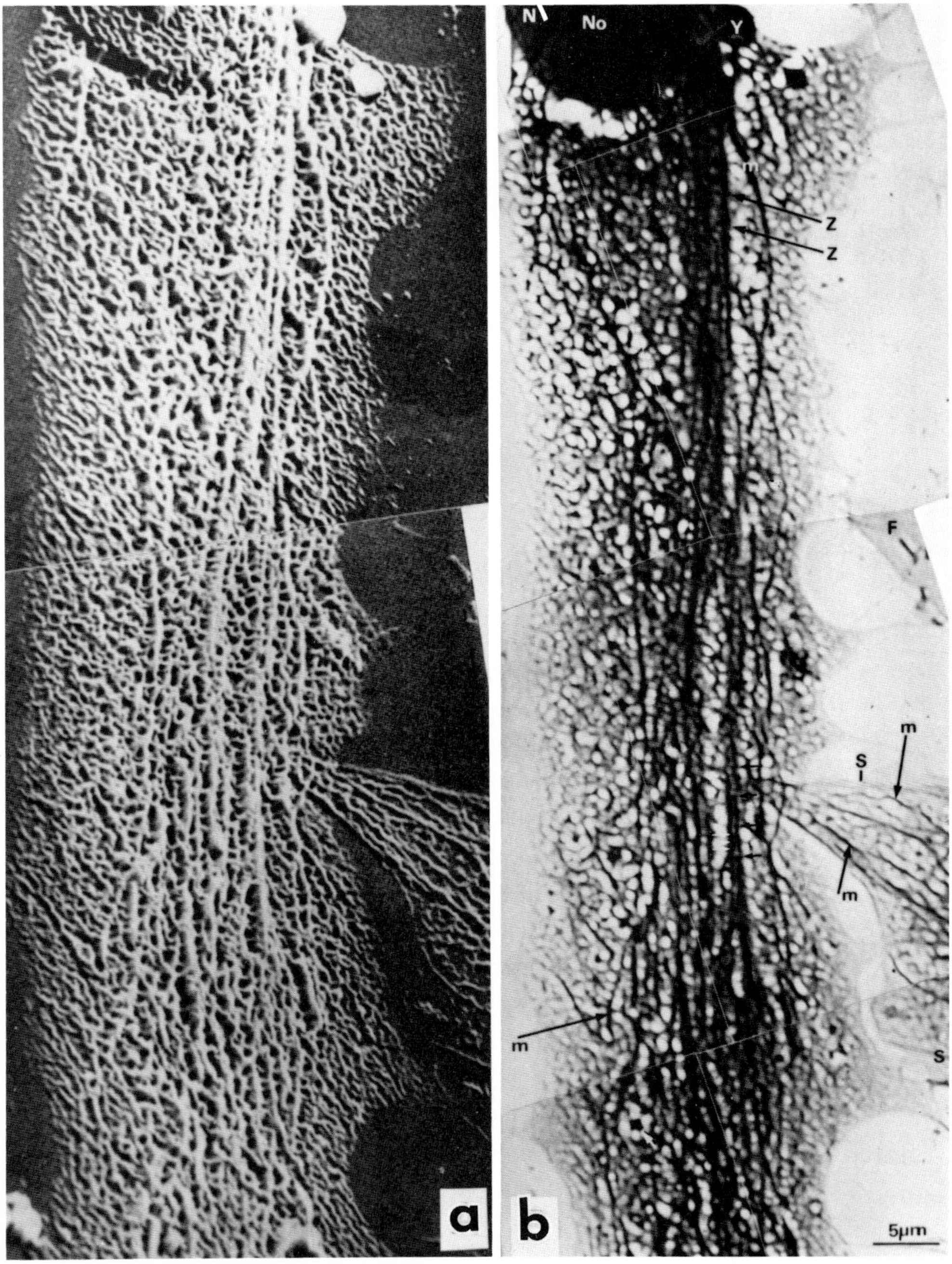

Figure 6.8. SEM (a) and TEM (b) magnified x-ray contact images of *Xenopus laevis* myotomal cell.

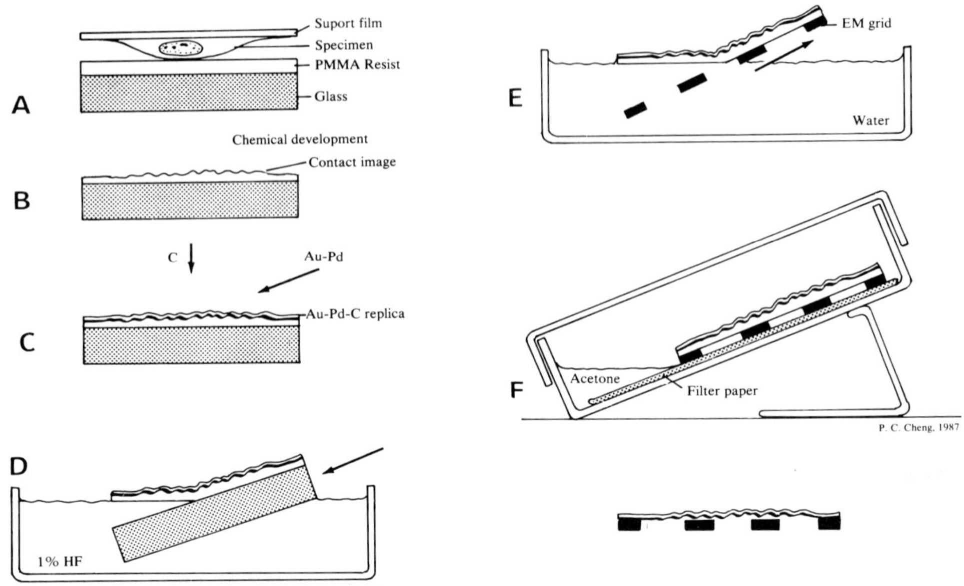

Figure 6.9. Surface replica obtained from the contact image provides the highest attainable resolution. A thin layer (~50Å) of Au-Pd is evaporated at a low angle (~10°), and a thicker layer (~500Å) of carbon followed at normal incidence. Then, the resist is lift off from the glass substrate with 1% HF solution, and the resist-metal-carbon sandwich is picked up on a 100 mash EM grid. The grid is then place on a filter paper wetted with acetone for 24hr. Care must be taken to ensure that the acetone does not dry out. Then the grid is carefully removed from the filter paper and readily for TEM examination.

is often used. If vacuum evaporation is used, the specimen should be rotated so all surfaces are coated. In all methods of coating, care should be taken to prevent overheating the resist surface during coating.

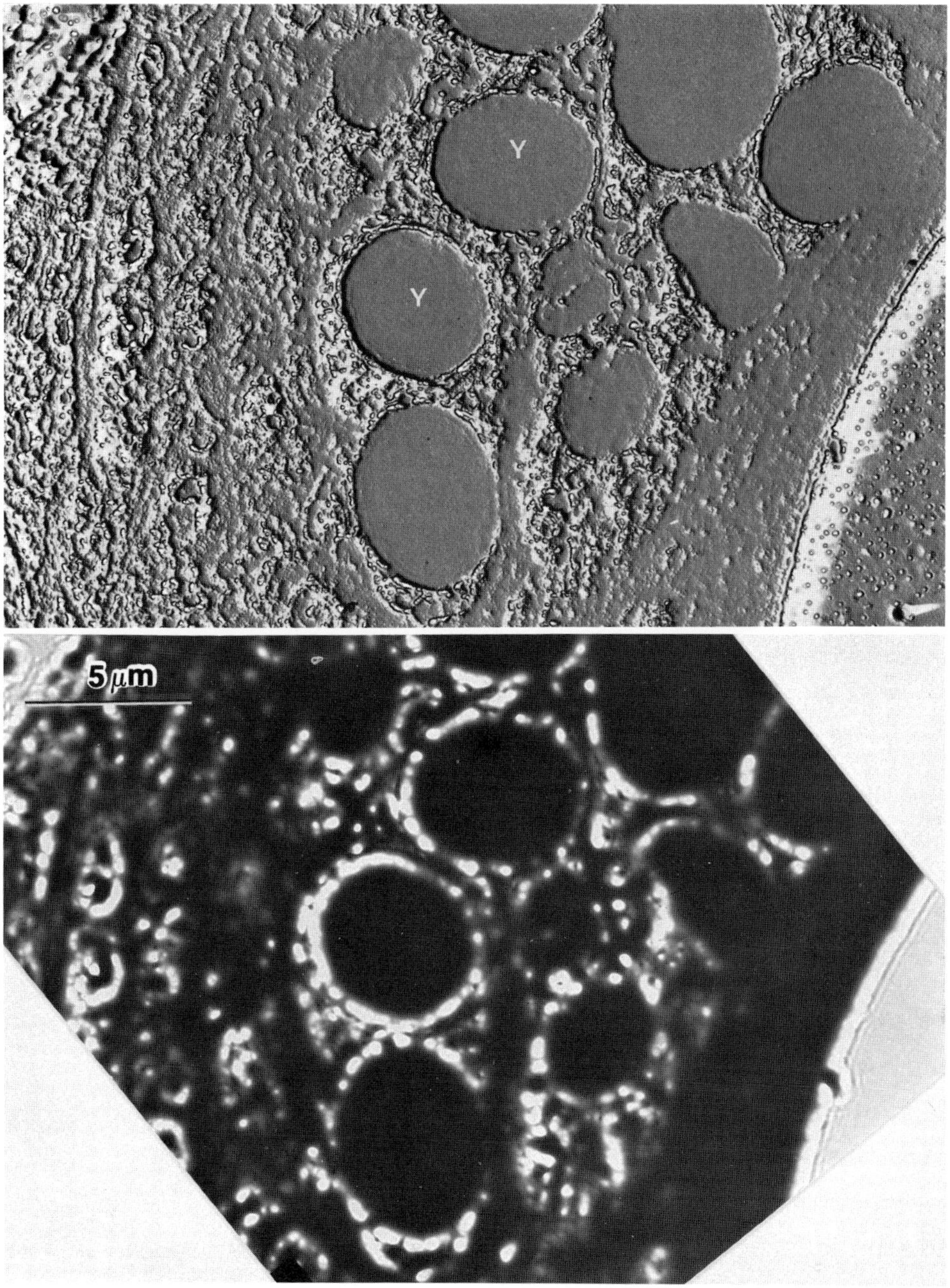

Figure 6.10. Cytoplasm of *Xenopus laevis* myotomal cell. (a): Contact image obtained by surface replica technique. (MIBK/IPA, 1:1 v/v, 45 sec.). (b): Contact image directly magnified by transmission electron microscopy. (MIBK/IPA, 1:1, 5 sec.). Y: york granules, S: stress fiber bundle.

If the resist is spun onto a thick silicon wafer or glass cover slip, the coated surface can be examined in a SEM directly (Figure 6.8). An electrically conducting path to ground should be attached to the metal coated surface (for example with silver paint) to allow charge to bleed off during examination. For typical exposures and developments, the surface relief will be relatively light and the contrast in the SEM (using collected secondary electrons) will be low. The tilt of the specimen (which determines the signal collected), and the gain and contrast of the signal should be optimized to observe the low contrast image. The radiation damage artifacts should be noted particularly at high magnification before image interpretation is attempted.

The observed image is not a direct shadowgraph of the original specimen. The bright/dark contrast in the SEM image is a relative measure of the secondary electron signal strength and not a direct measure of the local thickness of the resist. For example, differences in tilt of the surface of the resist with respect to the incident electron beam will produce differences in signal strength. Hence the observed contrast of the same structure in the surface will depend on the orientation of the surface tilt axis with respect to the electron beam. Hence the SEM image brightness is not necessarily an accurate record of the resist surface height.

Transmission electron microscopy provides a method of mapping the thickness of the resist directly. The resist can be supported on a thin Si_3N_4 membrane, and after exposure and development, the point to point thickness of the resist plus substrate is a map of the absorbed x-ray dose. The resist can be examined in an orientation exactly normal to the electron beam, and the thickness contrast (the scattered electron intensity is related to the mass thickness) can in principle be related to the mass thickness of the resist. There are a number of examples of TEM examination of soft x-ray images stored in resist surfaces (Feder *et al.*, 1981; Cheng *et al.*, 1982; 1984; 1985; Cheng, 1984) (6.7b), but direct observation of the resist , even when coated with a conducting layer, inevitably results in electron beam damage to the resist. The high resolution examination of the resist image is therefore fundamentally limited by this.

Two of the problems which are presently being addressed in contact imaging are (1) the accessibility of quantitative information from the image and (2) the recording of small, low contrast structures. The first is best approached through the direct measurement of the point by point thickness of the resist using normal incidence transmission electron microscopy or scanning transmission electron microscopy (using for example, the annular dark field signal as described in the accompanying paper by Shinozaki and Robertson). The second problem is best approached by using a metal shadowed carbon replica of the resist surface. Such replicas are stable in the electron beam and with care can be made with lateral resolutions limited by the grain size of the shadowing metal. In addition, the light surface relief inherent in a lightly developed (high resolution) resist image can be artificially enhanced by shadowing with the metal at a shallow angle with respect to the resist surface. The enhanced contrast and the insensitivity to the electron irradiation in the microscope means that the absolute resolution is not radiation damage limited as is the case in the direct observation of polymer surfaces. A more detailed discussion of this problem of high resolution examination of radiation sensitive polymers is given in the accompanying paper.

One procedure for making metal shadowed carbon replicas is shown in Figure 6.9. Figure 6.10 shows the comparison between replica image and directly magnified resist image. It should be noted that these methods are not new, and in fact have been used since the early practical applications of electron microscopy to the study of materials. Alternative methods may be found in standard electron microscopy textbooks and in the literature on polymer microstructure studies in materials science.

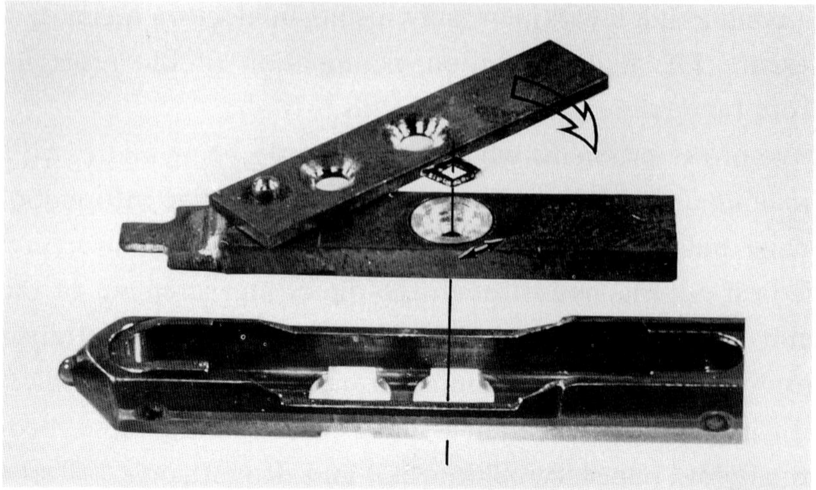

Figure 6.11. Window holding device for TEM observation. The window can be rotated by the small rotating table (light metal part) and by the goniometer of the microscope specimen stage (lower part of the photo). This holder is designed for the JEOL 100 series TEM equipped with goniometer stage

Silicon Nitride Window Holders for TEM

Silicon nitride windows are about 100nm thick, with a much thicker silicon supporting frame. They are inherently brittle materials, but being thin, can support quite large uniform loads, and one can observe elastic bending of the membrane under such conditions. However any inhomogeneous loading causes catastrophic fracture in the silicon nitride at small applied loads. The rectangular geometry of the crystallographically etched windows accentuates the inhomogeneous distribution of load under apparently innocuous simple loading conditions. Hence the windows must be handled carefully at all stages of the imaging process. If examined directly in the TEM, the holder used must be mechanically designed to minimize non-uniform loading on the window. The typical TEM specimen in biology or materials science is attached to a TEM grid or has a much thinner outside frame (in jet polished metal specimens for example), and in many cases is not brittle. Hence common commercial TEM holders are often not suitable to examine silicon nitride windows. Figure 6.11 shows a window holding device which was designed to be used in JEOL 100CX transmission electron microscope. The device consists of a rotatable holder in which a Si_3N_4 window can be placed. The holder can be rotated in order to align the window to a specific orientation (*i.e.*, for 3-D imaging). The whole device is then mounted on a JEOL 100CX specimen holder (Figure 6.11, lower part of the photo) which can be inserted into the goniometer stage of the microscope. Rotation about the specimen rod axis is achieved by the goniometer, and rotation about the axis of the window is achieved by the rotatable holders. It should be noted that different electron microscopes have different holder designs.

Three Dimensional Imaging with Soft X-rays

When compared to electron microscopy of biological materials, one of the perceived hypothetical advantages of soft x-ray imaging is the ability to examine thick samples. In very thin sections, the resolution of electron microscopy exceeds that of soft x-ray methods. However as the specimen thickness increases, multiple scattering and increased absorption of electrons of a given incident energy not only reduces the absolute resolution (due to increased chromatic aberration) but also reduces the transmitted intensity drastically. With x-rays, the wavelength can be decreased to optimize the resolution and the transmitted intensity through specimens too thick to examine with electrons. Synchrotron radiation is therefore ideal for this purpose.

The two major interests in the study of thick specimens is (1) many structures of interest in biology are larger than the thickness of TEM specimens, and sectioning methods often distort or destroy the

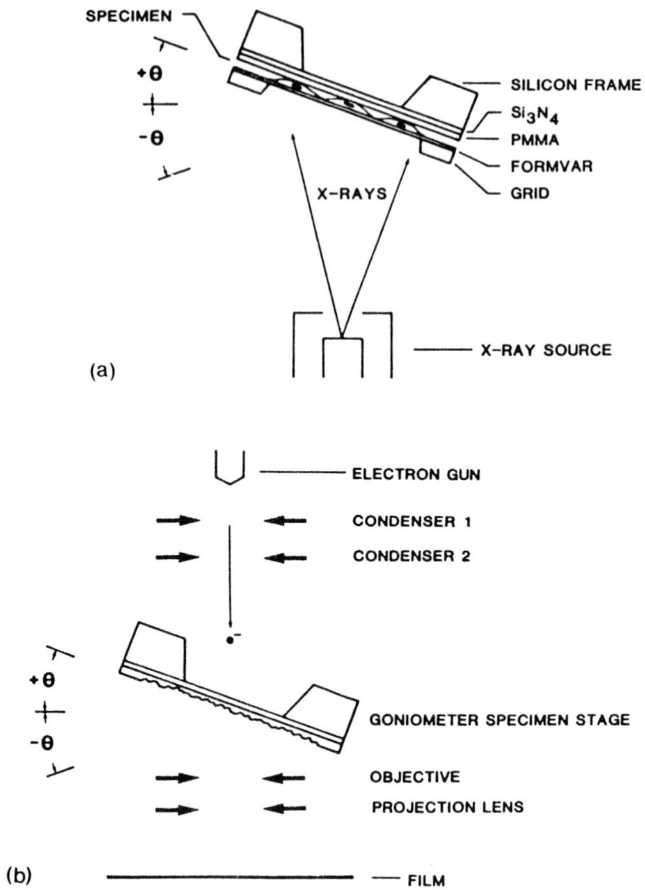

Figure 6.12. (a): Diagrammatic representation of a 3D imaging set-up. The resist-coated window is placed in intimate contact with a cell bearing grid. Exposures are made while the specimen-window complex is tilted at specific angles (*e.g.*, $\pm 20°$). TEM viewing of the contact image is done with the resist tilted at the same angle as it has been replicated.

structures, (2) unrelaxed three dimensional structures can be examined. The normal way to obtain three dimensional information from biological specimens is to serially section a cell (ultra-microtomy), or if the cell is sufficiently thin (*e.g.*, cultured fibroblast), to record stereo pairs directly in the TEM. Feder *et al.* (1980) have shown x-ray con-

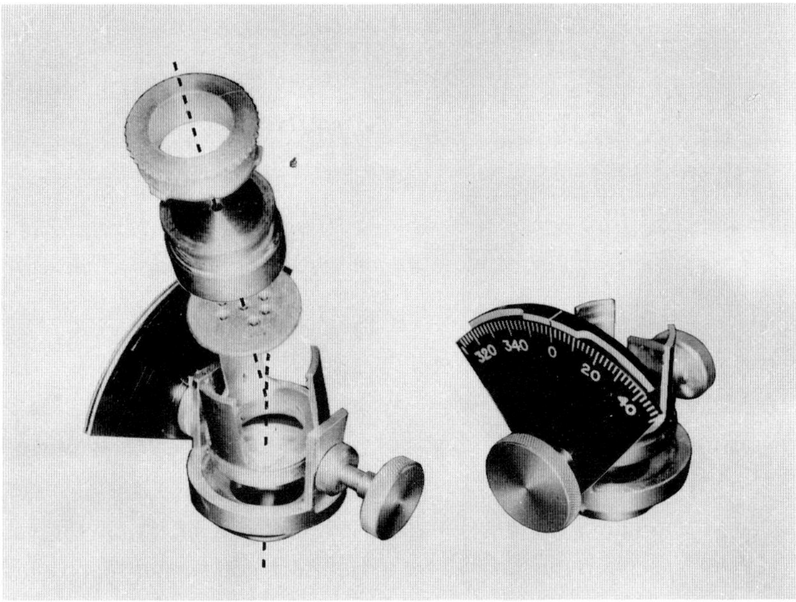

Figure 6.13. Window holding device for 3D imaging. The specimen-resist complex can be tilted at proper angles.

tact images of *Drosophila* giant chromosomes in stereoscopic image pairs. Their work showed that it is possible to produce 3D image-pairs by contact microscopy. Recently, further demonstrations of soft X-ray stereo pairs have been made of mouse 3T3 fibroblast and *Xenopus laevis* muscle cell (Cheng *et al.*, 1985)

The principle of stereo imaging is straightforward; two images of one area of the specimen are recorded. The two images are taken at different orientations, with the tilt angle and tilt axis being noted. The subsequent viewing of the images produces a synthetic parallax which reveals the three dimensional structure. Since the soft x-ray contact imaging process involves multiple steps before the final positive image is viewed, some intermediate steps are involved. A cell-bearing grid is placed in intimate contact with a Si_3N_4 -supported x-ray resist, and the exposure is made with the x-ray source positioned at a certain angle (*e.g.*, 30°) with respect to the axis of the grid (Figure 6.12). The

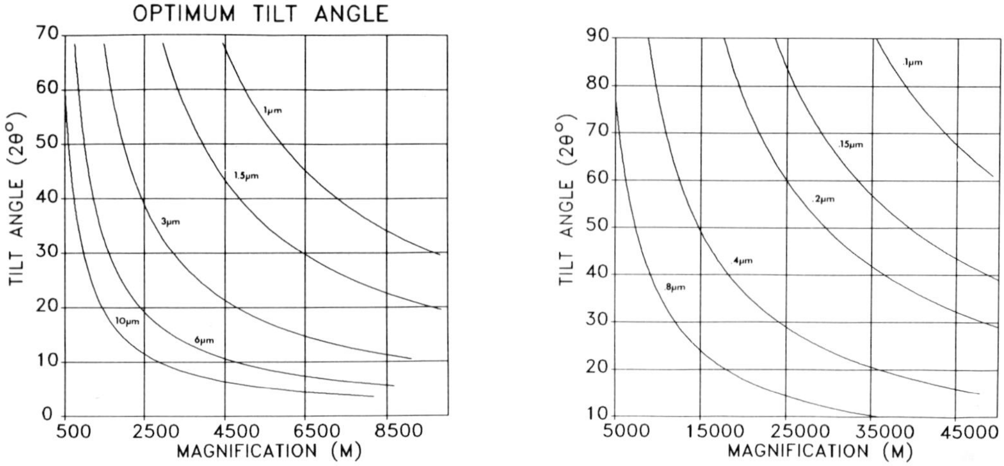

Figure 6.14. Optimum tilt angle plotted against magnification for various specimen thicknesses from 1μm to 10μm (upper graph) and 0.1 to 0.8μm (lower graph). The magnification (M) represents the final magnification of the viewing images, therefore, generally, a microscope magnification of M/2 to M/5 is recommended.

specimen is reloaded with a new resist and exposed at a different angle (*e.g.*, -30°). Figure 6.13 shows a specimen-window complex holding device used in this study.

It is essential to estimate the final magnification and specimen thickness when the imaging is carried out, because the optimum results are dependent on a proper tilt angle. The required relative tilt angles can be calculated from the work of Hudson and Makin (1970).

$$Y = 2\,h\,M \sin 2\theta \qquad (6.1)$$

where Y is the parallax (the difference in the distance between a corresponding point in the two images; for optimum results, Y is generally fixed at 0.3-0.5cm), M is the final image magnification, θ is half of the angle of tilt, and h is the vertical separation of points in the object at

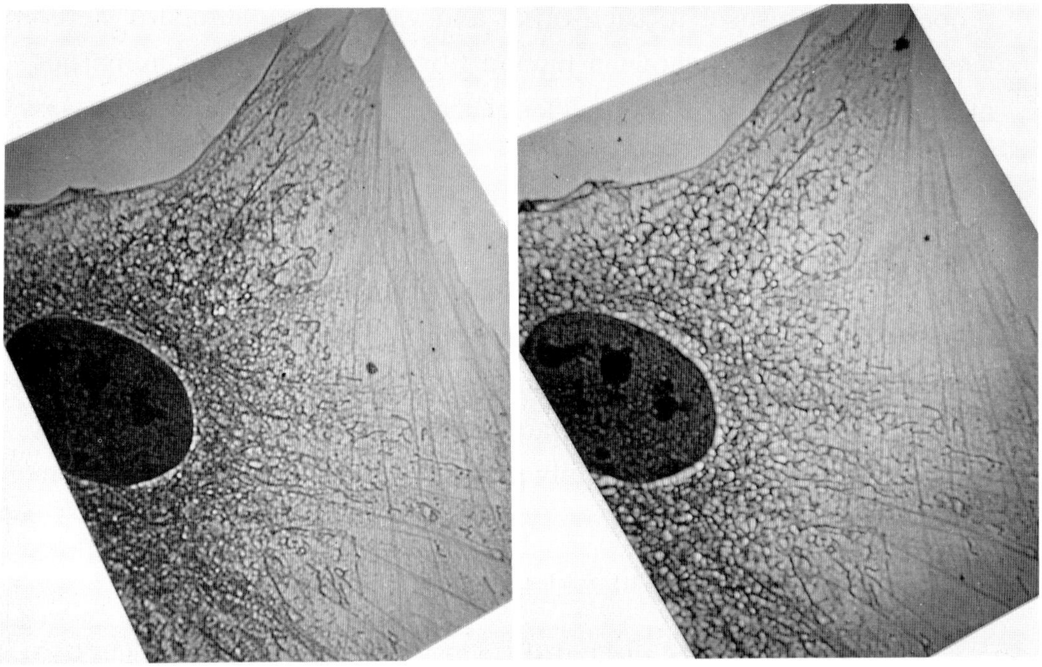

Figure 6.15. 3-D image of a human fibroblast taken with C-K radiation. The cell was cultured on Formvar-coated index grid, fixed in 1% glutaraldehyde, dehydrated in EtOH and critical point dried.

the angle midway between the two tilts. Figure 6.14 shows optimum tilt angle plotted against magnification for various specimen thicknesses from $1\mu m$ to $10\mu m$ ($Y=0.5cm$).

After their exposure, the two resists with images from different relative orientations are developed identically and viewed under a TEM. A microscope equipped with a goniometer stage is required. A specially made window holder is used (Figure 6.11). The TEM projection images of the x-ray replicas are obtained with the replica positioned at the proper tilt angle which corresponds to the angle at which the replicas were exposed (*i.e.*, projection images at $\pm$ 30°) (Figure 6.12). The positive prints of the images are then aligned properly with respect to each other and viewed in a standard stereo viewer. Figure 6.15 shows a human fibroblast. Note the numerous nucleoli suspended in the nucleoplasm. The tilt angle used was ± 30°.

One potential problem involved in the 3D stereo pair imaging arises from the requirement of intimate contact between the specimen and the resist. The mechanical contact between the resist and the specimen can result in a displacement or distortion of the specimen to produce slightly different objects in the two images. If this is a serious problem, it is necessary to use spacers between the specimen and resist so that the resist does not directly press on on the biological specimen.

Current Status in Wet Cell Imaging Experiments

As has been pointed out before, a second major area in which soft x-ray imaging can outperform electron microscopy is in the study of wet biological material (Robinson, 1982; Kirz and Sayre, 1980; Sayre *et al.,* 1977). This is related to the presence of the "water window" in the absorption spectrum in the soft x-ray region (Figure 6.16). The application of soft x-ray imaging to this problem has been relatively slow, restricted largely for experimental reasons. Based on the absorption characteristic of water and major biological materials (Figure 6.17), wet cell imaging should be possible.

The ideal wavelength for wet cell imaging therefore lies in the soft x-ray region. However the presence of water in the cell allows physical motion of the components of the cell and therefore long exposures of moving structures are not possible. This problem has been addressed by examining the possibility of using a short pulse of illuminating x-rays to essentially freeze the image. Since the resist has relatively low sensitivity, and the dose required to form an x-ray image in the resist is quite large, the dose rate must be high in such an experiment. Apparently the ideal sources are therefore the high intensity pulsed

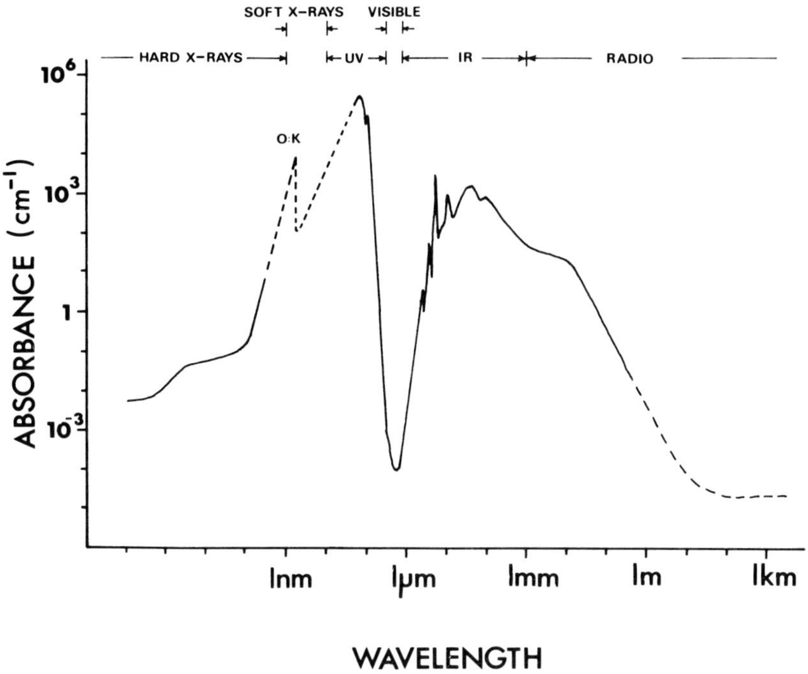

Figure 6.16. (a) Electromagnetic wave absorption of water. Two major high transmission windows exist in water, one of which is located in the visible light spectrum, and the other is in the soft x-ray spectrum. Although the absorption coefficient of water at 2.4nm wavelength is 6 orders of magnitude higher than at 600nm wavelength (visible light), the degree of transparency still allows a significant amount of x-rays to pass, provided that the specimen is relatively thin (*e.g.*, 10μm).

sources: laser generated soft x-ray sources (Ginter, 1984), and pinched gas plasmas. To see small structures, the Rose criterion shows that a minimum intensity is needed. Hence the wet cell problem reduces to having a very high flux of soft x- rays incident on the specimen / resist sandwich for a very short time. The absorption of energy in the specimen then becomes one limiting problem. Solem (1984) points out that if the pulse is very short, 10^{-12} seconds the image can

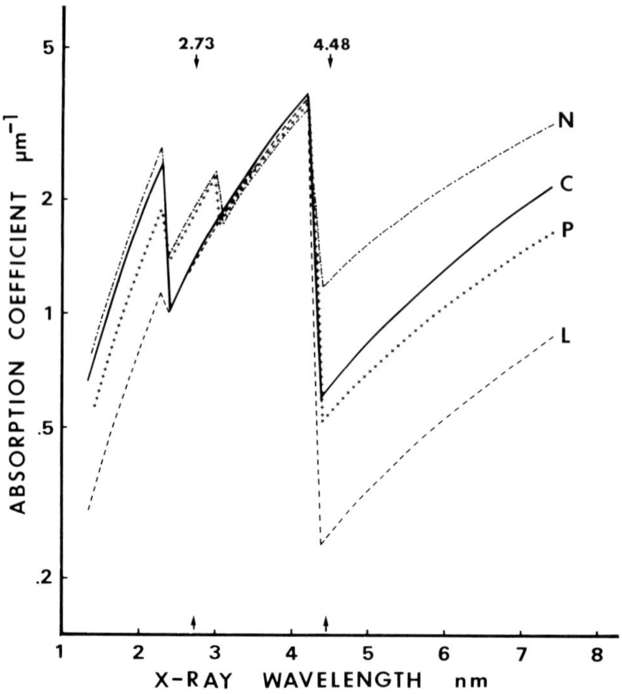

Figure 6.17. Calculated absorption coefficient of various biological materials. N: nucleic acid, L: lipids, P: proteins, C:carbohydrates, and water. Note the water film possesses an x- ray translucent window just below the oxygen K absorption edge (μ=0.1). However, the absorption of other biological materials is considerably higher (*i.e.*, protein: μ=1.3). Therefore, significant image contrast can be obtained by tuning the x-ray energy just below the oxygen K edge (~ 2.4nm).

be recorded before the specimen evaporates under the incident dose. It should be noted here that the problem of absorption of energy in the resist has not been considered. Rates of energy deposition which evaporate the biological specimen will likewise evaporate the hydrocarbon resist unless the dose transmitted by the biological specimen is carefully controlled. Any radiation which is incident on the resist after the biological specimen disappears will tend to overexpose the resist (analogous to overexposing a photographic film) making the image much more difficult to see.

The major questions in the study of wet biological cells are therefore: (1) the selection of the optimum soft x-ray source properties for imaging and (2) the proper design of a wet cell chamber. For both laser generated and pinched gas plasma sources, the spectral output of the source should be carefully matched to the requirements of the wet cell imaging process. Questions related to the UV content of the radiation, the debris emanating from the source, and the time structure of the pulse are important. In addition reproducibility of the output should be confirmed in each experiment.

The basic chamber design requires the following features: (a) it must to be able to maintain atmospheric pressure against the high vacuum environment of the x-ray source; (b) it must have a window which allows significant transmission of x-rays at the desired energies (*i.e.*, at the water window); and (c) the chamber design must allow for fast loading and unloading. Since the main objective of this chamber is to be able to image specimens in the presence of water, it is mandatory that the window material allows for a significant amount of useful radiation (*i.e.*, radiation with energies below the oxygen L absorption edge (2.4nm) to pass through, and maintains proper physiological condition for the survival of the specimen.

There are a number of published attempts at wet specimen imaging using contact soft x-ray methods (Panessa *et al.*, 1981; Feder *et al.*, 1985). There are two major experimental questions which remain in the light of these preliminary observations. These are related to the discussion above. Firstly, future experiments must ensure that the biological specimen used is both wet and alive. The overall size of the specimen (3mm diameter x a few microns thick) makes it very difficult to visually confirm that the specimen remains wet after the holder is exposed to vacuum. One method for evaluating this qualitatively is to examine the image for evidence of precipitation of salt crystals as it dries. Since in a low salt culture medium, formation of salt crystal is difficult to observe. Therefore, future experiments should involve a measurement of the water content of the specimen before and after the exposure. This requires the shielding of the specimen from destruction during pulsed exposures (due to debris) or some method of dynamically measuring the water content up to the moment of exposure. Such holders are presently being designed.

The question of whether the cell is alive is more difficult, and, in addition to temperature, is related to the pH, and the ionic concentration of the medium surrounding the cell, which is strongly dependent on the oxygen and carbon dioxide content of the medium. The physical size of the environmental chamber post difficulties on maintaining a proper physiological conditions. Further, the size of the specimen and thin window holder makes measurement of the culture medium experimentally difficult. Trypan Blue may be a useful stain for assessing if the cells under study remain viable. The principle of the technique is based on the ability for viable cells to exclude the dye. Further, one has to bear in mind that an improper environment may not kill a cell, but dramatically alter the cellular structures. In the absence of a direct measurement technique, the loading of a living cell must be handled very carefully, and relatively quickly to increase the chances of imaging a truly living cell. The lack of proper cell handling apparatus near all existing pulsed soft x-ray sources is a severe handicap in fulfilling this requirement. The final image obtained should be carefully compared to existing electron and optical microscope images to determine whether the image shows the expected differences.

The second major experimental question which remains in wet cell imaging is the spectral content of the imaging radiation. Calibrated filters should be used in front of the specimen to ensure that UV radiation or other extraneous radiation do not contribute to the image. In many cases, there is a large UV component to the radiation which, of course, causes exposure of the resist.

The role of diffraction in contact imaging

Ideally, a contact image is a true shadowgraph of the specimen microstructure. However in practice, the use of monochromatic soft x-rays (necessary for a variety of quantitative imaging processes) produces severe diffraction effects (Figure 6.18) which obscure the true geometric shadow of the microstructure. Examples of their effect on the observed image are shown in Figures 6.19 and 6.20.

Figure 6.20 is an image taken with 77Å monochromatic light, shows a typical contact image with pronounced Fresnel fringes. Images taken with the stationary target x-ray source or polychromatic synchrotron radiation characteristically show a deeply etched channel around the edges of objects, likely due to diffraction effects.

The presence of such diffraction patterns obscures the microstructure of the specimen, and being on a relatively large scale, is a severe limitation on the attainable resolution. Furthermore, methods requiring more than one image from the same area of the specimen, as in microchemical analysis (difference images above and below absorption edges), or stereo imaging, will be adversely affected, since the exact scale of the diffraction pattern depends on the exposure / development geometry.

The biological interpretation of the contact image must therefore be tempered in the light of these kinds of diffraction effects. For example, an x-ray transparent zone commonly found around cell nuclei of whole mount cultured cells may represent only the diffraction pattern (Figure 6.20)

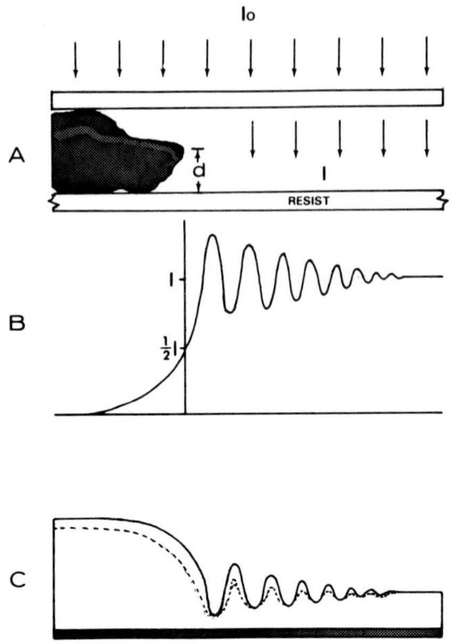

Figure 6.18. Fresnel type diffraction in contact microscopy. A: An x- ray opaque object is supported by a supporting film. The object and resist distance is d. The intensity of the incident light is Io, and the light transmitted through the supporting film is I. B: Fresnel diffraction fringes. Note the local intensity maxima have intensities greater than I. The edge of the object has 1/2 of I but not zero intensity. C: Resist surface profile. Note the local intensity maxima have lower resist profiles than the background.

Self-development and chemical development of x-ray resist

When an x-ray resist is subjected to a high dosage of x-ray irradiation, the image is observed before any chemical development. This phenomenon has been referred as "self- development" of the x-ray resist. This self-developed image can be seen under phase contrast microscope as shown in Figure 6.21a. Figures 6.21b and 6.21c show the progress of chemical development. These images are generally

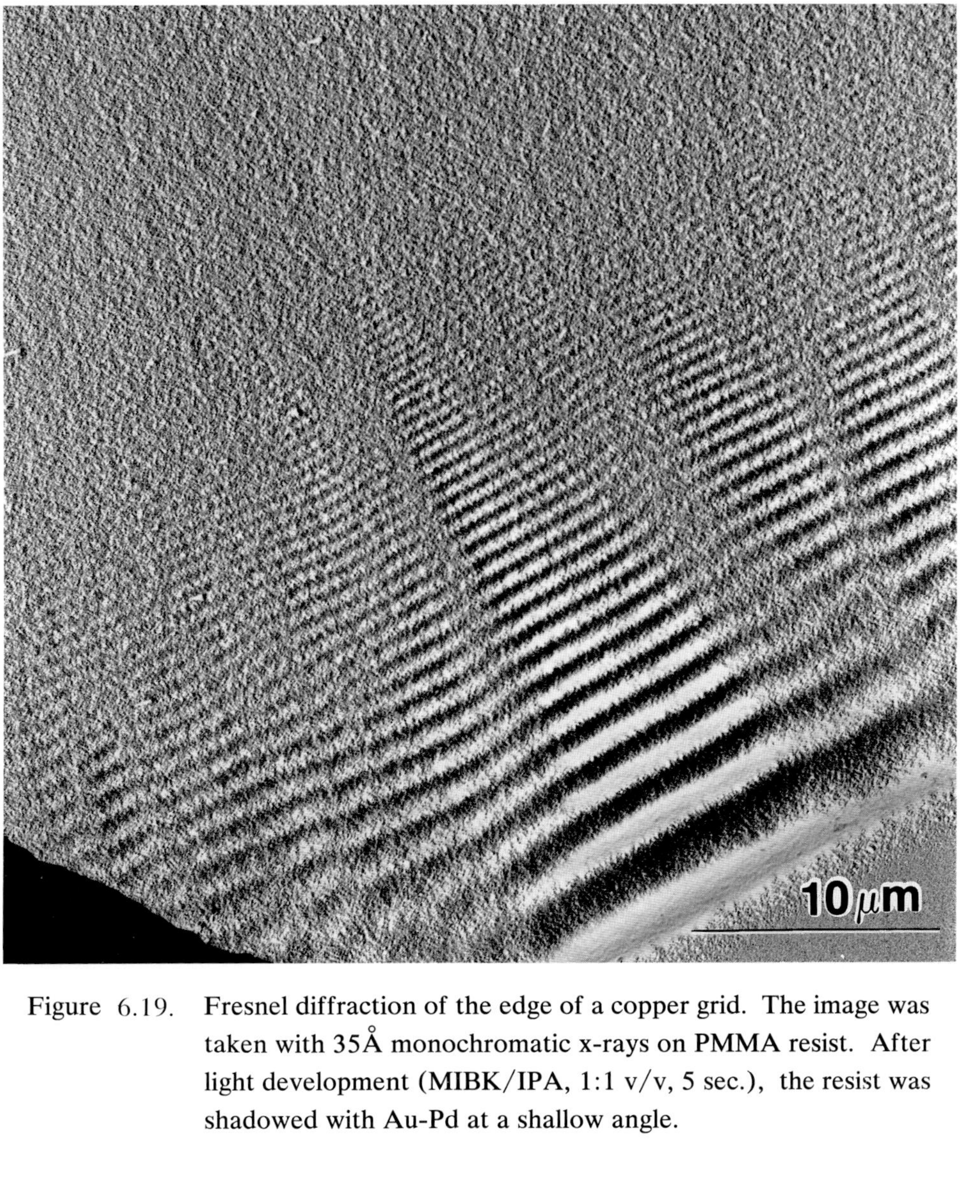

Figure 6.19. Fresnel diffraction of the edge of a copper grid. The image was taken with 35Å monochromatic x-rays on PMMA resist. After light development (MIBK/IPA, 1:1 v/v, 5 sec.), the resist was shadowed with Au-Pd at a shallow angle.

quite low contrast, and difficult to observe. The evaporation of the resist in the high dose areas is a result of the loss of side-groups and low molecular weight components produced in the polymer resist by soft x-ray irradiation.

The normal development processes produce different rates of dissolution in regions of the resist which suffer different amounts of exposure. For two adjacent regions with different doses, during development, the contrast increases with increasing development because of the different rates of dissolution. However, there is also a finite lateral development rate which quickly obliterates small surface features in the resist (Figure 6.22). Thus for coarse features, heavy development increases the visibility of the feature while for small features, it erases them from the resist surface. This, of course, has serious implications for the conditions of exposure and development required for ultra-high resolution imaging (Sayre and Feder, 1979).

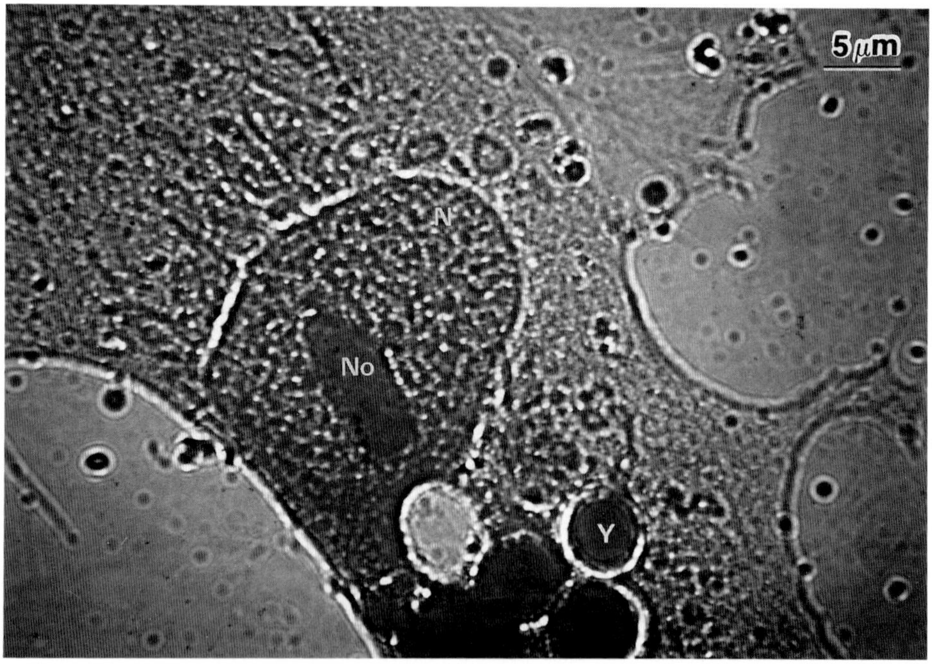

Figure 6.20. X-ray contact image of *Xenopus laevis* fibroblast. Note the x-ray transparent zone around the cell nucleus.

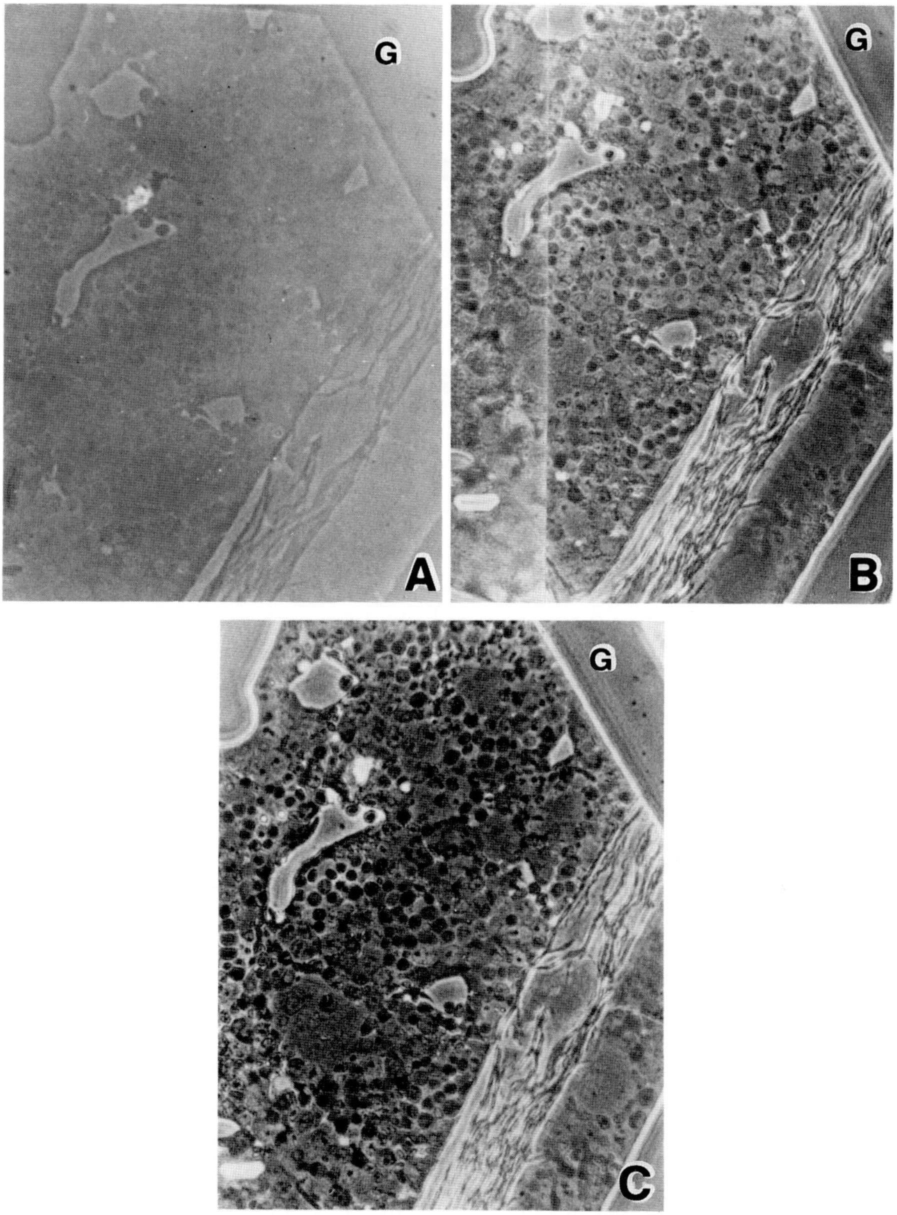

Figure 6.21. Self-development and chemical development. A: A self-developed contact image as viewed by a phase contrast microscope. The image is a rat bone marrow tissue (PEG section). B: The same resist as shown in A but chemically developed in MIBK/IPA for 14 sec. C: Chemically developed in MIBK/IPA for 24 sec. Note the various types of cells, including a large megakaryocyte (center of the picture). G: grid.

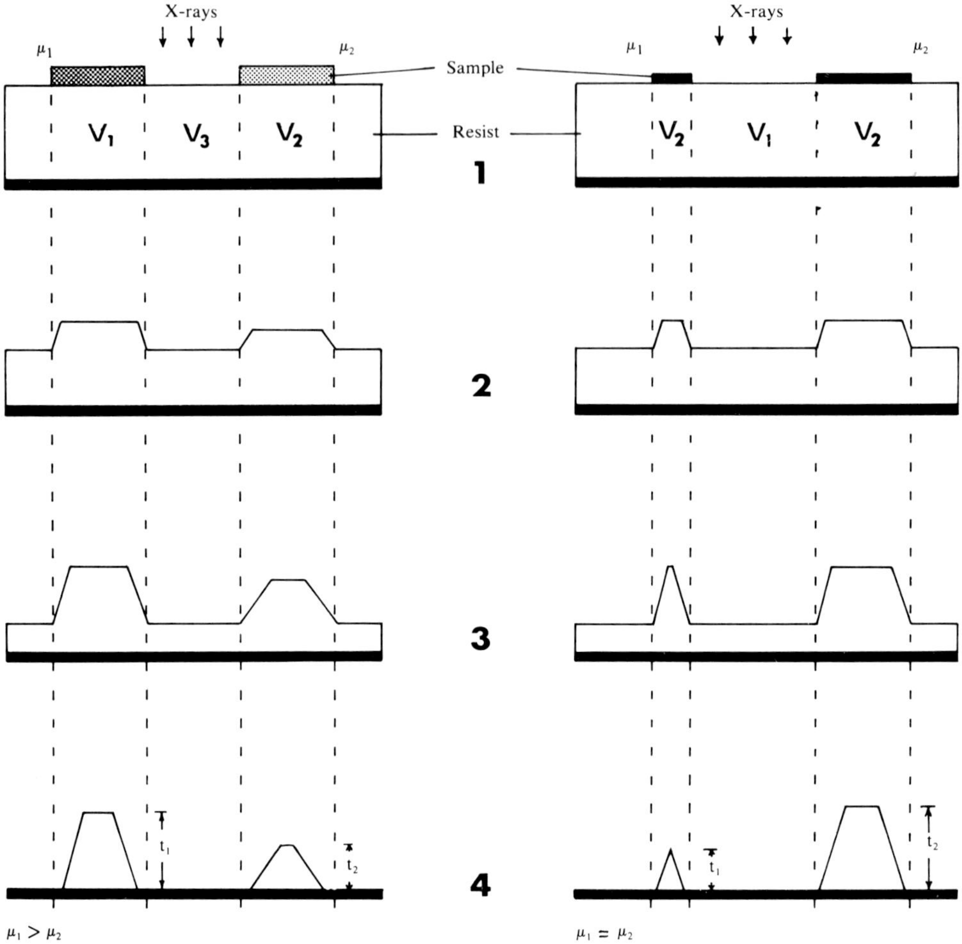

Figure 6.22. Effect of development on the surface profiles of resist. (a): Two object with identical dimension but different x-ray density ($\mu_1 > \mu_2$). X-ray exposure cause the resist to change from its initial dissolution rate to V_3,V_2 and V_1. Sequence of chemical development is shown in 2, 3 and 4. Note the feature with lower x-ray density has higher lateral dissolution rate.

(b): Two x-ray opaque "masks" with dimensions of d_1 and d_2 are placed on the surface of the resist. X-ray exposure causes the resist to change from its initial dissolution rate V_1 to V_2. Sequence of chemical development is shown in 2, 3, and 4. Note the features on the resist change as the development progress.

A further complicating aspect to the problem arises for long wavelength irradiation, in which the incident beam is strongly attenuated in the resist. In this case the absorbed dose varies with depth into the resist and the development rates also vary with depth.

Self-development is the result of resist mass loss due to the damage caused by the incident x-rays. This mass loss phenomena can be easily observed by x-ray spectroscopy. Figure 6.23 shows a set of x-ray absorption spectra of PMMA resist damaged by increasing x-ray dose. The damage is characterized by the thinning of the PMMA film. The roughness of the PMMA surface shows no significant changes in the self-developed resist. However, significant increases of surface

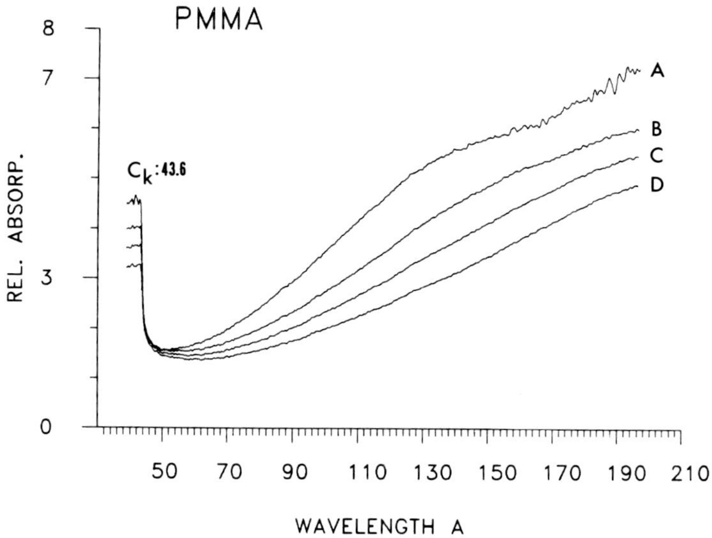

Figure 6.23. Relative absorption spectra of PMMA resist. The uppermost spectrum was obtained from a nondamaged resist; spectrum B was obtained from the same resist but has been damaged by zeroth order synchrotron radiation from the monochromator (1650 units). Spectrum C and D received 3250 and 6185 Units of radiation. The change of PMMA absorption spectra is due to radiation damage in the form of mass loss.

roughness is observed in x-ray exposed, chemical-developed resist (Shinozaki *et al.*, 1986)(Figure 6.24).

Typical Examples of Electron Beam induced damage in the resist

The method of viewing the final resist image also affects what is apparently seen. A detailed discussion of these electron microscope

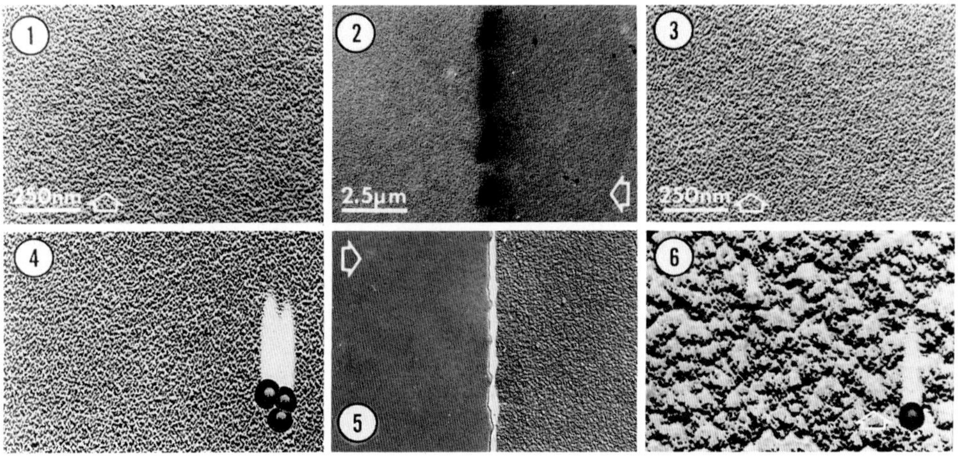

Figure 6.24. Soft x-ray induced surface roughness on PMMA resist. (1): High magnification view of an unexposed and undeveloped resist. (2): low magnification view shows the unexposed (left) area and self-developed (right) area; the resist has not been chemically developed. (3): high magnification view of the self-develop area. (4): High magnification view of an unexposed but chemically developed resist. (5): Low magnification view of unexposed (left) and exposed (right) area after chemical development (IPA, 5 sec.). (6): High magnification view of an exposed and chemical developed resist surface. Note the rough surface. Large arrows: direction of metal shadowing. The plastic spheres were spray onto the surface of the resist as internal standard. The length of the shadow provides information on the angle of the metal shadowing. (Adopted from Shinozaki *et al.*, 1986)

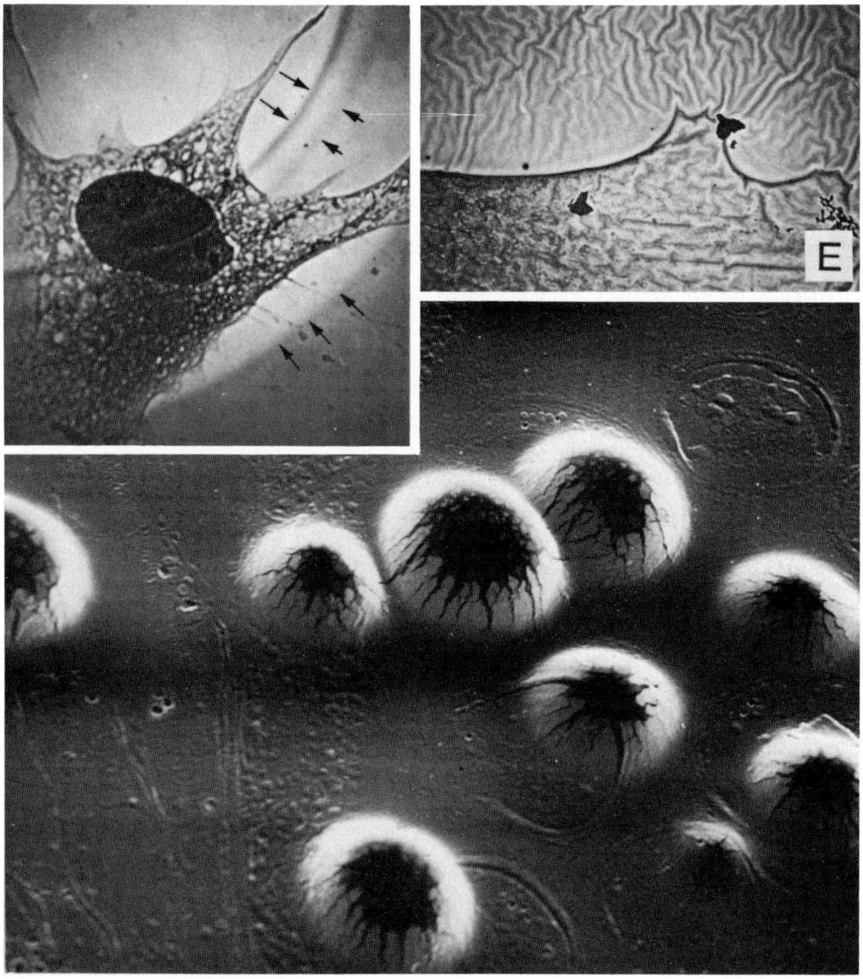

Figure 6.25. Electron beam-induced resist damage. A. TEM view of a Si_3N_4 window-supported resist. Note the electron beam-induced damage on the resist as thinning of the resist and formation of concentric rings (arrows). The darker concentric rings are the result of mass relocation. B. TEM view of a resist suffering from electron beam damage. Note the wrinkles on the resist. C. SEM view of a resist surface damaged by the electron beam of a scanning electron microscope. Note bubbles formed on the damaged resist. The bubble has a dark center and a bright peripheral area. This is due to the fact that the center of the bubble has no metal coating.

induced artifacts is given elsewhere (Cheng, 1985; Cheng *et al.*, 1986), but typical examples are shown in Figures 6.25.

Figure 6.25a shows electron beam induced mass loss of PMMA resist. With a thicker resist (>1.5μm), the resist begins to form wrinkles on electron microscopic examination (Figure 6.25b). This usually occurs in the thick regions of the resist such as those areas under the specimen grid or under dense specimens. It appears that resists with thicknesses of less than 1μm seldom show this type of damage.

After a heavy electron irradiation of a thick resist formation of bubbles is observed, as shown in the SEM image (Figure 6.25c), the bubbles usually have a dark center and a bright peripheral area, both of which are caused by cracking of the metal-coated resist surface. The area which has no surface coating appears dark because of the low secondary electron emission. Under a transmission electron microscope, this type of damage is characterized by the formation of clear spots with radiating rays.

Conclusions

A selected set of experimental results has been used to illustrate the present state of contact soft x-ray imaging. As more observations are recorded, the problems which remain to be solved become clearer. The literature contains a variety of predictions and observations which require further investigation.

Acknowledgements

Parts of this work have been supported in various ways by NSERC and MRC (Canada); NRC (Canada); R. Feder, P. Ho, D. Sayre (IBM); and J. Kirz (SUNY).

7. The Examination of Topographic Images in Resist Surfaces

D. M. Shinozaki[1] and B. W. Robertson[2]

[1]Faculty of Engineering Science, University of Western Ontario, London, Canada, N6A 5B9.

[2]Department of Materials Science and Engineering, McMaster University, Hamilton, Canada, L8S 4M1.

Introduction

The high resolution examination of the microstructure of biological materials using soft x-rays has become possible because of the recent advances in long wavelength x-ray optics and in very large scale integrated circuit fabrication technology. As all microscopists are interested in the limits of their technique, the question of resolution is a topic of central interest. A comparison of the different methods of imaging with soft x-rays reveals that the contact method, called in the earlier literature microradiography, has the highest claimed resolution (5nm) (Spiller *et al.*, 1979; Feder *et al.*, 1977; Panessa-Warren, 1984). However, this is at variance with Flanders' experiments in attempting to use carbon K α radiation to replicate lithographically tungsten lines using a mask which consisted of 30nm thick evaporated tungsten lines on a thin carbon support film (Flanders, 1980). The lines and spaces of the mask were examined in an electron microscope and were as small as 2nm in lateral width. Flanders found that the minimum linewidth possible to replicate with soft x-rays was 15nm. The height of this exposed and developed PMMA line was 60nm. At linewidths

X-ray Microscopy
Ed. by P.C. Cheng and G.J. Jan

narrower than this, the lines on the PMMA appeared to disintegrate. There are similar resolution limitations in fabricating small lines and spaces in PMMA using electron beam writing, for which Mackie and Beaumont report that it is possible to produce 10nm wide lines by lift-off (Mackie and Beaument, 1985). However the physical reasons for this apparent fundamental limit to resolution in PMMA (which is similar in both electron and soft x-ray lithography) are not well understood. It should be noted that the requirements for successful soft x-ray lithography in PMMA are somewhat different from those for useful x-ray imaging. Lithography requires a high contrast image while imaging requires a smoothly varying gray scale. The practical application of lithographic processes in manufacturing microelectronic circuits requires, in addition, a thick enough layer of resist to allow processing steps subsequent to the development of the image. However, the highest resolution images are obtained in the thin layer near the surface of the resist (of the order of 25nm) (Kirz and Sayre, 1980).

It is therefore of particular interest to examine the feasibility of obtaining the 5nm claimed resolution using soft x-rays in a reproducible fashion. From the viewpoint of microscopy of biological and synthetic microstructures, there exist a wide variety of structures at this level which it would be useful to image with soft x-rays. For the microscopist, the difference between resolving a 50 and a 5nm structure is often crucial for a number of applications. For example, in synthetic materials, it would be of practical importance to examine light element distributions in metallic thin films (prepared by thinning from bulk specimens or by evaporation) at grain boundaries or in the region near precipitates. The inability of soft x-ray imaging to achieve the extremely high resolutions of transmission electron microscopy may be counterbalanced by the ability to examine relatively thick specimens and to obtain 3 dimensional information. For the practical microscopist in biology or in materials science, the useable resolution in these thick specimens is a critical question.

It is clear from Flanders work that small structures in PMMA will also be quite low contrast structures, the height of the narrow line imaged in PMMA being only 60nm or less. In general, structures in PMMA which are small in the lateral direction (in the image plane) will be quite thin in the direction of viewing. For example, it is not expected to be possible to find a 5nm structure in the PMMA surface which stands 500nm high. Rather, one might expect structures to have an aspect ratio much less than 1:10. In many cases of soft x-ray contact imaging, the PMMA is supported by about 100nm of silicon nitride. During subsequent examination in an electron microscope, the effective contrast of 5nm structures in PMMA is then really quite small, especially if a light development is used (typically, the PMMA thickness before development is about 1000nm). The contrast of the topographic relief image in the resist surface is an important parameter in limiting the resolution at low doses in the electron microscope (as discussed by a variety of sources (Hren *et al*, 1979; Reimer, 1984; Thomas and Goringe, 1979) applying the Rose criterion for visibility (Rose, 1948).

One can consider the method of microradiography step by step, and examine the limitations of each step on the ultimate resolution. This would include a discussion of Fresnel diffraction, photoelectron path length in the resist, penumbral blurring, lateral development during dissolution in the solvent, and the effect experimental conditions such as specimen-to-resist distance, incident wavelength *etc.* (Beese *et al.*, 1986). The purpose of this paper is to consider the final step in the process: namely electron microscopic examination of the image in the resist, a necessary step in high resolution contact imaging since the resist image is not itself a magnified image of the specimen.

A typical high resolution resist such as poly(methylmethacrylate) (PMMA) has a high molecular weight of the order of 4.5 x 10^5. The molecules are in the form of long chains, with repeat units:

$$\begin{array}{c} CH_3 \\ -\ C\ -\ CH_2\ - \\ CO_2CH_3 \end{array}$$

The primary interaction of the incident soft x-ray radiation involves the production of photoelectrons in the resist which in turn interact with the surrounding molecular segments. The absorption of energy by the polymer molecule can involve immediate breakage of chemical bonds or transfer of energy to other parts of the molecule or to other molecules. For a positive resist such as PMMA, the local reduction in molecular weight makes this region more readily soluble in a solvent developer such as methylisobutylketone. The exact mechanism of molecular degradation may be inferred from the studies on radiation damage of PMMA irradiated with gamma rays (Dole, 1972). The disintegration of the molecule involved side-group evolution as well as main chain scission. Similar results have been reported by Hiraoka (Hiraoka, 1977). These observations show that there are a number of chemical effects related to the resist molecule which are distinct from the physics of the energy absorption process. The local dissolution properties are dependent on chemical effects coupled to the physical effects of the boundary layer fluid mechanics in the solvent. The developed contours in the surface of resist exposed in the contact imaging mode cannot be quantitatively modelled using our present understanding of the physics and chemistry of the exposure-development process at the limit of resolution of the contact imaging method. However, there have been a number of attempts to model lithography development processes at lower resolutions (Betz *et al.*, 1986). If one wishes to relate the depth of development in the exposed resist to the incident soft x-ray dose (which allows one to map the absorption in the original specimen), an empirical relationship between the two parameters must be obtained at the scale of interest (in this case at the limit of resolution of the contact imaging process).

In order to obtain quantitative information from the soft x- ray contact image in the surface of the resist,the topographic contours of the exposed and developed resist must be mapped with sufficient spatial resolution. The only method to achieve the resolution claimed by the contact method (of the order of 50nm or better) is electron microscopy. (There is a possibility of using a surface mapping technique such as scanning tunnelling microscopy, but this has not yet been accomplished.) In most cases, the resist itself has been examined in the electron microscope. Unfortunately, positive polymer resists such as PMMA are inherently sensitive to electrons, and this leads to some experimental limitations to the recordable resolution in resists.

Image Forming in the Electron Microscope

A number of modern references deal with the subject of electron microscopy (Scanning (SEM), Transmission (TEM), and Scanning Transmission (STEM)) (Hren *et al.*, 1979; Reimer, 1984; Thomas and Goringe, 1979). Image forming processes in the electron microscope primarily involve scattering of electrons in the specimen from the incident beam. If no such interaction occurs, no contrast arises and no image is observed. Images of various kinds are formed from the contrast generated by elastic and inelastic scattering (see Figure 7.1), and by the x- rays emitted. For elastic interactions with small scattering angles, little energy is lost from the incident beam. However for larger scattering angles (for example in backscattering), a significant amount of energy is transferred to the nucleus of at least one atom in the specimen. For inelastically scattered electrons, the energy of the primary electron is partially absorbed in electron excitations in the specimen, principally in the creation of plasmons (collective electron oscillations which decay into photons or phonons), or in ionization (ejection of an atomic electron). The vacant site left after ionization of an inner atomic shell is filled by electrons from a higher energy state with the consequent emission of a photon or an Auger electron. The images which can be formed in the electron microscope utilize these various electrons and photons. The essential observation is that, except for the primary electrons scattered elastically through small angles, all other contrast mechanisms involve ionization or transfer of energy to the specimen. Furthermore, even if only small angle elastically scattered electrons are used to form the image, inevitably the other processes deposit energy into the specimen. The deposited energy which does not appear as emitted electrons or photons will lead to bond rearrangements and to loss of material from the specimen or will be stored in the specimen as heat.

Resist Heating in the Electron Beam

For thin, lightly absorbing specimens examined in transmission, the specimen temperature stabilizes when the steady state condition is reached (when the rate of heat generated by electron-specimen interactions is equal to the rate of heat lost to the surrounding environment by radiation and conduction). If the temperature rise in the specimen is not too large, the radiative heat loss is small because the excitance or emittance for typical materials is small and because the radiative losses increase as T^4. Under these conditions, Reimer has calculated the maximum temperature T_{max} in thin foils for small beam radii and for uniform illumination to be (Reimer, 1984):

$$T_{max} = T_o + \frac{j\rho}{2e\lambda} \bullet \frac{\Delta Q}{\Delta x} r_o^2 \ln \frac{R}{r_o} \qquad (7.1)$$

for a focussed spot in a TEM, and

$$T_{max} = T_o + \frac{j\rho}{4e\lambda} \bullet \frac{\Delta Q}{\Delta x} R^2 \qquad (7.2)$$

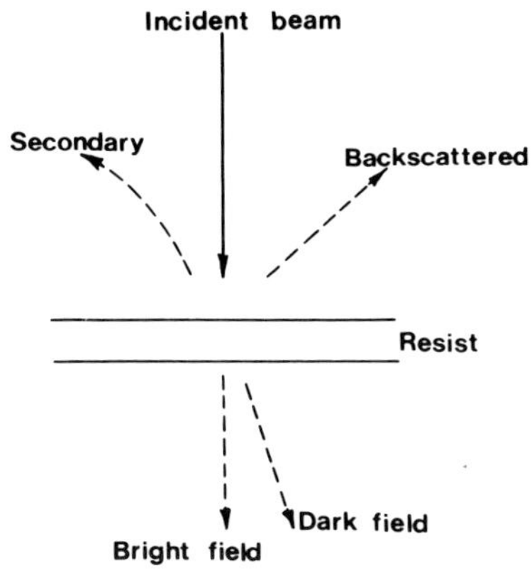

Figure 7.1. Electron microscope images are formed by collecting and recording the secondary or backscattered electrons in an SEM; or the transmitted electrons in a TEM or a STEM (bright field and dark field). The signal to noise ratio in the image depends on the strength of the interaction of the incident beam with the molecules of the resist.

for a uniform broad illumination, where $\Delta Q / \Delta x$ is the contribution to specimen heating of one electron per unit mass thickness; j is the current density in the electron beam; ρ and λ are the foil density and thermal conductivity respectively; and e is the electronic charge. This assumes that a high thermal conductivity material such as a grid bar (at a distance R from the center of the irradiated area) is at ambient temperature T_o. Schematically, the temperature variation across a uniform specimen is shown in Figure 7.2.

It should be noted that T_{max} depends on specimen thickness only through the term $\Delta Q / \Delta x$ which increases for thicker specimens due to multiple scattering and for lower electron energies due to the larger probability for inelastic scattering.

From the viewpoint of the operation of an electron microscope, an interesting aspect of Figure 7.2 is that small area illumination with much higher current density can result in a smaller maximum temper-

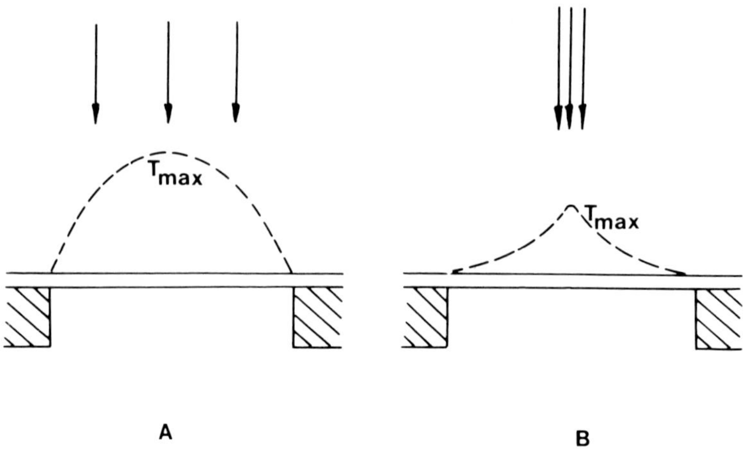

Figure 7.2. The temperature variation in a polymer film supported on a TEM grid and examined in a TEM: (A) Uniform irradiation of the resist as in ordinary imaging, (B) localized irradiation for a focussed beam. (The quantitative description of the temperature distribution is found in (Reimer, 1984).

ature than large area illumination because the heat flow from the centre of the illuminated area is predominantly by conduction in the unilluminated regions of the thin film to the grid bar.

For the smaller beams found in a STEM, the temperature rise in the irradiated region of a PMMA film supported on a thin silicon nitride membrane is of the order of 10 to 40°C (for a 1nA beam of electron energy 100keV). For thicker specimens and the lower electron energies usually employed in the SEM, a larger fraction of the incident electron beam power is deposited as heat in the resist by the incident flux. The quantitative comparison with the case of thin transparent specimens is difficult since the heat loss by conduction is a three dimensional problem in the SEM case.

Other estimates of the beam heating effects include those by Hobbs (found in reference Hren *et al.*, 1979) and Fisher, 1970). These authors come to the conclusion that the maximum temperature reached in the resist can be minimized by using small beam currents, since the temperature rise is proportional to the beam current. This runs counter to the attempt to obtain a large signal to noise ratio in the image, which requires a large electron dose unless extended recording times are employed (in this case, specimen drift may then be a factor limiting resolution).

Since the glass transition temperature for PMMA is around 114°C, local temperatures of this magnitude will result in distortion of the surface features in the resist (*i.e.* the x-ray image). If higher temperatures are reached with the beam currents which are within the range of typical operating conditions for SEM, TEM or STEM operation, the specimen can locally melt to produce complete destruction of the resist as shown in Figure 7.3. These kinds of images have been reported earlier by Cheng (1985) and Cheng *et al.* (1986). In addition to direct heating effects in the the resist, the electrons also produce significant amounts of radiation damage which produces other changes in the resist surface.

Radiation Damage in PMMA

PMMA can be used as an electron beam resist for high resolution lithography, and is therefore highly sensitive to electron damage. Hiraoka has studied the radiation chemistry of a variety of poly (methacrylates) using UV, gamma rays, and electron beams (Hiraoka, 1977). Even very low voltage electron beam irradiation (less than 2.5keV) produced significant mass loss with a relatively high yield of hydrogen. At temperatures near 373°K, a variety of monomeric compounds also were detected evolving from the PMMA under irradiation. When compared to the radiation damage mechanisms suffered under UV or gamma irradiation, the main difference with electrons was the high yield of hydrogen, particularly at low electron energies. This was apparently related to the high electron affinity of the free radicals generated upon irradiation.

Figure 7.3. Bright field STEM image of PMMA. At successively higher magnifications, the rastered area is reduced in size and the dose rate increases (A), (B), (C). The resist is most transparent (thinnest) at (C) because of severe mass loss due to irradiation and heating. Even modest exposures to the electron beam, corresponding to short times of examination result in a visible change in the transparency of the resist (A).

The Measurement of Mass Loss of PMMA Examined in a TEM

The relative rate of mass loss can be measured by recording the annular dark field current in a STEM as a function of time as in Figure 7.4. In this example, the original specimen was a uniform layer of PMMA about 140nm thick on a 100nm silicon nitride window (provided by R. Feder). The electron beam in the VG-HB5 was spread to irradiate a relatively large area (in normal use the beam is focussed to a spot smaller than 10nm.). The annular dark field current and the virtual objective aperture current were measured simultaneously (the virtual objective aperture current monitored the incident beam current which is time dependent for a room temperature field emission gun). For specimens of this mass thickness, the electron current scattered from the primary beam into the annular detector is related to the mass thickness of the specimen. In this case the annular dark field signal represented only a relative measure of the mass thickness of the specimen. The irradiated area not only becomes more transparent to electrons as the resist evaporates, but in the boundary region between the irradiated and unirradiated areas, the resist becomes thicker (Figure 7.5). This suggests that the electron irradiation of the PMMA produces a significant mass transport to the boundary region, giving an appearance similar to that of contamination build-up. The relative mass loss rates are similar to those measured by Reimer on PMMA using 60keV electrons (Reimer, 1984). He observed mass losses of the order of 38% to 52%. He also observed that the rate of mass loss increases with temperature. In both his observations and in the present work, the typical doses which result in most of the mass loss are reached in a matter of seconds in the electron microscope unless special precautions are taken to minimize the dose (using low dose imaging procedures). An alternative method to obtain the mass loss with electron beam irradiation is to measure the zero loss peak in the energy loss spectrum as suggested by Egerton and Cheng (1985).

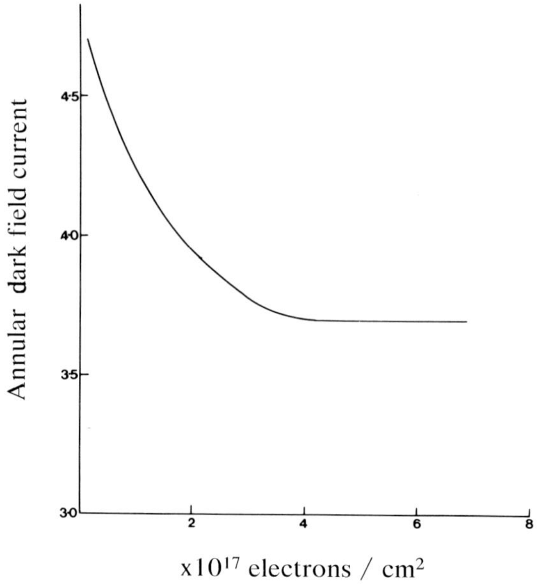

Figure 7.4. Annular dark field decay with electron dose in PMMA (the electron beam was 13.8 micrometers in diameter) measured in a VG- HB5. The dark field current scale is measured in arbitrary units.

Resolution and Electron Beam Damage in Resists

For bright field TEM images recorded with perfect detection efficiency, in two adjacent resolution elements the contrast (C) is related to the difference in numbers of electrons transmitted by each element as:

$$C = \frac{\Delta N X^2}{N_o X^2} \tag{7.3}$$

where N_o is the number of electrons per unit area, and ΔN is the difference in number of electrons per unit area transmitted through

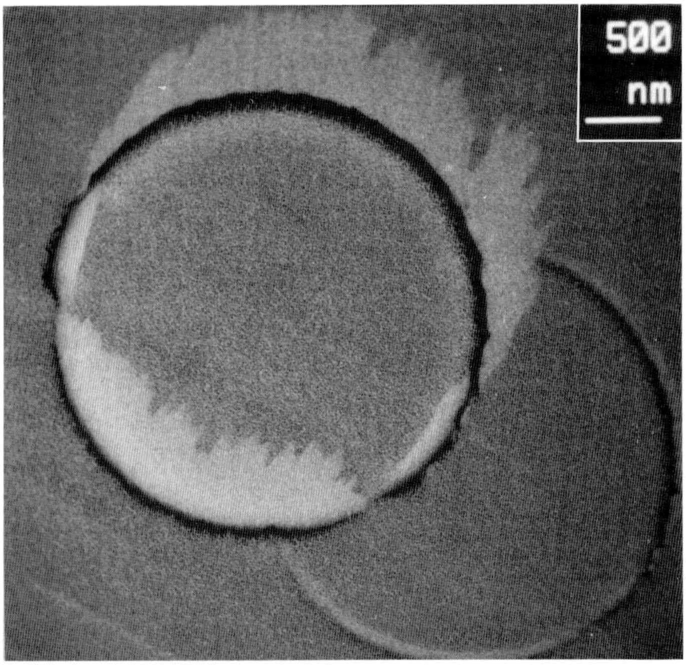

Figure 5. Circular area on PMMA irradiated as for Figure 7.5, then removed from the microscope and shadowed with AuPd. The shadow is cast on both sides of the irradiated crater, suggesting a raised "lip" to the crater. Part of the mass lost from the center of the crater redeposits in the rim. This mass transport is enough to cause distortion or obliteration of fine topographic features in the resist surface.

neighbouring elements of area X^2 (Figure 7.6). The background noise assuming Poisson statistics is:

$$\sqrt{N_o X^2}$$

and the signal to noise ratio is:

$$\frac{N_o X^2 C}{\sqrt{N_o X^2}}$$

which then gives a form of the Rose criterion (contast C detectible with a signal to noise ratio $\geq$ 5 (Rose, 1948)):

$$\sqrt{N_o} \geq \frac{5}{XC} \qquad X \geq \frac{5}{C\sqrt{N_o}}$$

Substituting the dose 10^{-4}C/cm^2. (equivalent to 6 electrons/nm^2) for ~1% mass loss from a thin PMMA specimen in the TEM and assuming a local contrast of 0.1, the resolution element (X) is about 20nm. (for X = 5nm, the dose per unit area must be raised, leading to a mass loss of about 16%).

Now the local contrast must be much greater than the local mass loss during observation in the TEM if the contrast in the resist image is to be recorded correctly. This condition is necessary since mass loss will determine the resist surface topography. In a contact x-ray image in the resist, the contrast is much lower than 0.1. A 20nm structure will be of the order of 20nm higher than the surrounding resist. Given the starting thickness of the resist can be as great as 1000nm and that the silicon nitride substrate is about 100nm thick, the inherent contrast of a lightly developed resist will be less than 2% (To retain 20nm topographic features in the resist surface, the depth of development must be small). Hence at the same electron dose (10 C/cm^2) for an average of 1% mass loss from the resist, the observed resolution (X) of the image will be greater than 100nm.

There are two additional points to note for electron microscopic examination of resists. The human eye is able to detect light intensity variations of about 5%. Firstly, the contrast in the photographic film must itself be greater than this, emphasizing the need for proper exposure control in the electron microscope to observe small structures in the resist with a minimum electron dose. The TEM magnification must be set low enough to match X to the film resolution element (typically about 30 micrometers) in order to achieve sufficient optical

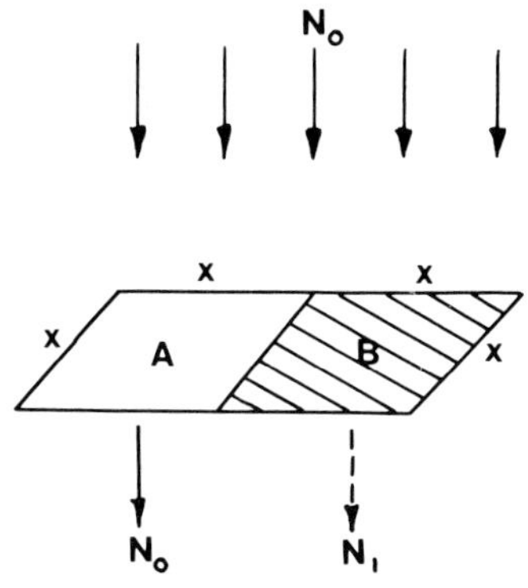

Figure 7.6. Two resolution elements (A) and (B) in the resist have different electron transparencies.: To resolve an element of a size "x", a minimum electron dose is required. A higher dose is required to see a smaller "x".

density that the contrast of the electron signal will be detectable reliably despite the influence of granularity of the emulsion.

Secondly, the numbers suggested by these estimates are really very low imaging doses, when one considers typical beam currents in a TEM of 0.01A/cm^2 (6.24 x 10^{16} electrons.cm^{-2}.sec^{-1}) at 10,000 times magnification. Complete destruction of the PMMA corresponds to 0.01 C/cm^2 *i.e.* 1 second at 0.01A/cm^2. An image must therefore be recorded in the first fraction of a second of exposure to the electron beam. This precludes the leisurely examination of the resist on the phosphor screen in a TEM, and means that an approximate focus must be fixed while viewing an adjacent specimen region before the resist is exposed to the electron beam and a photograph then recorded without prior examination on the screen. For this reason, as well as for minimising the temperature rise, the illuminated region should be no larger than the region to be observed if unintended exposure before recording is to be avoided.

It is obvious therefore, that for practical purposes, in a transmission or scanning transmission electron microscope aligned in the standard way to examine beam insensitive materials, any attempt to resolve structures as small as 20nm or less in PMMA will inevitably result in significant mass loss in the first few seconds of examination. The attendant temperature rise mentioned earlier only exacerbates the problem. The contrast of the image changes, and any attempt to correlate the electron microscopic image (stored electronically or on film) with the thickness of the resist, as one might attempt in microchemical analysis, must take the resist mass loss into account.

Furthermore, the rapid loss of mass during examination in the electron microscope is likely to occur non-uniformly at the scale of 20nm or less, near the resolution limits being discussed here. This scale is close to the typical molecular dimensions of the PMMA molecule (of high molecular weight) in a randomly coiled conformation. Any topographic features of this size in the resist will therefore consist of almost isolated molecules, and the volatility of the chain segments resulting from chain scission is likely to be different from that of molecules imbedded in the bulk. These aspects of the problem have not yet been studied either experimentally or theoretically. Any attempt to resolve structures near the limit of resolution (20nm or less) in the resist surface must therefore examine this problem first, since it is quite possible that small features in the resist surface will evaporate first upon exposure to the electron beam.

Scanning Electron Microscopic Examination of Resists

The use of transmission or scanning transmission electron microscopy yields the highest potential resolution, but in many cases, in particular in the study of the quality of structures produced in resists for lithography, the SEM is commonly used. In almost all cases, the lithographic structures are about an order of magnitude larger (about 100nm) than those examined in imaging (about 10nm). Although the typical SEM uses lower energy electrons (15 to 50keV) than does the

TEM or STEM, the radiation damage mechanisms are similar, and mass loss will be observed with quite short observation times at moderate magnifications (X10000) unless extraordinary precautions are taken to minimize the dose. There are, in addition, other considerations which must be taken into account when attempting to image extremely small structures in the resist surface with an SEM.

In an SEM, a finely focussed primary electron beam is rastered across the specimen, and the secondary or backscattered electrons are used to produce the image in most cases relevant to resist examination (Goldstein and Yakowitz, 1976). The resolution depends on the size of the beam, the volume of interaction in the specimen, and the signal to noise ratio at the detector.

In using the secondary electrons to form the SEM image, the resolution for a given dose depends on the electron beam diameter and the contrast of the structure being examined. The secondary electrons are ejected from a thin layer near the surface, typically less than 10nm thick, by primary beam electrons and backscattered electrons.

The fraction of electrons which are backscattered falls with decreasing substrate atomic number and with decreasing substrate thickness (if the thickness is less than the range of incident electrons). Therefore, provided that a resist is coated with a thin, heavy element metal layer to enhance secondary electron emission and is supported by a low atomic number substrate (preferably a thin one), the resolution of the secondary electron image will be affected only slightly by beam broadening processes in the resist and substrate. In this case, the absolute limit is fixed by the diameter of the primary electron beam or the granularity of the metal coating. For a typical SEM with a heated tungsten filament as an electron source, the minimum beam diameter will be about 5nm, while for a LaB_6 electron source it can be as small as 2nm. This will be the smallest structure resolvable if the contrast of the structure is high (greater than 25%). If the contrast is low, as expected for small resist structures, the effective resolution would be closer to 20nm.

The backscattered electrons have a lateral distribution with an intense component of width a few tens of nanometer on a very broad background (≥ 1 micrometer). Thus, using backscattered electrons to form the image reduces the resolution and contrast somewhat. Most importantly, however, since the backscattered electrons originate from a greater depth in the specimen, they convey little information about the resist surface.

Practical Considerations in Using the SEM

A resist surface is normally coated with a metal to eliminate electrical charging and to increase the signal from the surface. The coating must cover all the areas to be examined, and is usually deposited by sputter coating or vacuum evaporation. In the latter case the specimen should be rotated during evaporation to eliminate uncoated regions in the shadowed areas. Some care must be taken in the method to ensure a uniform fine-grained coating, and to avoid surface damage during sputter coating and sample heating during vacuum evaporation. The thickness should not be too great since the small topographic features can be obscured. In general, most of the primary beam will penetrate into the resist and radiation damage there will be a problem, since the evolution of gases or volatile components can distort the surface metal layer.

The Rose criterion shows that a minimum incident dose per resolution element is needed, and as the resolution element size decreases, the number of electrons per unit area increases. Typical operating conditions used to examine metals or beam insensitive materials at high magnifications will leave observable imprints of the rastered area in the surface of coated PMMA, a manifestation of radiation damage mass loss. This has been observed extensively in studies of polymer microstructure using electron microscopy (Grubb, 1970). The practical use of an SEM to study resist surfaces should include a recognition of the following problems.

Firstly, because topographic features which are small in the lateral direction (in the plane of the resist surface) are also likely to be small in the direction normal to the resist surface, the detectability of the feature will depend on the tilt of the specimen in the SEM. (Figure 7.7). Furthermore, small features with a slowly varying contour will be difficult to see under conditions which limit the exposure dose. This is because the topographic contrast depends on the relative signal originating from two adjacent resolution elements, and a 1° change in surface orientation produces typically only a 1.8% contrast. Thus the development of the resist should be directed towards producing sharp changes in topography in the resolution element of interest. This suggests that the examination of the x-ray image in the resist using an optical microscope may be of limited use in optimizing the visibility of very small structures in the resist surface. It is a common practice to check the development of the surface relief image by alternately immersing the resist in the developer and examining the image optically, proceeding with development until a high contrast image is seen optically. The depth of development is so great under these circumstances, that the small structures of interest (say 20nm.) are likely to be obliterated by sideways development or at least smoothed out considerably (reducing the contrast). Finally, to observe low contrast structures of this kind, the electron dose should be increased. Hence there exists the trade-off between radiation damage and visibility of the small structures.

Two instrumental changes can improve the situation in examining resist surfaces using an SEM. Firstly, the gun brightness can be increased and the beam diameter reduced to perhaps 2.5nm (with field emission tips). Secondly, the accelerating voltage can be reduced to restrict the penetration of the primary beam to the metal layer (Joy,

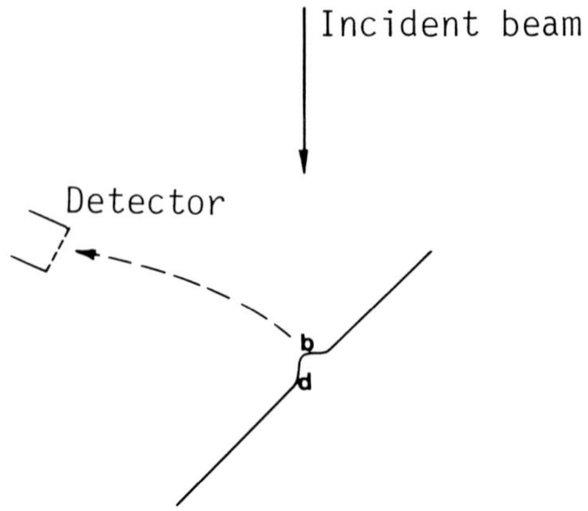

Figure 7.7. The detectability of the topographic feature in the surface depends on the relative signal strength from neighbouring bright and dark regions (b) and (d).

1984). However, this results in a large beam diameter (perhaps 0.1 to 0.5 microns) unless a field emission gun is used (Boyes, 1984). It should be noted that presently, these kinds of instruments are not normally used by groups examining soft x-ray images in resist surfaces.

Conclusions

The purpose of this paper has been to discuss the feasibility of using standard electron microscopy to examine surface relief images in polymer resists. The examination of the resist itself in a typical transmission or scanning transmission electron microscope will produce radiation damage in the inherently electron sensitive resist. To observe low contrast soft x-ray images at the limits of the resist resolution, the electron dose must be so large that some changes in the thickness of the resist must result. No experiments have yet proven that small topographic features in the surface of the resist remain unchanged as mass is lost during electron irradiation. As a result, extreme care must be taken in the interpretation of high magnification images in resists.

Acknowledgements

The support of CSRF, NSERC, MRC (Canada) and of the Centre of Chemical Physics has been used in this work. One of the authors (DMS) was on sabbatical leave at the IBM T. J. Watson Research Center (supported by P. Ho). Discussions on various aspects of this work with P. C. Cheng, L. Beese, R. Feder, and D. Sayre are gratefully acknowledged.

Note added in proof: Similar observations and conclusions have recently been reported in the scanning electron microscopy examination of resist used for lithography, and in microfabrication (Erasmus, 1987).

8. The Shadow Projection Type of X-ray Microscope

Sterling P. Newberry

CBI Labs, Box 11, Rd. 6, South Westcott Road, Schenectady, New York, 12306, USA

Introduction

The shadow projection method is the most convenient of all approaches to x-ray microscopy providing its resolution limit in the neighborhood of the optical microscope is adequate for the problem at hand. See for example, The History of X-Ray Microscopy, else where in this volume. The purpose of this paper is to present a working knowledge of the shadow method, its supporting equipment, capabilities and future possibilities. The instrument developed by the author at General Electric is the type equipment currently used in his laboratory and is therefore the model used here for illustration of the method. For recent references on shadow projection microscope, readers should turn to those papers by Horn and Waltinger (1978), Fletcher *et al.* (1983), and Yada and Ishikawa (1980).

Shadow Projection X-ray Microscope

The basic structure is shown schematically in Figure 8.1.a The equipment logically and physically divides into two regions, the electron beam path in vacuum and the x-ray path in either atmosphere or vacuum. The two regions are separated by a thin window of beryllium or other appropriate material which is coated on the vacuum side with a thin x-ray target material. In some cases the target is able to act as window as well as target, an example is aluminum foil 6

X-ray Microscopy
Ed. by P.C. Cheng and G.J. Jan

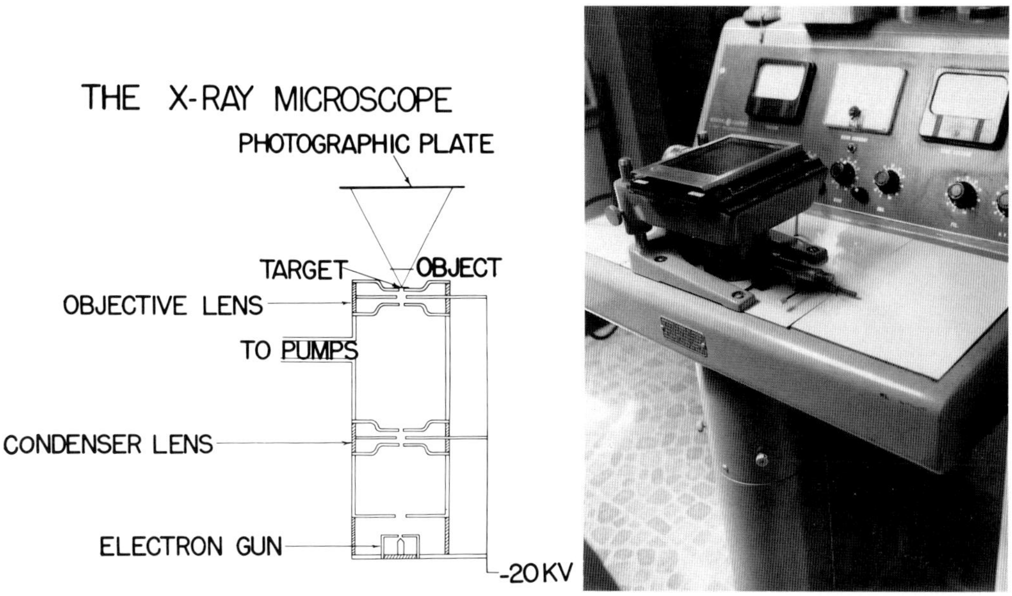

Figure 8.1. (a) Schematic diagram of shadow projection x-ray microscope construction. (b) Upper front of x-ray microscope showing electrical controls, rear panel, x-ray camera front, and aperture controls right of center at table top level. Beam centering screws may be seen lower center.

micrometers thick. The requirement in the electron beam path is to form the smallest, high current density electron probe possible at the precise location where the target crosses the optic axis. The small spot is accomplished by essentially reversing the beam direction in a transmission electron microscope or considered another way by using scanning microscope optics (without need for the scanning coils) with a very strong final lens. To give practical exposure times the final lens should have a focal length less than 3mm. As is well known, electrons may be focused by either magnetic or electrostatic lenses. The instrument described here uses the electrostatic Einzel lens. With this lens, to obtain the required short focal length, it is necessary to use the target as the final element of the lens which in turn requires that the target must be axially symmetrical and smooth of surface. For this penalty one buys a very comfortable space for the x-ray region compared with the space available in the narrow bore of a magnetic lens and at the same time improves the optical performance of the Einzel lens to the same neighborhood as a magnetic lens of the same focal length. Good microscopes have been produced using both basic lens types. The x-ray properties are the same for either system and the description below, with very little change, fits the magnetic lens system as well as the authors machine.

Continuing with Figure 8.1, let us now concentrate on the x-ray region, and the x-ray requirements of the target which separates the two regions. The targets are usually composed of two layers, one a window to support up to an atmosphere pressure differential and the other a much thinner target material to optimize x-ray production in the desired wave length region. It is important to note that targets are slowly damaged by the electron beam and thus should be movable perpendicular to the axis by a small amount. A variety of target elements are useful which mandates that the targets should be readily interchangeable. The present machine uses the simple expedient of interchanging the entire target window assembly as explained below.

The x-ray enlargement is by simple geometrical projection of the specimen shadow on to a more distant photographic film or fluorescent screen. It is very difficult to realize how sharp and clear these shadow images can be until one sees a representative collection of photographs. We are all conditioned by the relative softness of ordinary shadows where the source is usually large or if the source size is reduced Fresnel diffraction destroys the sharpness.

The "projection of shadow" as opposed to "contact shadow imaging", substantially enhances the convenience of x-ray microscopy. One can obtain survey pictures with instant film in less than one minute and can, with an image intensifying tube obtain a useful fluorescent

Figure 8.2. Two target plates, x-ray side on the the left, electron side on the right. Plates are 6mm thick by 100mm diameter. Window at center usually 20 micrometer Beryllium.

image. Thick specimens are as sharp as thin ones and stereo views are easy to record and view. The x-ray space is of convenient size which is determined by the properties of projection shadow imaging and the properties of available photographic emulsions. The photographic film distance generally is from 2 to 5cm. Since photographic enlargement

Figure 8.3. Electron beam focusing elements. The target plate of Figure 2 is necessary to complete the final lens (objective lens).

is preferred to be in the range of 3 to 20 X maximum, one can only obtain the full resolution of the instrument by bringing the specimen quite close to the target. Thus the transmission type of target is employed as shown rather than a solid target which would have to be recessed a considerable distance from the window and thus would require very long camera distances and very long exposures to obtain the needed magnification. The need for small camera distance and the desire to visually focus at top magnification, dictate a window thickness of 50 micrometers or less. A representative value is 20 microns for a beryllium window with a 0.2 micrometer thickness of heavy metal for the target. Even with the most favorable specimen, magnification at the plate is generally set at 200 X or below with the full magnification which may be as high as 2,000 X obtained through photographic enlargement. Visual focus is only done occasionally, to optimize the electron spot, *e.g.* when a filament is changed in the electron gun. The instrument is then in focus for the entire normal range of accelerating potential camera distance and specimen position. One simply puts the specimen in place and Photographs it.

The implementation of this basic design is shown in Figures 8.1 through 8.3. Figure 8.1a shows the upper front of the instrument. One notes the use of a standard photographic view camera film holder of 4x5 inch size (approximately 9 x 12cm). The control panel has a thermocouple type vacuum gage on the left side, a beam current meter with overload relay in the center and a kilovoltage meter on the right side. Knobs below adjust condenser and objective lens focus, bias voltage, filament temperature and acceleration potential. Lenses can be turned off individually during beam alignment and test. Inside the camera is a mechanical stage with X, Y, and Z motions which are controlled by calibrated screws through x-ray and light tight seals. The thinner parts of the camera casting have been filled with lead putty for additional x-ray shielding and the fluorescent screen is coated on lead glass used for 100kV x-ray installations. The camera is interlocked so that the beam will instantly turn off if any part of the protection is removed. Figure 8.2 shows two of the target plates, one right side up (on the left) the other with target side (also electron beam side) upper most. These target plates fit to an "O" ring seal on the end of the electron beam vacuum housing. The vacuum housing is 100mm diameter and the plates are 1mm less in diameter to permit the necessary slight lateral adjustment required to occasionally locate new target areas. This adjustment is made using 4 small screws which slide the target plate over the "O" ring seal. The thin window sits in a small recess in the plate. It is sealed into the plate by epoxy glue, then the target material is evaporated on to the window. Target material may be renewed as required by light polishing and subsequent evaporation of new target material of the same species. A damaged window may be removed for replacement by heating the epoxy to destruction on a hot plate. Figure 8.3 shows an exploded view of the electron beam focusing elements. The electron gun uses a simple hair pin wolfram cathode with a 0.5mm wehnelt opening. Its insulator is an epoxy casting with inert filler. The metal parts are mild steel and brass. The condenser lens is a symmetrical Einzel lens design. The central plate is cast in a molded epoxy insulator. This lens may be centered by external micrometer threaded screws. The Objective lens is the first two elements of an asymmetrical einzel lens which is completed by the target plate. The nominal separation between the center element and the target is 2mm, but this can be adjusted for extra high voltage operation or extra low voltage in the soft x-ray region by use of spacers. The 2mm spacing can cover the range from 30 kilovolts to 1.5 kilovolts with out need to refocus the beam. However a ten fold increase of x-ray intensity at the low end can be obtained by the simple expedient of placing a 1mm step at the outer edge of the target plate which will be used for the soft rays. This step moves the target to 1mm from the center lens element and when additional focus voltage is applied the lens has approximately 1/3 its former focal length, thus the gain in intensity. This 1mm spacing can only be used in the range up to 12.5 kilovolts however. The objective lens insulator is machine turned and must be held to 1/4 milliradian parallelism to avoid coma. None-the-less the objective elements can be easily dismantled for cleaning and replaced without fear of losing centering or parallelism. A moderately skilled operator can completely dismantle the electron beam column, clean it, assemble and realign it in two hours. The vacuum housing is made of mild steel with flash nickel plating to prevent rust. Since the target plate is of the same material the beam is totally enclosed in a magnetic shield. An additional mu-metal shield is placed inside between the condenser lens and the gun insulator thus affording additional shielding for the very sensitive electron source region. The final beam column element is a sector shaped multiple aperture mechanism seen in Figure 8.3. The click stop mechanism for changing from one aperture to another is best seen in Figure 8.1b just under the camera and on the right hand side of the visible part of the electron beam housing right at desk top level.

Once the electron beam column is understood, it becomes a less significant part of the technology. Its relevance to the microscopes operation recedes to a background chore when replacing gun filaments or cleaning apertures. The important considerations are the operation as an x-ray microscope. The projection shadow method of imaging has the following interesting characteristics:

1. It is a fixed focus method, the entire image space is in focus from the source to the photographic plate with equal resolution except for a small difference in the onset of Fresnel diffraction as the specimen is placed nearer the source.
2. The geometrical distortion due to the flat specimen is nearly compensated by the distortion due to the flat recording field out to approximately an angular field of 50 degrees total angle.
3. The magnification varies linearly with the ratio of "source to plate" over "source to specimen" distance. Thus a thick speci-

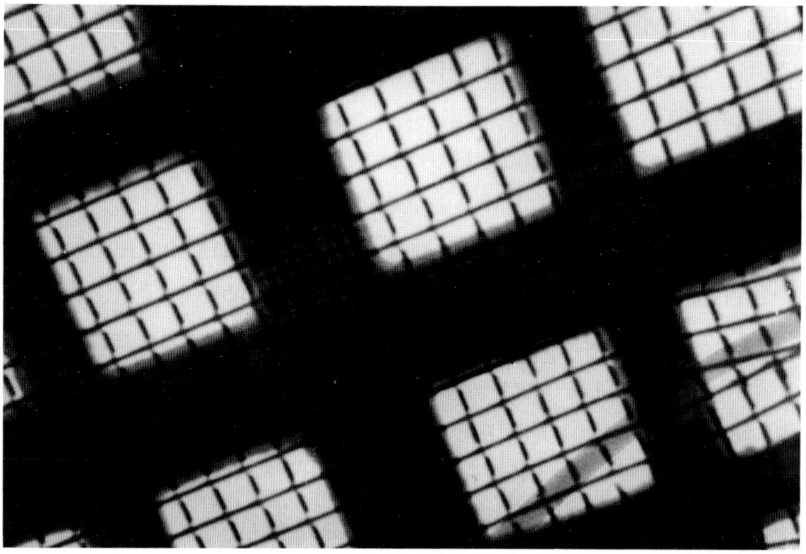

Figure 8.4. Silver grid 1500 mesh on 200 mesh "Anthene" copper grid. Silver bars are 3.5μm at widest point.

men has a noticeable variation in magnification from front to back giving an effect of enhanced perspective. For accurate measurements a known specimen shift may be used in a double exposure to give a distance marker at each position in the image.

4. Stereo pairs are readily produced by a small lateral displacement between exposures equal to 63 mm divided by the viewing magnification.
5. Compared with the contact shadow method a new variable is added, namely, the effect of scattering. It is primarily observed as a haze which lowers contrast with very soft radiation, but as pointed out below it could be used to enhance contrast.

In all other respects shadow projection images are the same as any other x-ray images.

These images can be appreciated by a few examples. In Figures 8.4 to 8.10 we show some representative pictures which are briefly described in the list of captions.

One of the useful applications has been time lapse motion pictures of stress corrosion. We also took time lapse movies of small insects at three frames per second in which one could see the muscles in action which are connected with breathing and locomotion. The camera was removed and replaced by a 16mm film transport driven by a speed controlled motor. More refined set-ups were made later. By placing a double deflection unit in the region just before the objective lens we were able to produce three dimensional movies of crystal growth which were shown at the 1958 EMSA meeting in Los Angles.

The design was planned to easily accept a vacuum camera. The authors version of the vacuum camera is shown in Figure 8.11. It has

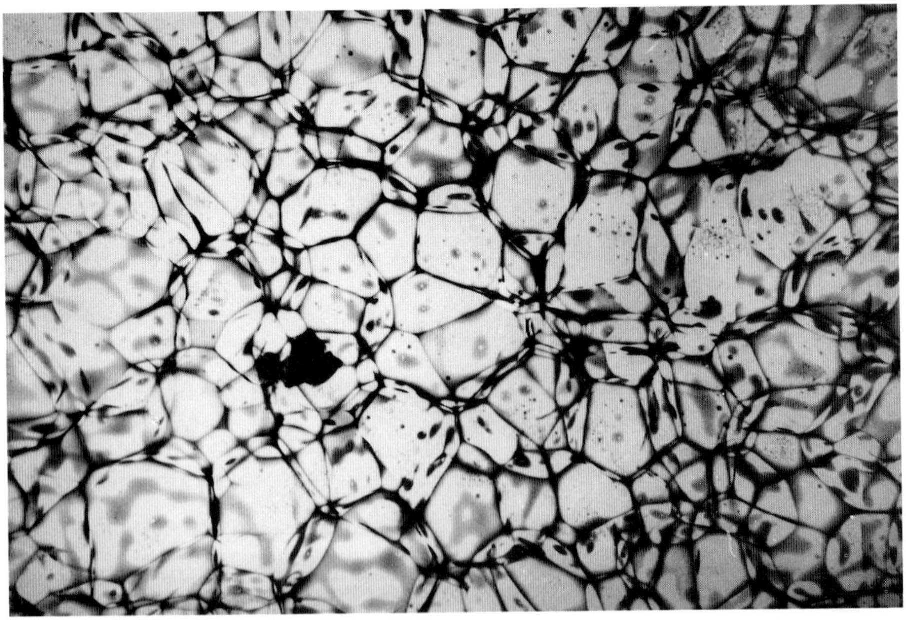

Figure 8.5. X-ray micrograph of Aluminum 5% Sn alloy, 1/2mm thick. The image was taken with a Walfram target at 20kV with 3min exposure on Kodak Lantern Slide medium contrast plate. magnification on plate was appox. 20X.

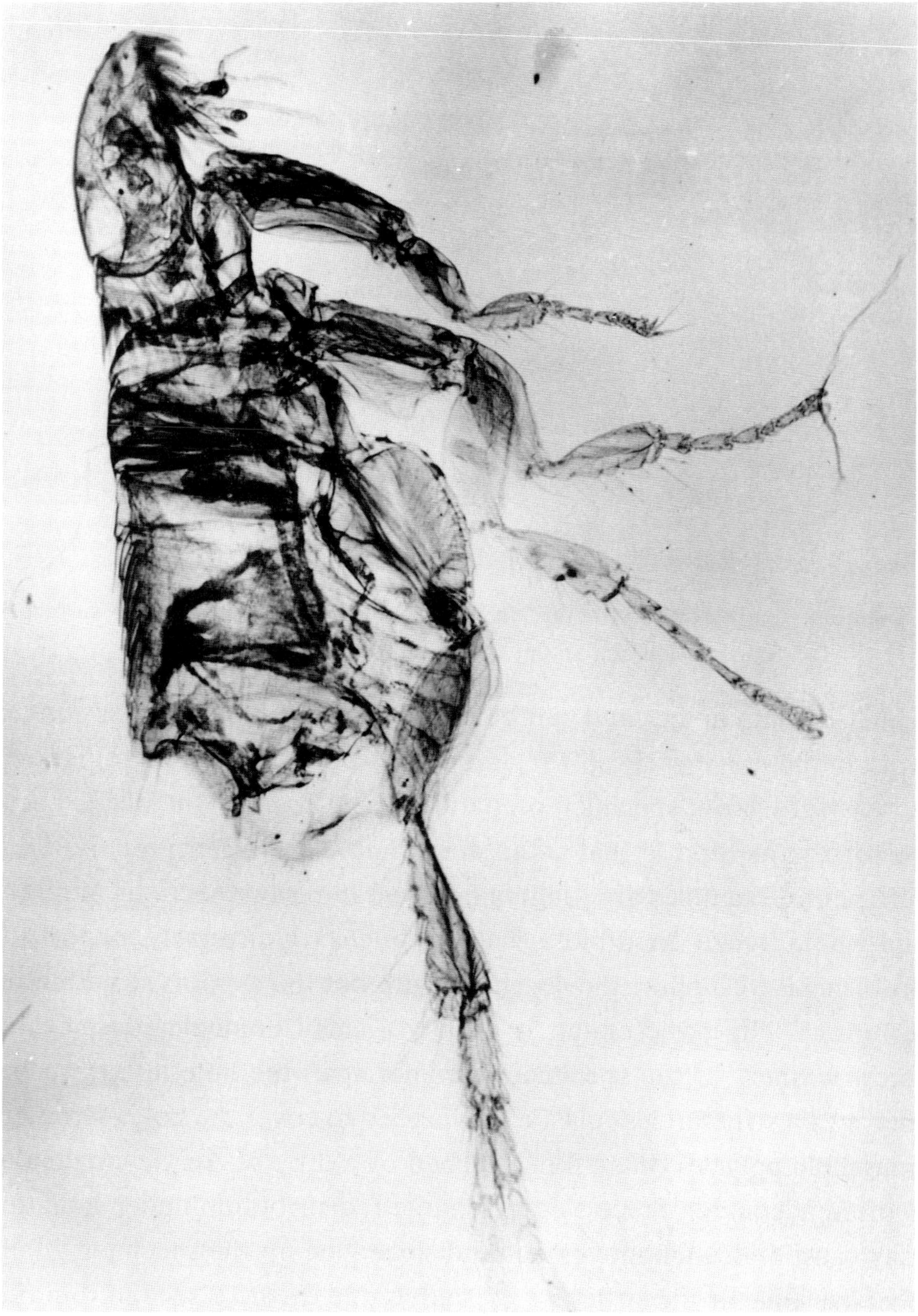

Figure 8.6. X-ray shadow projection micrograph of a flea. The specimen was freeze dried preparation. No fixatives or contrast media used. Note a cotton fiber caught in tarsus.

Figure 8.7. Dendritic structure in a Niobium alloy sheet showing surface corrosion and an unexpected third phase.

a built in plate holder and can be used for either projection or contact work in the moderately soft region of the spectrum (*e.g.* Al radiation).

A word about specimen preparation is in order. One of the beauties of the method is that often very simple specimen preparation is sufficient. Examples are roughly polished thin sawn sections of metal, and whole freeze dried biological specimens. Conversely, one of the problems is to support the specimen with out the support structure interfering. The basic mount is usually a small nonmagnetic machine screw washer. If the specimen does not span the hole in the washer then a thin structureless plastic film is used to cover the hole. Ordinary household polycarbonate film attached by a drop of Acrylcyanate glue is often adequate. Since electric charge redistribution under radiation may cause specimen movement and since glue on the specimen would obscure some of the structure it is advisable to add a second thin film stretched over the top of the specimen. For thin sections the household film is too thick and one must cast his own by one of the electron microscope methods. The availability of relatively simple methods to produce silicon nitride films on silicon frames may offer a universal support membrane by using two frames one above the other below the specimen. Finally the use of contrast enhancement has been relatively primitive up to this time.

More needs to be said about the use of the vacuum camera. It came as a surprise that vacuum exposures of five micrometer sections of mouse tibia were inferior in contrast to exposures in air. The problem is that the very soft radiation (*e.g.* 8.3Å from the Aluminum target) scatters appreciably in both the specimen and the air but the specimen scatter fuzzes out the sharp boundaries where as the air scattering produces a uniform background increase which is less dele-

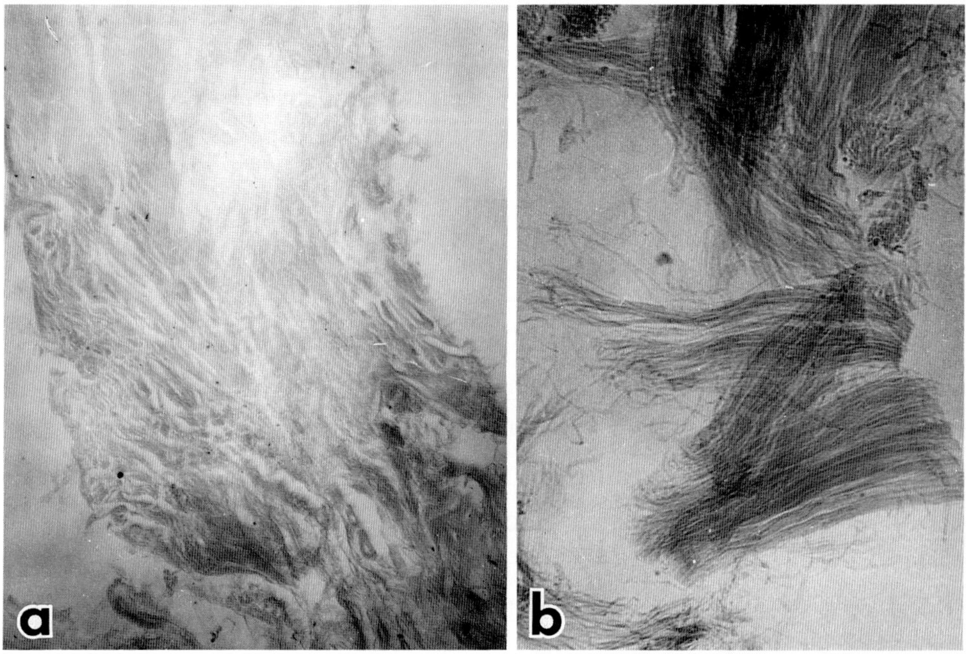

Figure 8.8. (a) Twenty five micrometer thick section of fetal mouse tibia taken with vacuum camera with an Aluminium target at 5kV. (b) Reconstituted collagen fibers stained with plutonium. Sample courtesy of Dr. Reuben Borasky while at Handford plant. Chromium target.

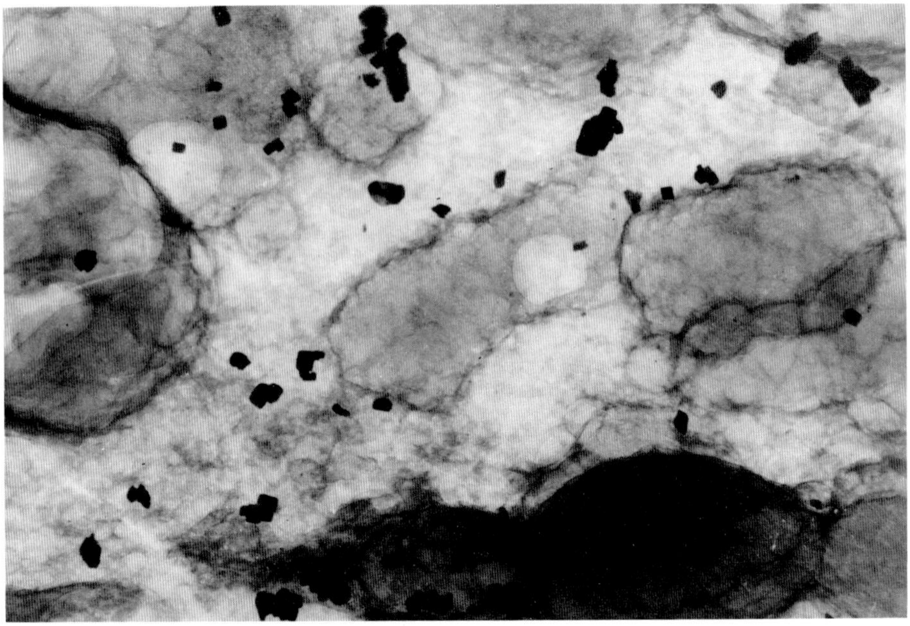

Figure 8.9. X-ray projection image of potato chip (also called potato crisp). Thin potato slices deep fried. Note sodium chloride crystals are more x-ray opaque than starch and cell debris.

terious to contrast. The air also absorbs the longer wave length modified rays preferentially thus acting as a filter as well as contrast diluter. The scattering problem was avoided by using the projection instrument as a convenient source for contact radiography see for example Saunders (1969) or Butler *et al.* (1969). It would be better to restore the advantages of the projection method by, developing a vacuum filter based on the well known Bucky grid (alsc know as Potter-Bucky Grid) of medical radiology see for example Selman (1972) page 332. For the microscope the grid should be a lamella of a sphere concentric with the source and imposed between the specimen and the photographic plate.

The design of this microscope dates back to 1954. A great many improvements in supporting technology have been made during the

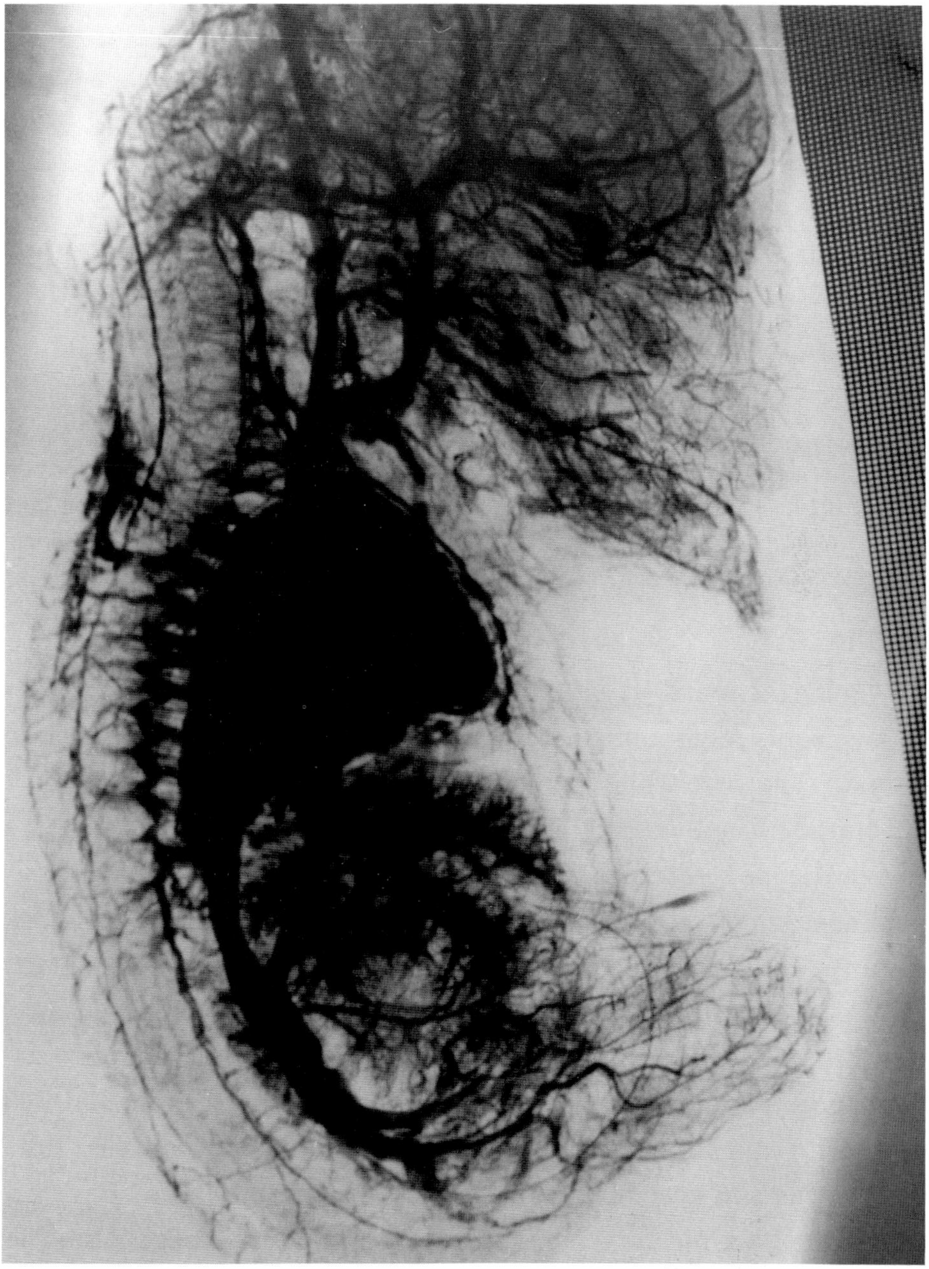

Figure 8.10. Paraffin embedded rat embryo (5mm thick whole mount). The rat embryo was perfused with x-ray opaque solution. The objective is to check the orientation of embedding prior to section for proper orientation. Sample courtesy of Dr. Eric L. Effman of Duke Medical School.

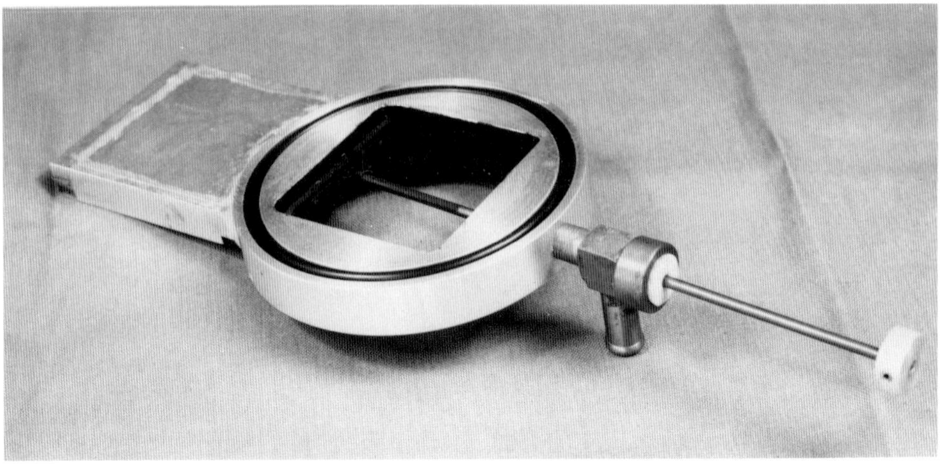

Figure 8.11. Another version of vacuum camera which rests directly on the target plate and has own light tight compartment for a 5cm square plate.

intervening 32 years. It is therefore appropriate to consider how these improvements might enhance the usefulness of the shadow projection method. The first which comes to mind, is the considerable gain of convenience if one adds a digital electronic image capturing system to the microscope. Alternatively, if one has a separate image processing equipment which can work from the x-ray negatives, then one needs only a very high gain image intensifier tube to assist in visual alignment and focus. To obtain the same visual capability without the image intensifier tube one must sit in total darkness for one half hour to dark adapt the eyes and then operate the controls using a faint red light. Ernest F. Fullam Inc. loaned us one of their image intensifying tubes. Visual focus and specimen positioning then became more convenient than taking Polaroid survey exposures. With out the image intensifier, visual focus and alignment is rarely used. Davies *et al.* (1986) have recently described another image intensifier for their shadow projection version of the microscope.

Another obvious improvement would be the substitution of a turbomolecular pump or other clean pumping system for the diffusion pump. The cleaning of apertures, a major nuisance, could then be a rare event. Another obvious improvement would be to change to one of the modern high brightness cathodes such as lanthanum hexaboride (LaB_6, which would probably require addition of a small ion pump). It is probably not yet practical to put a field emission gun in the microscope. The considerations for this step are complex and need to be addressed in a separate paper.

Summary

In summary the shadow projection method fills a very valuable technological niche in the array of equipments for delineation of microstructure. A few microscopes are still available for cooperative research. If more of these microscopes are to be made available the hard ware needs to be updated to take advantage of the many advances in supporting technology since the design was set 32 years ago.

Editors' note: following references are recent works on x-ray shadow projection microscopy: Hall *et al.* (1972); Horn and Waltinger (1978); Saunders *et al.* (1971); Yada and Ishikawa (1980); Takahashi *et al.* (1983).

9. The Application of Synchrotron Radiation to X-ray Imaging

P. J. Duke

SERC Daresbury Laboratory, Daresbury, Warrington WA4 4AD, UK

Introduction

Many phenomena in chemistry and biology occur at interfaces where two matrices interact in a variety of ways across the boundary. The list of such processes is almost endless. Semiconductors, catalysts, corrosive action, adhesion and electrolyte processes all depend on the physico- chemical properties of surfaces and interfaces. Life at the cellular level depends on the steady intake of nutrients across the cell membrane and the expulsion of cellular products. These processes depend in turn on the properties of the membranes which act as a boundary between the cell and its immediate environment. The breakdown of cellular life can be induced by entities such as viruses which penetrate the boundary, with devastating consequences not merely for the life of the cell but often also for the entire organism.

The imaging of phenomena such as these, in real space, and in real time, so that the observer is presented with a detailed picture, offers a tremendous challenge to chemists and biologists and demands the development of new techniques of observation. All imaging techniques are limited by spatial resolution (the ability to distinguish one adjacent point from another), by contrast (the ability to distinguish one physical or chemical region from another) and by time resolution (the ability to observe temporal change as it takes place). Each of these limitations is dependent on the physical characteristics of the specimen

X-ray Microscopy
Ed. by P. C. Cheng and G. J. Jan

(depth, density, crystallinity *etc.*), on its chemical composition (light or heavy elements for example) and on the rate of change of its properties with time. The highest spatial resolutions have been achieved with electron microscopy with contrast down to the atomic or molecular level in the case of high energy transmission electron microscopy through thin inorganic specimens with well defined structural patterns. Biological materials give poorer contrast which must be enhanced by staining techniques, complimented by freeze drying and slicing to submicron thicknesses. All of these methods are essentially static and dynamic life processes must be arrested before observation can take place. At the other end of the scale, optical microscopy can view dynamic processes directly, with large volume specimens (provided they are transparent) but with resolution limited by the shortest wavelength (200-300nm) which can be used for direct observation.

Imaging using x-rays is limited, in general, to resolution levels inferior to those obtainable with electrons in the case of specimens <100nn in thickness although useful complementary information may be obtained. For thicker specimens (either frozen or in their natural state) the greater penetrating power of x-rays may lead to comparable resolution values. The contrast enhancement techniques applied to specimens prepared for electron microscopy may be carried over to x-ray microscopy but the latter also offers the possibility of natural contrast enhancement by means of differential absorption of x-rays in the vicinity of absorption edges. The application of dynamic methods to electron microscopy is limited both by specimen preparation requirements and by the need to place the specimens in an ultra high vacuum. The use of x-rays to image kinetic phenomena is now beginning - the present limitation is the intensity of the sources available and once this limitation is removed, as is quite likely over the next few years, then kinetic work using x-rays could be an important new field of investigation. This highlights the importance of developing x-ray imaging techniques now and applying them to static problems so that they can be applied to dynamic situations once higher x-ray intensities are available.

Synchrotron Radiation as an X-ray Source

Synchrotron Radiation (SR) provides an excellent radiation source for a wide variety of scientific applications, including x-ray imaging. The radiation is produced from a beam of electrons stored between the poles of the dipolar magnets of an electron storage ring. Such an arrangement is shown diagrammatically in Figure 9.1. The total radiated power (P) in watts from the dipole magnets integrated over the entire spectrum is given by Eq. (9.1).

$$P = 88.5\, E^4\, I/R = 26.5\, E^3\, IB \qquad (9.1)$$

Where E is the electron beam energy in GeV, R is the radius of the electron beam trajectory in metres in a magnetic field of strength B Tesla and I is the beam current in milliamps.

The spectrum itself is shown in Figure 9.2 for the case of the synchrotron radiation source (SRS) at the Daresbury Laboratory, for which E = 2 GeV. The ordinate axis of Figure 9.2 shows the photon flux (ϕ) integrated over all vertical angles in units of photons/(sec•mrad) into 0.1% bandwidth ($\Delta\lambda/\lambda$), when I=100mA. Flux curves are shown for the dipole magnets (λ_c = 0.4nm) and for a 5T 3-pole wiggler magnet (λ_c = 0.1nm). The range of vertical angles over which the radiation is emitted is given approximately by Eq. (9.2).

$$\sigma_{Ph} = \frac{0.57}{\gamma}\left(\frac{\lambda}{\lambda_c}\right)^{0.43} \text{radians} \qquad (9.2)$$

$$\phi(\lambda = \lambda_c) = 1.6 \times 10^{10}\, EI \text{ photons/ sec} \bullet \text{mrad)} \qquad (9.3)$$

$$\lambda_c = 0.56\, R/E^3 = 1.86/(BE^2) \text{ nm} \qquad (9.4)$$

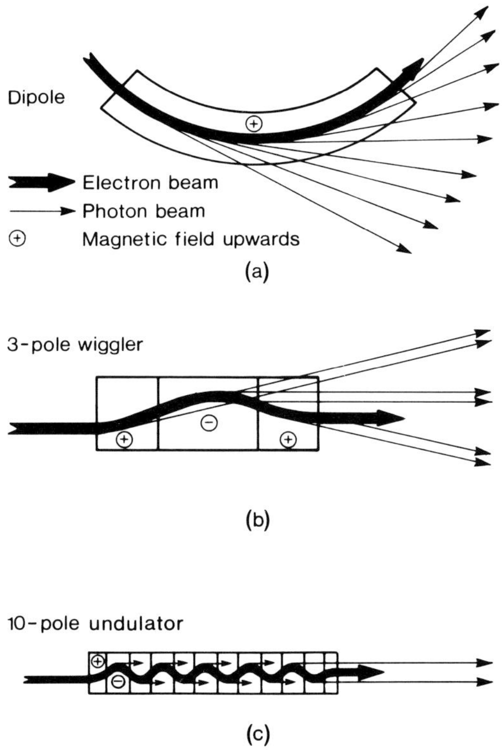

Figure 9.1. Production of synchrotron radiation: (a) by deflection of an electron beam in the magnetic field of a dipole magnet: (b) by successive large amplitude deflections in a wiggler magnet and (c) by successive small amplitude deflections in an undulator magnet.

in which $\gamma = E/m_0c^2$ and m_0c^2 is the electron rest mass in energy units (0.511MeV). σ_{Ph} can be regarded as the standard deviation of a Gaussian distribution (at least when $\lambda \lesssim \lambda_c$). The quantity λ_c is a scaling parameter (often called the characteristic wavelength) so that all storage rings give the same photon flux at the same value of λ/λ_c, for the same value of the product EI. When $\lambda = \lambda_c$ the photon flux is given by Eq. (9.3). Flux calculations for any value of λ/λ_c can be carried out using the detailed theory of SR production (Krinsky *et al*, 1983). These show that the spectrum exhibits a broad maximum whose peak is at $\lambda/\lambda_c \simeq 10$ and that useful flux is obtained even when $\lambda/\lambda_c \simeq 10^{-1}$ although the usefulness depends on the experiment.

In an actual SR source account must be taken of the finite dimensions and divergence angles of the electron beam which emits the radiation. These quantities are different for each installation and, for a given storage ring, depend on the position of the radiating region around the circumference of the electron orbit. It is quite common for the electron beam divergence to be greater than σ_{Ph} as given by Eq. (9.2), particularly in the x-ray region when $\lambda \lesssim \lambda_c$ and $\sigma_{Ph} \lesssim \gamma^{-1}$ It is useful to define a quantity often called the source brightness:

$$\text{Brightness} = \frac{\text{No. of photons / sec}}{(\text{mrad})^2} \text{ into } 0.1\% \text{ bandwidth} \quad (9.5)$$

The source area can be taken into account by defining

$$\text{Brilliance} = \frac{\text{No. of photons / sec}}{(\text{mrad})^2 \bullet (\text{mm})^2} \text{ into } 0.1\% \text{ bandwidth} \quad (9.6)$$

A second point, which is very important to note, is that the brilliance of the storage ring defined by Eq. 9.6 cannot be increased by any optical system. At best, according to Liouville's theorem, this quantity is conserved. In practice the brilliance of the radiation received at an experiment is reduced by inefficiencies in the mirror/monochromator system and in particular by the inability of most imaging systems to accept all the available photons.

For this reason the development of SR sources has been towards the design of sources with high intrinsic brilliance and has followed two main directions:

(1) the construction of sources with low electron beam dimensions and

(2) the use of multipole insertion devices to enhance the SR output in a particular region of the spectrum.

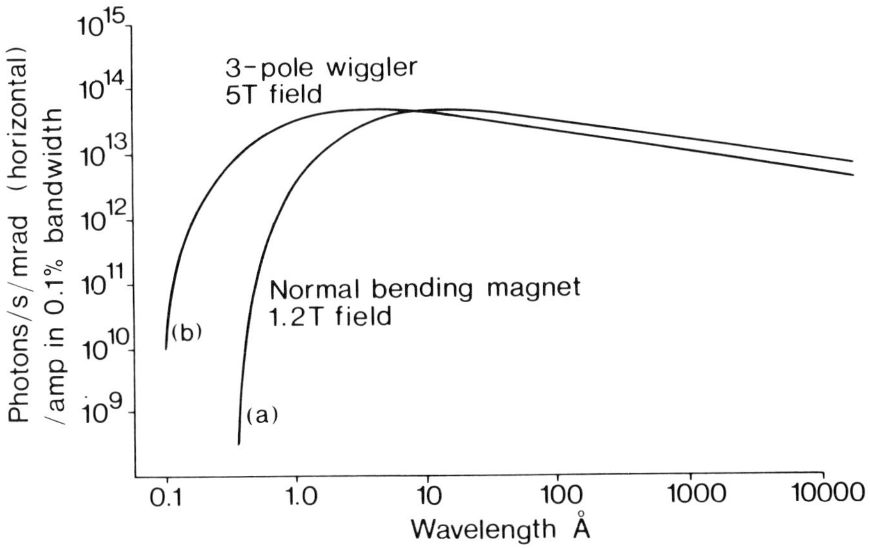

Figure 9.2. Flux spectrum from the Daresbury SRS, from the 1.2 T dipole deflecting magnets and from the 3-pole 5T superconducting wiggler magnet.

Insertion devices, which can be placed in the drift-space between successive dipole magnets, are illustrated in Figure 9.1(b) and 9.1(c). There are two types of insertion device, the multipole wiggler, where the electron beam deflection angle $\alpha >> \sigma_{Ph}$ and the undulator for which $\alpha \lesssim \sigma_{Ph}$. The former device acts as a wavelength shifter in which $\lambda_c(w) = \lambda_c(D) \bullet B_w/B_D$ where D and w refer to the dipolar and wiggler magnets respectively and B is the magnetic field strength in each case. In these devices $B_w >> B_D$ and the brilliance is enhanced at short wavelengths mainly on account of the flux gain resulting from the low value of $\lambda_c(w)$ as shown in Figure 9.2. In the undulator, on the other hand, $B_u << B_D$ and an observer close to the axis of the device will see an N-fold increase in brilliance at long wavelengths since in this case the low magnetic field makes $\lambda_c(u) << \lambda_c(D)$. The increased brilliance comes from the superposition of the sequence of sources generated by the N successive periods of the undulator, so that, at $\lambda = \lambda_c$, the photon intensity, $J(\lambda = \lambda_c)$, is given approximately by equation 9.7 in which $K = \alpha\gamma = 0.0934\ \lambda_u B_u$ and λ_u is the undulator period in mm.

$$J(\lambda = \lambda_c) \;=\; 3.3 \times 10^{10} \times \text{NKI photons / sec} \qquad (9.7)$$

$$\lambda_c \;=\; \frac{0.27\ \lambda_u}{(KE^2)}\ \text{nm} \qquad (9.8)$$

A photograph of the undulator installed in the Daresbury SRS is shown in Figure 9.3. Equation 9.7 is valid only when K>1 and the actual behavior of an undulator is much more complex. It has been shown (Krinsky *et al.*, 1983) that, in addition to the source superposition referred to above, constructive interference of the radiation from the sequence of sources will also generate a harmonic series of intense radiation maxima at wavelengths λ_i given by Equation 9.9.

$$\lambda_i \;=\; \frac{\lambda_u}{2\gamma^2 i}\left(1 + \frac{K^2}{2} + \gamma^2\theta^2\right) \qquad (9.9)$$

in which i is the harmonic number and θ is the angle between the undulator axis and the direction of observation. An undulator can be designed to give enhanced output at those wavelengths which will be of most value to the experimenter. In practice, however, because of the finite size and divergence of the electron beam, the conditions for an ideal undulator are never realized and the experimentalist must be content with a photon brilliance limited by the electron beam parameters of the storage ring. Table 9.1 shows the expected brilliance in the soft x-ray region of several undulator magnets installed or planned for installation. It can be seen that future x-ray sources can be very brilliant indeed and offer quite new potentialities for x-ray imaging. Figure 9.4 shows the spectrum from the SRS 10-pole undulator compared with that from the dipole magnets. An enhancement factor $\gtrsim 10$ is apparent in the flux from the undulator. The irregularities in the undulator spectrum correspond to the harmonic series of Eq. (9.9) with $\lambda_1 = 19$nm. The development of x-ray microscopy using such sources is already beginning at the NSLS, Brookhaven (Jacobsen and Rarback, 1986) and at the Daresbury SRS (Poole *et al.*, 1983), and some of these developments are described elsewhere in these proceedings. The remainder of this brief review will show how the less brilliant radiation from the dipole magnets of the SRS is being used to image crystalline objects by means of diffraction imaging and non-crystalline objects by means of x-ray absorption.

Table 9.1

Source	Location	Status	Energy	Brilliance[a]
BESSY, Dipole	Berlin, FRG	Operating	800MeV	2×10^{13}
SRS, Undulator	Daresbury, UK	Operating	2GeV	2×10^{14}
NSLS, Dipole	Brookhaven, USA	Operating	700MeV	6×10^{12}
NSLS, Undulator	Brookhaven, USA	Operating	2.5GeV	1×10^{16}
1-2GeV, Undulator	Berkeley, USA	Proposed	1.5GeV	5×10^{18}
ESRF, Undulator	Grenoble, France	Proposed	6GeV	1×10^{18}

[a]Brilliance is defined as photons/($mm^2 mrad^2$ sec) into 0.1% bandwidth at an x-ray wavelength of 4.4nm and a circulating beam current of 100mA.

X-ray Diffraction as an Imaging Method

One form of x-ray imaging, known as topography, is being used as a tool in solid-state chemistry to image chemical changes taking place at the boundary between different regions of a crystalline material. The method relies on x-ray diffraction (Bragg reflection) from crystal planes in order to map the 3-dimensional volume of the crystal onto the 2-dimensional surface of a detector such as a high resolution photographic film. Contrast results from defects in the crystal lattice or from internal changes (lattice spacing or orientation changes for example), which occur naturally or which can be induced by changes in the external environment. In transmission topography, illustrated in Figure 9.5, the observer sees a Laue diffraction pattern in which each Laue spot provides a map of the components of the strain field within the crystal. This topograph was taken using the Daresbury SRS (Herres and Lang, 1983) and illustrates the use of an almost parallel beam of x-radiation with a broad energy band width to provide a whole series of topographs from a single orientation of the crystal. This technique can be extended to the study of dynamic phenomena

Figure 9.3. The permanent magnet undulator at the Daresbury SRS. The gap between the poles of the magnet can be adjusted to provide a range of field values up to 0.34T.

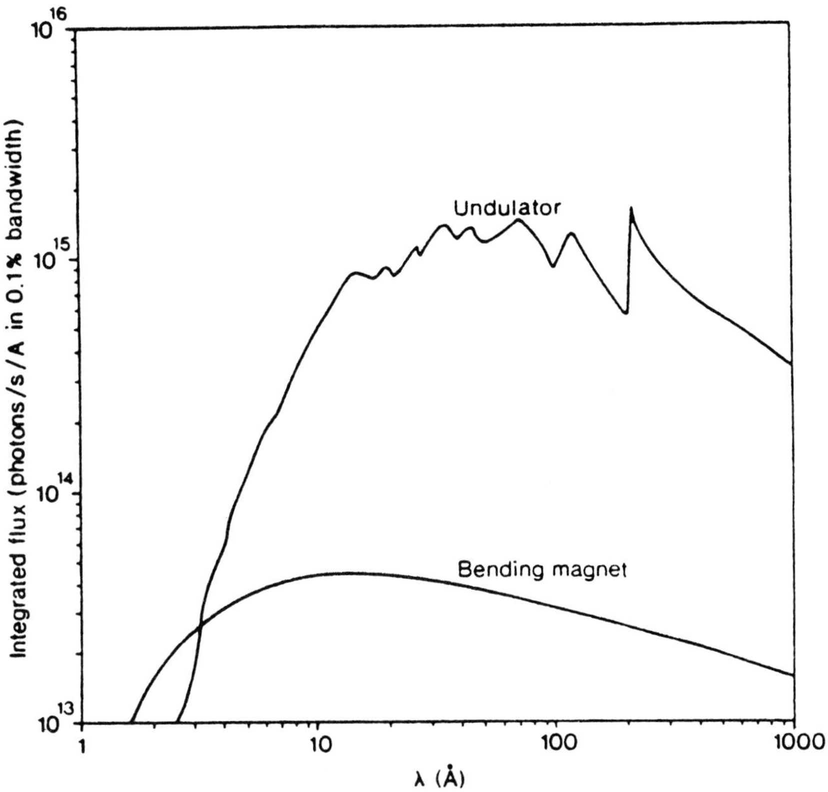

Figure 9.4. The flux spectrum from the Daresbury SRS undulator compared with that from the dipole magnets.

either by the use of electronic imaging detectors or by using the pulsed time structure of the radiation as a strobe to present a 'frozen' topograph of a regular recurrent phenomenon. Figure 9.6 shows a topograph of s surface wave traveling across a lithium niobate crystal (Whatmore *et al.*, 1982). The transducer which stimulated the wave was pulsed in phase with the arrival of the x-ray pulse from the synchrotron. The time width of the x-ray pulse was about 300ps with a repetition frequency of 320ns. The peaks and troughs of the travelling wave can be clearly seen. It is interesting to note the interaction of the wave with imperfections in the surface structure and to observe that there are phase changes presumably induced by surface changes which could be of a physical or chemical nature. Similar work at other synchrotron centres has demonstrated that bulk progressive waves can also be revealed and standing wave patterns could also be photographed by this method. These are all essentially physical phenomena but the possibility of using wave phenomena in a crystal as a chemical probe should not be overlooked. Although the method is applicable only to regular crystalline solids nevertheless a wide variety of materials covering a broad density range. Chemical composition and crystal volume can be studied under many different physical conditions and combining both spatial and temporal resolution with good, naturally occurring, contrast.

Imaging by X-ray Absorption

The dominant absorption process for x-rays of wavelength >1nm (energy $\lesssim 10$keV) is the photoelectric effect in which an x-ray photon is absorbed by an atom with the emission of a photoelectron. This process is described in terms of a mass absorption coefficient, μ and by an absorption law $I = I_o e^{(-\mu\rho t)}$, where I_o is the incident x-ray intensity, I the transmitted intensity, t is the thickness and ρ density of the material. The product $\mu\rho$ is termed the linear absorption coefficient. Values of μ for compounds can be calculated approximately as the weighted sum of the mass absorption coefficients so that $\mu_t = \sum_j \lambda_j \omega_j \mu_j$, where ω_j is the atomic mass number divided by the molecular weight and λ_j is the number of atoms of type j in a single molecule. Figure 9.7 shows values of the linear absorption coefficient for several elements and compounds over a range of x-ray wavelengths. In general $\mu\rho$ varies smoothly except at certain well defined points on the energy scale where absorption edges occur. These correspond (in the case of the K-series of absorption edges) to the energy required to eject an electron from the innermost electronic shell which, in the case of the carbon atom, occurs at 4.4nm (282eV). Water, on the other hand, will continue to be relatively transparent until an x-ray energy (corresponding to the oxygen K-edge) of 532eV is reached. This difference in x-ray opacity is the basis of the lithographic method of x-ray microscopy illustrated in Figure 9.8. A biological object, enclosed in its native environment, can be placed on an x-ray sensitive surface. A

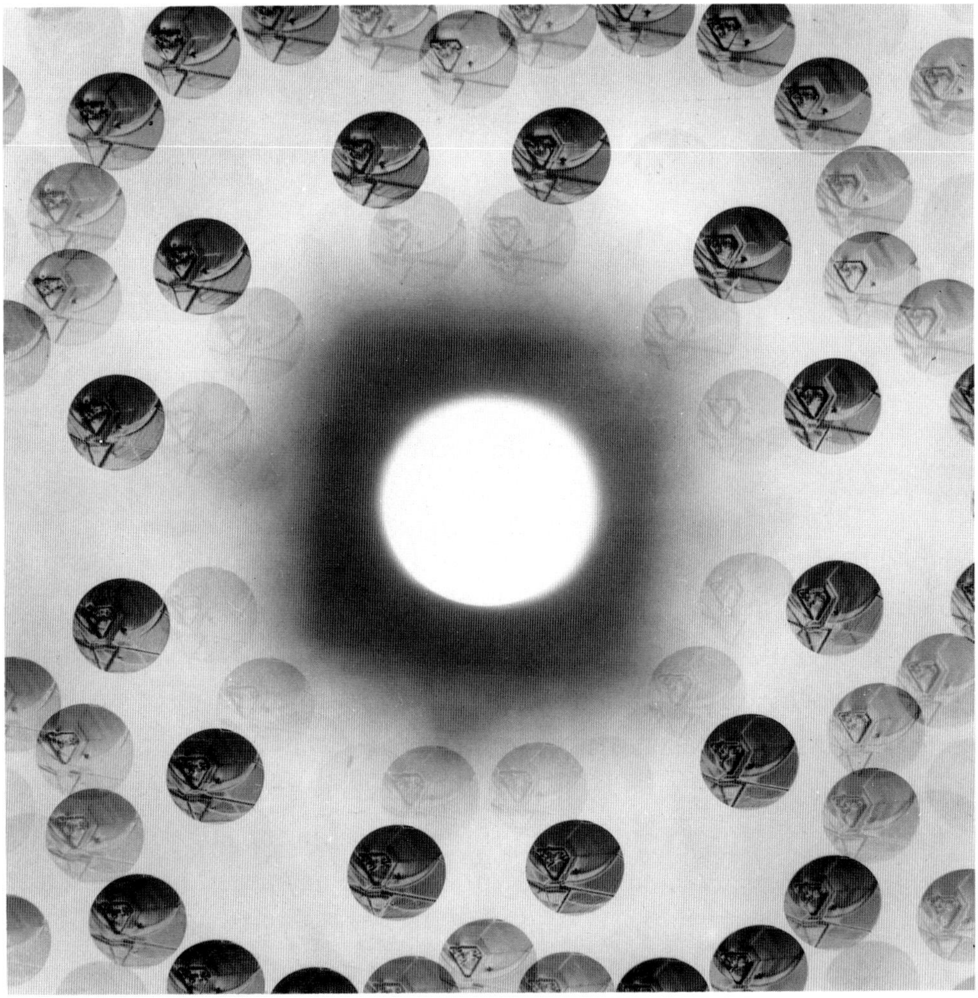

Figure 9.5. Synchrotron radiation x-ray topograph of a beryl crystal showing the Laue pattern of topographic images.

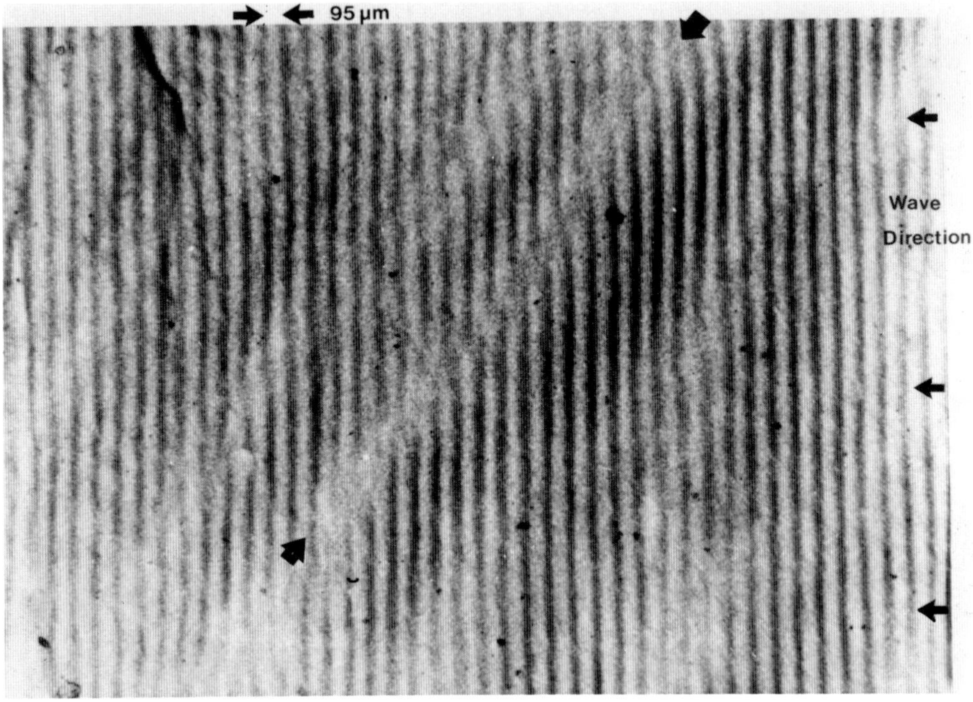

Figure 9.6. Surface acoustic waves travelling across a lithium niobate crystal. The heavy black arrows indicate a surface feature which is inducing a 90° phase change to the propagating wave.

parallel beam of x-rays will be absorbed as it passes through the specimen, so that an intensity profile is projected on to the surface. The latter is usually a layer of photoresist material (PMMA polymer is often used) which retains the intensity profile in the form of radiation damage. After exposure the specimen is removed and the photoresist is chemically etched (the etch rate depends on the degree of radiation damage) to reveal the intensity profile as a contour map of x-ray absorption in the specimen. This profile can be examined in an electron microscope. The method combines the attractive features of x-ray microscopy (penetration through several microns, wet specimens and natural contrast via differences in linear absorption coefficient) with those of electron microscopy (high spatial resolution and convenience of examination under standard conditions). Figure 9.9 shows an example of some recent work (Sylvia *et al.*, 1986) at the Daresbury Laboratory carried out by Dr. Sylvia Richards (Keele University) and colleagues from Daresbury. It is part of a programe to study the take-up sites of lead in common earthworms. Figure 9.9(a) shows a TEM micrograph of a thin section of earthworm tissue taken from a worm which had matured in a lead environment. This is compared with an SEM micrograph (Figure 9.9(b)) of a photo resist sheet exposed to the same area of tissue. The contact image shows a very clear replication of the lead inclusions, some of which are <100nm wide. Figures 9.9(c) and 9.9(d) show similar micrographs from normal earthworms in which the lead inclusion features are absent. The contact micrographs show a remarkable degree of detail and indicate that this method of microscopy can be used to supplement the direct electron microscopy of thin sections. The spatial resolution (Bowen, 1982) which can be obtained is determined by geometry and diffraction as shown in Figure 9.10. The resolution r is given by the quadratic sum of the individual contributions:

$$r^2 = (sb/a)^2 + (b\lambda) + r_{det}^2 \qquad (9.10)$$

The first term in Eq. (9.10) (R_g in Figure 9.10) is the geometric contribution and is the source size s multiplied by a geometry factor which is the ratio of the specimen to detector distance b and the specimen-to-source distance a. For Use synchrotron at Daresbury, s is <1mm in the horizontal direction, a is typically 10-15m so that if b is ~10μm then the contribution from this term is $\lesssim$1nm. The second term (R_d in Figure 9.10) describes the effect of Fresnel diffraction at the edge of the sample which, at an x-ray wavelength of 4.4nm, (corresponding to the carbon absorption edge) and for feature within 1μm of the detector surface, will be <100nm. The third term is the intrinsic resolution of the detector which for a high quality photographic plate will be ~250nm and for photoresist material developed for the electronics industry can be ≤10nm. This method of x-ray imaging (using photographic film as the detector) has been in use from within a few months of the actual discovery of x-radiation by Röntgen in 1895, and apart from the difference in geometry (b << a) it is little different from radiography as practiced in medicine and industry. What is new is that present day synchrotron sources offer high intensities which mean short exposure times (a few seconds) and the possibility of using monochromatic radiation to exploit the sharp changes in absorption coefficient which occur across absorption edges. The importance of high intensity is immediately apparent when the relationship between resolution and contrast is considered, because a tiny particle of dimensions comparable to the spatial resolution, r, will still not be observed unless its absorption compared with its environment is sufficient to create a detectable signal difference. Calculations using the absorption coefficients indicate that to detect a protein feature of 100nm dimensions in an aqueous environment 10μm in thickness would require an incident illumination of 10^{12} photons mm^{-2} on the specimen. Assuming that the synchrotron source is ~10m distant (determined by radiation shielding thickness and other practical considerations) then this translates to a required source brightness of ~10^{15} photons mrad^{-2} after allowance for beam loss in vacuum protection windows and the like. This is comparable to the calculated brightness of the SRS of 10^{15} photons/(s • mrad^2) into 10% wavelength or energy band width, and would indicate the synchrotron source is well matched for the requirements of x-ray lithographic microscopy with high resolution and good contrast.

X-ray Microprobe Imaging

The development of methods of direct high resolution x-ray microscopy is prevented by the difficulty of focusing x-ray beams. Unlike electro-magnetic radiation in the visible region of the spectrum, x-ray refractive indices differ negligibly from unity and x-ray lenses are an impossibility. Elsewhere in these proceedings, methods of x-ray focusing in the soft x-ray region are described using diffractive optics, and a comprehensive survey of this field is presented in Schmahl and

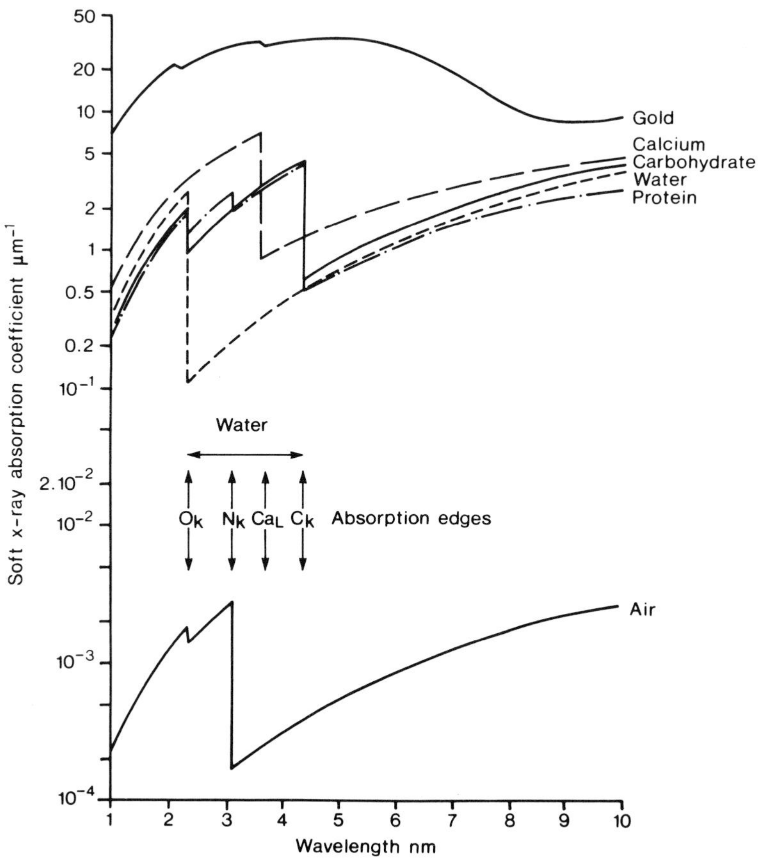

Figure 9.7. Linear absorption coefficients as a function of x-ray energy. The exact behavior at the absorption edge is shown diagrammatically only.

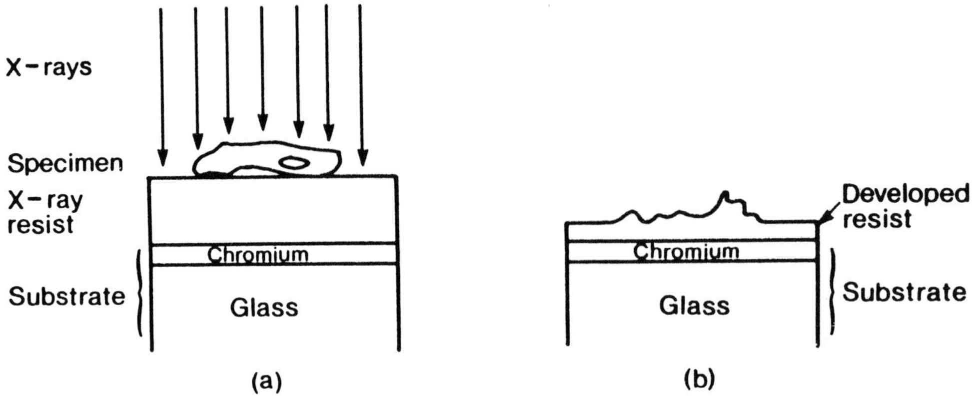

Figure 9.8. X-ray lithography geometry. (a) X-ray irradiation of the resist through the sample. (b) Resist after development.

Rudolph (1984). These methods can be applied only to soft x-rays, and harder radiation can only be focused using suitably figured mirrors from which x-rays are reflected at near grazing angles. However, the very high intensity of synchrotron radiation makes it possible to test methods of x-ray imaging using pinhole optics, first attempted by Horowitz and Howell (1972). Such a microprobe could be applied to surface studies with the detection of fluorescence x-rays or the detection of photon-emitted electrons. Both of these techniques could lead to high resolution mapping of surfaces by surface fluorescence or surface photoemission. A further possibility is that of using the x-ray microprobe to obtain a 3-dimensional reconstruction using the methods of computerized axial tomography (CAT) developed by Hounsfield (1973). Preliminary work by Elliott *et al.,* (1986) illustrated in Figure 9.11 which shows a tomographic slice of a bone chip from a human femur. The Haversian canals can be clearly seen. The pixel size in the reconstructed image shown is about 10μm. The principal difficulty of achieving higher resolution is the development of a rotating and scanning stage with a high degree of rotational stability.

Overcoming these problems could lead to 3-dimensional elemental mapping with sub-micron resolution.

Conclusion

The imaging of material properties using x-ray topography is opening new possibilities in solid state and surface chemistry and x-ray lithography can give radiographic images with resolution approaching 10nm. These results open up the use of new, natural contrast methods

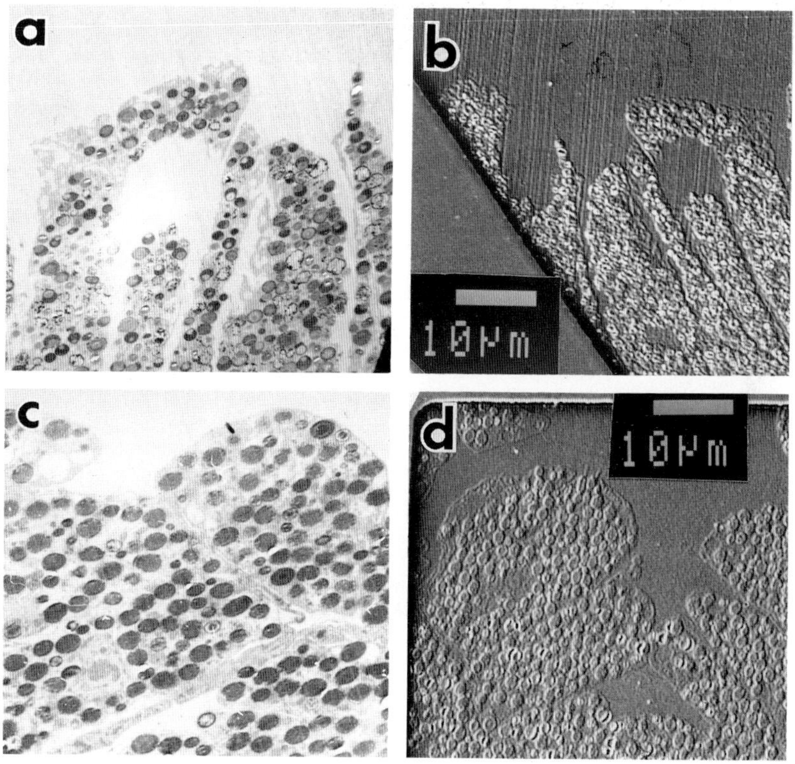

Figure 9.9. (a) TEM unstained section of lead contaminated earth worm cells (~x1300); (b) x-ray contact image of the same area; (c) as for (a) but control tissue; (d) x-ray contact image of control tissue adjacent section to (c).

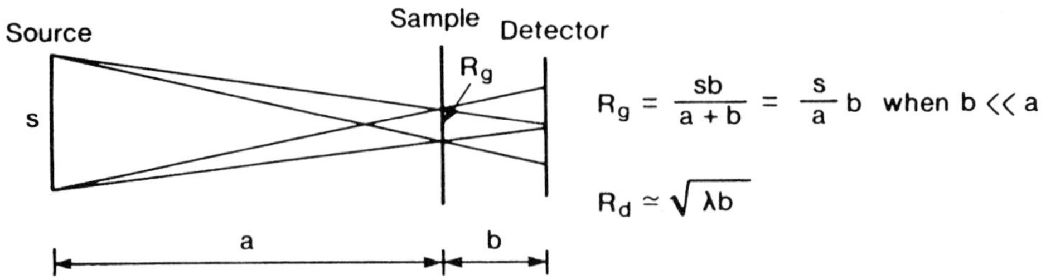

Figure 9.10. Contributions to resolution in contact imaging.

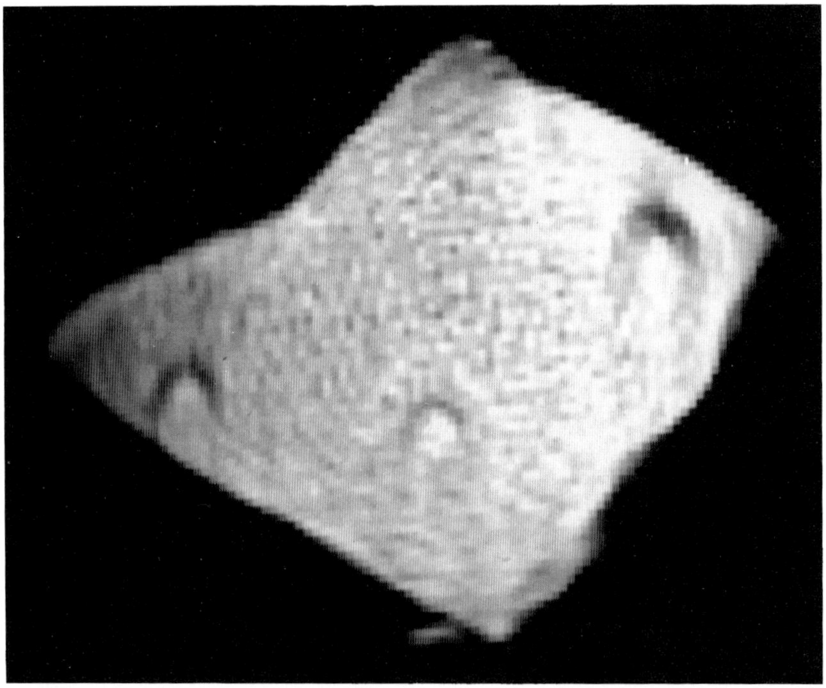

Figure 9.11. X-ray tomograph of a section of human femur obtained using x-rays from the Daresbury SRS.

for the detailed examination of materials on the micron and sub-micron scales.

Acknowledgements

I wish to thank Dr. D. K. Bowen (Warwick University), Prof. A. Lang (Bristol University), Dr. S. Richards (Keele University) and Dr. B. K. Tanner (Durham University) for the use of photographs and for their helpful advice. I am most grateful to my colleagues at the Daresbury Laboratory for their constant support.

10. Laser-produced Plasmas as Soft X-ray Sources

K. G. H. Baldwin[1] and R. J. Rosser[2]

[1]The Laser Physics Centre, Institute of Advanced Studies, Australian National University, GPO Box 4, Canberra City, ACT, 2600, Australia

[2]160 Blythe Road, London, W14 OHD United Kingdom

Introduction

A number of sources have been developed for the production of radiation in the soft x-ray region. While much work has been published on synchrotrons as continuous soft x-ray sources, recent experiments have shown that pulsed sources such as gas puff z-pinches and laser-produced plasmas can yield high x-ray fluxes in the 10 - 100 Angstrom region. This paper will outline the development of laser-produced plasmas as a soft x-ray source with particular application to x-ray microscopy.

Advantages of Laser-produced Plasmas

For the purposes of x-ray microscopy, an x-ray source must produce a sufficient flux of x-rays for the duration of the experiment to expose the image of the biological sample on the recording photoresist. However, this requirement must be balanced against the damage caused to the specimen by the incident x-ray flux. In the case of continuous, low flux devices such as synchrotron sources, the total x-ray dosage applied to the specimen is spread out over a long time scale such that the damage probability is low at any given instant over any given area. However, the total x-ray dose delivered may ultimately cause damage to the specimen.

X-ray Microscopy
Ed. by P. C. Cheng and G. J. Jan

In the case of pulsed x-ray sources, the aim is to expose the specimen to the entire flux instantaneously, such that any damage mechanisms which may alter the specimen's structure do not have time to be manifested before the image is recorded. Indications are that timescales for the x-ray pulse of the order of microseconds or less (Rosser, 1984) are sufficiently short that x-ray damage will not occur until the pulse is over.

Typically laser-produced plasmas produce x-ray emission on the nanosecond or even picosecond timescale. It is therefore possible using such a source to generate images of living biological samples, since the intense x-ray flux results in destruction of the specimen only after the image has been recorded. By careful choice of target materials and laser parameters it is possible to generate x-rays in the water "window" between 23 and 44 Angstroms. Therefore, x-ray microscopy of living biological specimens can be carried out in a hydrated environment.

Hence laser-produced plasmas have a considerable advantage over continuous sources which run the risk of damaging the specimen during the order of seconds exposure time required to produce the image. This is a result of the inherently low intensity of synchrotron sources compared to the high x-ray intensities generated by laser-produced plasmas (Figure 10.1, after Rumsby, 1985).

By comparison with the other major pulsed x-ray source (the gas puff z-pinch *e.g.* Feder *et al.*, 1984) the laser-produced plasma offers a number of significant advantages. Firstly, it reliably produces x-rays with every laser shot, which is not always true of the irreproducible z-pinch. Secondly, the position of the x-ray source is completely reproducible at the laser focus, whereas the gas puff x-ray emission originates from a dense plasma core which may move of the order of a millimetre from shot-to-shot. Thirdly, there is greater flexibility in the source to specimen distance using laser-produced plasmas, as the specimen may be located as near as ten millimetres from the laser tar-

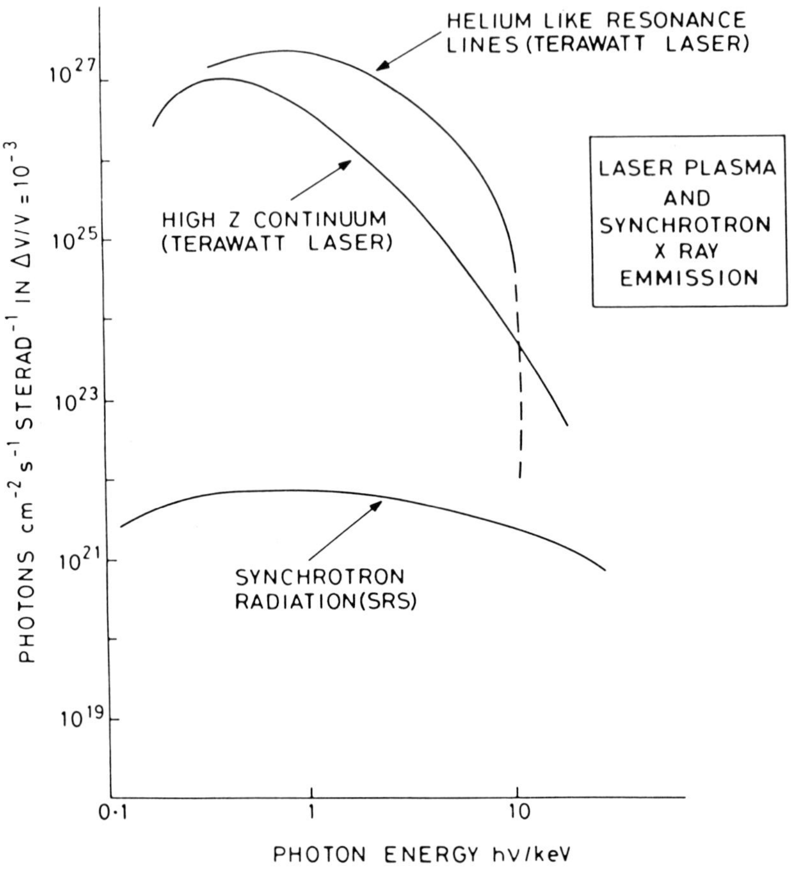

Figure 10.1. Comparison of instantaneous spectral brightness from line and continuum emission from laser-produced plasmas with that of the Daresbury synchrotron radiation source (Rumsby, 1985).

get (Rosser *et al.*, 1985a). Source to specimen distances of this size are difficult to achieve using gas puff z-pinches because of the inherently large initial diameter of the imploding gas and the size of the differentially pumped windows. Fourthly, the plasma emission region can be of the order of 10μm for laser produced plasmas compared with the larger (mm) emission regions in gas puff z-pinches (Bailey *et al.*,

1982). This can be an important factor if the microscopy resolution is source size limited due to penumbral shadowing.

Finally, the laser pulse may be split to produce multiple plasma x-ray sources on separate targets. This may be useful for producing sequential images of the time evolution of fast, microscopic processes using optical delays for laser pulses.

Laser-plasma X-ray Generation

The energy density of a focussed laser beam can be extremely high due not only to the large pulse energy, but also to the small focal spot sizes obtainable with high quality optics and low divergence laser beams. As an example, a single arm of the multi-beam Vulcan laser at the Rutherford Appleton Laboratory Central Laser Facility in the U.K. produces up to 200J per pulse with a pulse duration of the order of 1 nanosecond (>100 gigawatts), which when focussed to a $100\mu m$ spot yields intensities in excess of $10^{15}W/cm^2$. Even small laboratory sized lasers such as the 10J neodymium laser at the Australian National University Laser Physics Centre can, with careful attention to the laser optics, produce spots sizes of the order of a few microns and intensities ~$10^{18}W/cm^2$ (Burgess *et al.*, 1984).

However, such high intensities on target are not necessary or even desirable for x-ray microscopy applications. This can be understood by examining the mechanisms leading to x-ray production. The high laser intensity incident on the target causes rapid heating on the target surface, which in combination with other processes such as multiphoton absorption causes the material ablated from the target surface to be ionized. The resulting plasma then acts as a shield from the incoming radiation, which is either absorbed or reflected at or near a region known as the critical density surface. This is due to the shielding action of the electrons in the plasma which oscillate so as to cancel the high frequency electric field produced by the laser. As a first order approximation, the critical density region occurs at an electron density n_e given by equating the electron plasma frequency ($\omega_p = \sqrt{n_e e^2/\varepsilon_o m_e}$) to the laser frequency. For a 1.06μm neodymium laser the critical density surface corresponds to $n_e \sim 10^{21}cm^{-3}$.

The laser energy is absorbed by processes such as resonance absorption, parametric decay and ion-acoustic turbulence absorption at the critical density surface. The energy is transported inwards by a number of mechanisms e.g. via hot electrons generated by non-linear processes in the absorption region, which heat the high density material inside the expanding plasma plume. The upper limit on the plasma density is then given by the solid density near target surface ($\sim 5x10^{24}cm^{-3}$). Laser radiation is absorbed more efficiently at higher densities, and shorter wavelength lasers are therefore often employed which can penetrate to higher critical densities. Hence frequency doubling or even tripling is often used on laser systems to produce higher x-ray yields, even though the harmonic-conversion efficiency may be significantly less than unity (Yaakobi *et al.*, 1981). Depending on the laser wavelength and the focussed intensities used, plasma temperatures of 100-1000eV can he produced at densities of $10^{21} - 10^{24}cm^{-3}$. The presence of highly stripped ions in the plasma at these high temperatures results in the emission of energetic x-rays, since for example, the energy of hydrogen-like emission lines scales with Z^2 (Z being the nuclear charge).

X-ray yields of up to 20% of the laser pulse energy can be obtained from such plasmas (Mead *et al.*, 1981) in a pulse which roughly follows the pulse duration of the driving laser radiation. However, the x-rays may be distributed over a large range of wavelengths depending on the mechanism generating the x-ray emission. These processes may be non-thermal in origin, such as emission from inner shell x-ray transitions, or continuum emission from bremsstrahlung radiation generated by suprathermal electrons produced in the laser absorption region. Thermal x-ray emission also occurs in the high density regions, and may take the form of bound-bound emission caused by collisional electron excitation or post-recombination relaxation, and continuum emission due to recombination or bremsstrahlung radiation. An example of such a spectrum is shown in Figure 10.2.

For the purposes of contact x-ray microscopy, it is desirable (but not strictly necessary) that the emission be concentrated in the water window from 23-44 Angstroms. (Providing that there is sufficient contrast between the x-rays absorbed below the carbon K edge wavelength and the unabsorbed background radiation, x-rays of other wavelengths can be tolerated). Enhancement of x-ray generation in the water window can be achieved mainly by the variation of two parameters: the target material used to generate the plasma, and the intensity of the laser when focussed on the target.

An example of use of the first parameter is shown in Figure 10.3, where the thermal emission lines of hydrogen-like (CV) and helium-like (CVI) carbon (one and two electrons around a carbon nucleus) is shown. Emission from the higher carbon resonance lines in particular would be expected to fall in the carbon K shell absorption region, since the K shell ionization process is assisted by partial shielding from outer electrons which are not present in the hydrogen- and helium-like carbon. Hence carbon based target materials (graphite or Mylar) are often used in contact x-ray microscopy applications. This is particularly so since the carbon spectrum (like that of other low Z materials) can be dominated by line emission from highly stripped ions, with very little broadband continuum emission present to reduce the background contrast (Eason *et al.*, 1986; Gerritson *et al.*, 1986). The x-ray emission from a single line can in fact be as high as 1% of the incident laser energy (Matthews *et al.*, 1983). Materials other than carbon may also have strong x-ray emission in the water window, and studies have shown that high levels of continuum emission may fall in the 23-44 Angstrom region when high Z materials are irradiated (Gerritsen *et al.*, 1986).

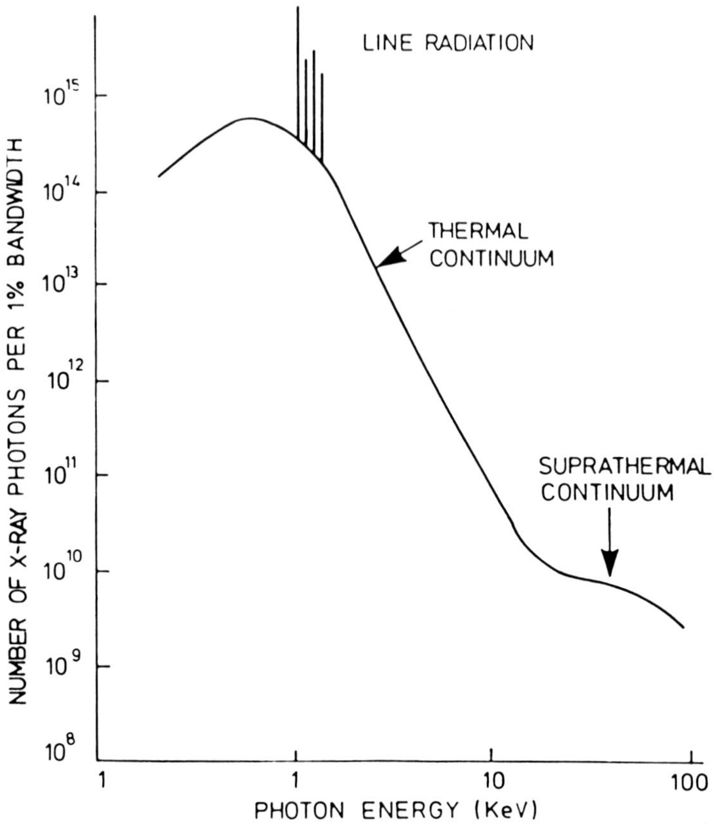

Figure 10.2. Typical x-ray emission from a laser-produced plasma (Rumsby, 1985).

Variation of the laser intensity focussed on the target can also be used to optimize the x-ray emission in the desired region. However, the intensity required is dependent upon the target material, as can be seen by referring to Figure 10.4. The temperature at which 50% of the plasma ions are hydrogen-like is shown as a function of atomic number (the rest of the ionic species being predominantly helium-like). The calculation is given for two densities (solid and 1.06μm critical) between which lies the density region from which most of the x-ray emission occurs in laser produced plasmas. It can be seen that in order to generate line emission from hydrogen- and helium-like carbon species, temperatures of the order of 100 eV are required. (Although the helium-like carbon ionization potential is 392eV, because of the greater number of ways of partitioning the energy in the hydrogen-like ion plus electron system c.f. the helium-like ion system, ionization can occur for a kinetic temperature that is only a fraction of the ionization potential. Similarly, the larger statistical weights of the higher excited states result in significant bound-bound x-ray emission at a temperature which is but a fraction of the transition energy).

However, if the plasma temperature were to be increased by increasing the intensity of the focussed laser on target, it would then "burn through" the helium- and hydrogen-like ion stages leaving only

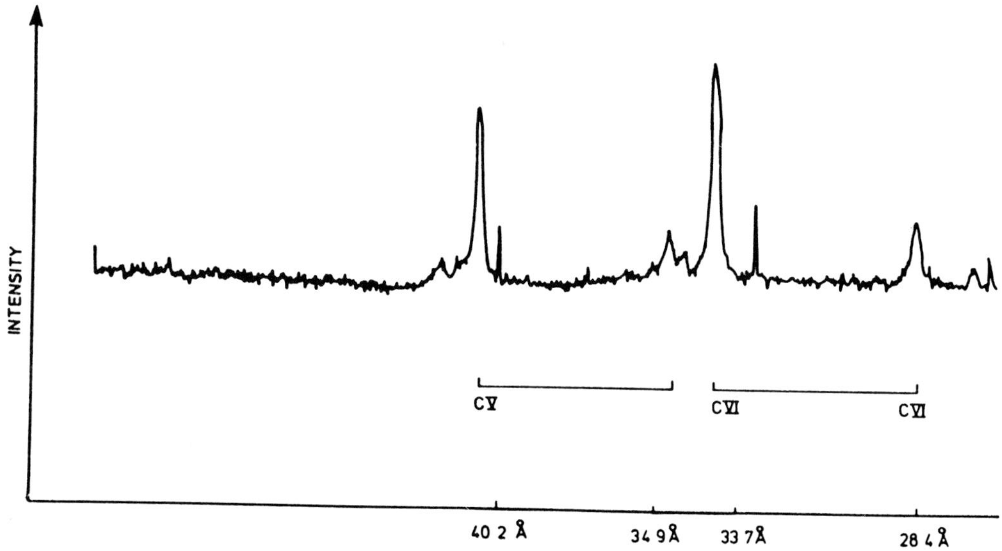

Figure 10.3. Spectrum of helium-like (CV) and hydrogen-like (CVI) carbon emission lines taken using a rocket spectrograph from illumination of a carbon target with a 0.53μm, lnsec pulse (2x 10^{13}W/cm^2) from the Rutherford Vulcan Laser. (The intensity zero is arbitrary as the background level contains a significant contribution from film fogging).

completely stripped carbon. As a result no line emission would be observed while the plasma is at these temperatures, and the resultant continuum x-ray emission would be spread over a broad band of wavelengths, much of it outside the water window. Hence it is not desirable to go to the highest intensities available with some of the larger lasers. In fact, intensities of around $10^{13}W/cm^2$ produce significant levels of x-ray emission between 23 and 44 Angstroms for

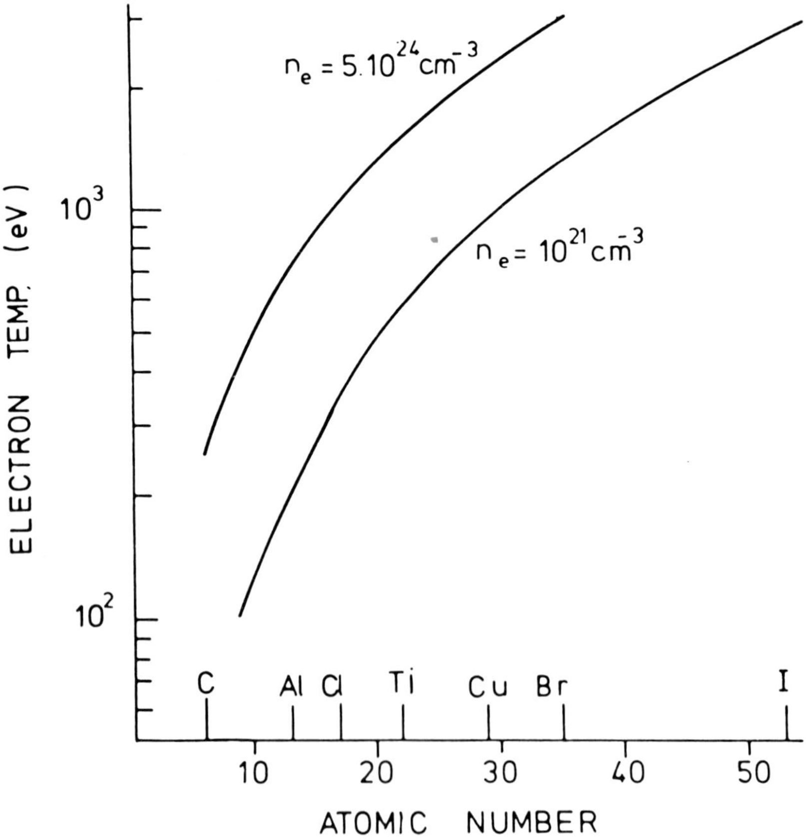

Figure 10.4. Plot of the electron temperature at which 50% of the ions of the species indicated are hydrogenic at the labelled electron densities (Rumsby, 1985).

x-ray microscopy applications (Rosser *et al.*, 1985a; Rosser *et al.*, 1985b; Eason *et al.*, 1986). Much of the x-ray flux from these carbon targets is found to be in the form of line emission in the water window region, and the relative line intensities (which are temperature dependent) can be altered by varying the focused laser intensity on target (Eason *et al.*, 1986a).

All of the above studies were carried out using the second harmonic (530nm) beam of a neodymium laser. However, similar results have been obtained using a KrF excimer laser (operating at 249nm) whose shorter wavelength couples the energy into the plasma more efficiently (Eason *et al.*, 1986a). The peak laser irradiance used was $5x10^{13}W/cm^2$ and produced a total x-ray conversion efficiency of up to 1% in the water window region using a thin (250μm) Mylar target. The x-ray flux was divided evenly between line and continuum emission and was emitted from a source approximately 200μm in diameter (roughly twice the focal spot size). Later experiments (Al-Hadithi *et al.*, 1986) using a solid carbon target yielded a five fold increase in the x-ray flux in the water window region using the same laser conditions. Conversion efficiencies in excess of 10% into the water window were obtained at lower (5J) laser energies using similar focal spot sizes. The decrease in conversion efficiency into the water window region at higher laser intensities may be due to burning through the CV and CVI ionization stages at higher temperatures.

A Short History of Laser-produced Plasmas as Sources for X-ray Microscopy

The following represents a brief summary of some of the historical developments that have led to the use of laser-produced plasmas as a source of x-rays for pulsed x-ray microscopy of living hydrated biological specimens. The list is not exhaustive.

Basov *et al.* (1966): First high temperature, high density laser-produced plasmas generated by focusing a Q-switched ruby laser onto a solid target. Interest in thermonuclear applications.

Burgess *et al.* (1967): XUV (20-200 Angstroms) emission spectra obtained by focusing a 500MW (5J in l0nsec) Q-switched ruby laser onto solid targets.

1973-1986, Numerous studies of small, laboratory size lasers used to generate laser-produced plasmas for XUV applications (e.g. Breton and Papoular, 1973; Carroll *et al.*, 1980; Nagel *et al.*, 1974, 1984; Ginter and McIlrath, 1986).

1979 Rutherford-Appleton Laboratory. Large Nd laser system (developed for investigation of inertial confinement fusion) used l00J 1nsec pulse focussed onto solid targets to generate ~ 8 Angstrom x-rays for backlighting microballoon implosions. Pinhole camera used for x-ray microscopy of targets with several micron resolution (see Rumsby, 1985).

Nagel *et al.* (1984): Repetitively pulsed laser plasma X-ray exposure of photo resists. Attempts at microscopy ruined by debris and destruction of sample.

Yaakobi *et al.* (1984): Single shot lithography at ~10 Angstroms on high sensitivity, low contrast negative resists using 10μm beryllium debris shield. Attempts to expose high contrast resist using a thin (2μm) Mylar shield inhibited by debris damage.

Rosser and Baldwin (1984): Attempts at laser x-ray microscopy using zone plates foiled by insufficient etching of the zone plate substrate.

Rosser *et al.* (1985): First successful single shot exposure of high contrast PMMA resist using a laser-plasma source. Contact microscopy of hydrated biological samples done. Debris problem overcome by using sacrificial debris shield of silicon nitride and ultrasonically cleaning exposed resist before developing.

Successful contact microscopy of a wide range of biological samples using both Nd and KrF laser systems, (Eason *et al.*, 1986b). Scaling experiments carried out to prove viability of contact x-ray microscopy on smaller lasers (Rosser *et al.*, 1986).

1985/86, Number of experiments on smaller lasers (UBC SMIRF, ANU Laser Group) show that contact x-ray microscopy is possible using small (few joule) single shot devices (Rosser *et al.*, 1985b).

The Future

The scaling experiments of Eason *et al.* (1986a) and the successful use of small laboratory size lasers for single shot contact x-ray microscopy (Rosser *et al.*, 1985b) have demonstrated the viability of a technique which is now feasible in many research establishments around the world. This "portability" is particularly important since the bigger laser facilities are often dedicated to large scale programs such as the development of inertial confinement fusion.

However, one area in which the larger laboratories may contribute significantly to future development of x-ray microscopy is in the realization of x-ray lasers. Already substantial gain has been achieved at ~206 Angstroms and ~155 Angstroms by irradiating selenium and yttrium foils using the large Novette laser facility at Livermore (Matthews *et al.*, 1985; Rosen *et al.*, 1985). The technique involves focusing the 500J, 450psec pulse into a thin line (10mm x 200μm) at a modest intensity (~10^{14}Wcm^{-2}) to produce a long, uniform plasma for the enhancement of amplified spontaneous emission. Other experiments using Z-pinch plasmas (Suckewer *et al.*, 1985) also demonstrated gain, this time in the 182 Angstrom CVI line, and efforts are underway to extend x-ray gain studies to other systems.

Although the Livermore experiments only produced x-ray fluxes of the order of 10^{10} photons per pulse, the spectral brightness (in terms of the units in Figure 10.1) for an x-ray microscopy experiment located 10mm from the x-ray laser is ~ 6x10^{26} photons/cm^2/s in a bandwidth $\Delta\upsilon/\upsilon \sim 10^{-3}$. This is much brighter than the synchrotron sources, and as bright at this stage of development as some line emission from present laser-produced plasma x-ray sources. However. the collimation of the x-ray laser radiation will allow greater source-specimen separations and will reduce penumbral shadowing effects. Further, the coherent nature of the x-ray laser opens up the possibility of interferometric and holographic x-ray microscopy (Solem and Baldwin, 1982).

Acknowledgement

We would like to acknowledge the contribution of Dr P.T. Rumsby, Rutherford Appleton Laboratory, for allowing the use of Figures 10.1, 10.2 and 10.4.

11. Single Shot Soft X-ray Contact Microscopy with Laboratory Laser Produced Plasmas

R. J. Rosser[1] and K. G. H. Baldwin[2]

[1]160 Blythe Road, London. W14 OHD, United Kingdom

[2]Laser Physics Centre, Australian National University, G.P.O. Box 4, Canberra, A.C.T. 2600, Australia

Introduction

The high intensities generated by laser produced plasmas in the water window (23-44Å) region of the x-ray spectrum make possible single shot contact x-ray microsocopy of living, hydrated biological samples (Baldwin and Rosser, 1986). The short (typically 1ns) laser pulses result in exposure of the specimen image onto the photoresist prior to structural alteration of the specimen by x-ray induced damage. This is a significant advantage over continuous x-ray sources such as synchrotrons, but is a common feature with other flash x-ray sources such as the gas puff z-pinch. However, the laser-plasma x-ray source offers a number of futher advantages:

1. It produces reproducible x-ray intensities with every laser shot.

2. The laser source position is reproducible to within the focussed laser spot size.

3. The small spot size at the laser focus ($<100\mu m$) reduces the effect of penumbral shadowing from a source of finite size.

4. The compact geometry of the laser-plasma source allows small ($<10mm$) source to specimen distances to enhance the x-ray intensity.

X-ray Microscopy
Ed. by P.C. Cheng and G.J. Jan

The usefulness of the laser plasma x-ray source for contact x-ray microscopy studies has already been demonstrated (Rosser *et al.*, 1985a). However, the source used in the original experiments was a large (>100J/pulse), multi-million pound research laser facility (the VULCAN Nd glass laser at the U.K. Rutherford Appleton Laboratory). Gas puff z-pinches (Bailey *et al.*, 1982) on the other hand, can be built at much lower cost in most well equipped laboratories. Therefore, there has been considerable interest in determining whether small laser systems could be used to generate x-rays for contact microscopy and retain the considerable advantages over gas puff z-pinches listed above. The progress in achieving this aim is summarized in this paper.

Laser Contact Microscopy - The Method

The process of x-ray contact microscopy has been well documented (*e.g.* Cheng *et al.*, 1986) and can be applied directly to laser-produced plasma sources without any direct modification to the laser-target system. Because the laser is focussed to high intensities on target, the target must be located in a vacuum to prevent ionization breakdown of the air, since the resulting plasma would prevent the laser from reaching the target in a tightly focussed spot. The vacuum environment also allows transmission of the x-rays produced, and hence the hydrated biological specimen must be located in a suitable environmental chamber.

An example of such a chamber (used by Rosser *et al.*, 1985a) is shown in Figure 11.1). It consists of an o-ring sealed outer housing with a window to allow transmission of the laser-generated x-rays. The window is made of Si_3N_4 etched from a silicon wafer. The window area is 1mm^2 and is 1000Å thick, sufficient to withstand an atmosphere pressure difference. Often the window is used with a periodic grid (typically 40μm period made of photoresist) in order that the shallow wells thus formed can hold the biological specimen in position. The specimen is then sandwiched between the window and a silicon wafer coated with photoresist. A spring is used to ensure close contact between the specimen and the photoresist surface in order to minimize the loss of resolution caused by penumbral shadowing and diffraction effects. The photoresists used are sensitive polymers such as PMMA, or more highly sensitive copolymers (such as P(MMA-MA)) and terpolymers.

The environmental chamber can be made very small (a few mm diameter) and shaped in such a way that it can be brought very close (<10mm) to the laser produced plasma without interfering with the incoming laser beam. A schematic of the experimental geometry used by Rosser *et al.* (1985a) is shown to illustrate the technique in Figure 11.2. Following illumination of the target surface by the incident laser pulse, x-rays emitted from the expanding plasma plume irradiate the specimen and record the contact image onto the photoresist. Some time later, debris from the plasma plume strikes the chamber window. Depending on the source distance, target material and the laser parameters, physical destruction of the window and the specimen can result. However, the pre-recorded image on the resist can be recovered by ultra-sonically cleaning the resist surface in ethyl alcohol to remove the debris. The resist can then be developed (usually in a 1:1 solution of methyl iso-butyl ketone (MIBK) in isopropyl alcohol (IPA) using a minimal development time to prevent lateral dissolution of the image. The image can then be viewed by a variety of methods: interference or phase constrast optical microscopy; or for high resolution: scanning electron microscopy following metallic shadowing of the resist surface, or transmission electron microscopy using carbon replicas of the resist.

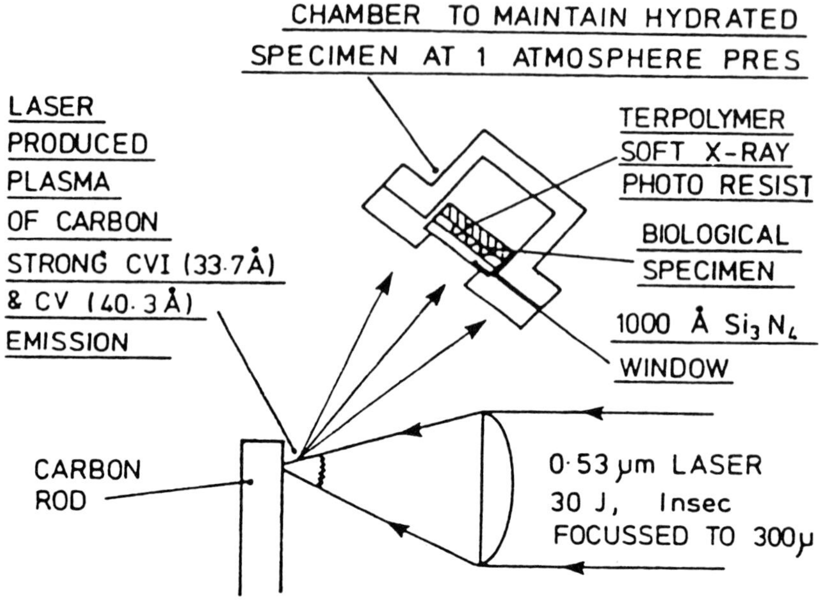

Figure 11.1. Experimental arrangement for laser generated soft x-ray microscopy.

Development of Contact Laser X-ray Microscopy

The first demonstration of a laser-produced plasma x-ray source for the imaging of biological specimens was carried out on the Rutherford Appleton Laboratory VULCAN laser system by Rosser *et al.* (1985a). The system used was a large ($<$100J/1ns pulse) Nd glass laser which was freguency doubled to produce ~ 30J of green (0.53μm) radiation to enhance the soft x-ray conversion efficiency. The laser was focussed to a 300μm spot on a carbon target as shown in Figure 11.2. The first specimens employed were diatoms whose gradated structure provided a useful test of the system resolution. Figure 11.3 shows the image of a diatom on the terpolymer photoresist as viewed using a scanning electron microscope, and indicates resolution of better than 1000Å.

Following this initial work, further experiments were carried out at the Rutherford Appleton Laboratory on a wide range of biological samples (Eason *et al.*, 1986). However, two different types of laser were used in these experiments which, in addition to the Nd:glass system already mentioned, employed a KrF excimer laser (SPRITE). The KrF output at 0.249μm produced 80J pulses of 50nsec duration, which were focussed more tightly (into an 80μm spot) to produce the same irradiance onto the carbon target (~ 3 x 10^{13}W/cm^2). Some examples of the results obtained are shown in Figure 11.4 using a source to specimen separation of ~ 60mm.

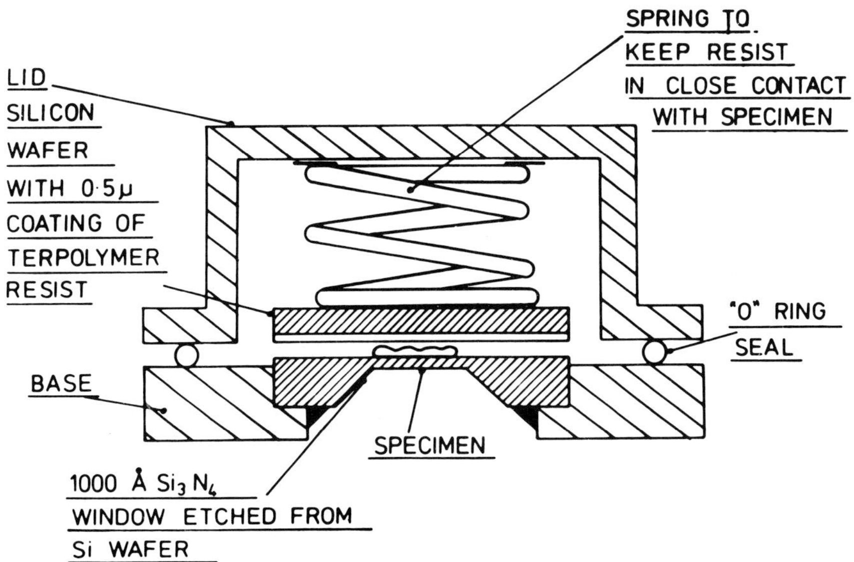

Figure 11.2. Environmental chamber for imaging of hydrated biological specimens.

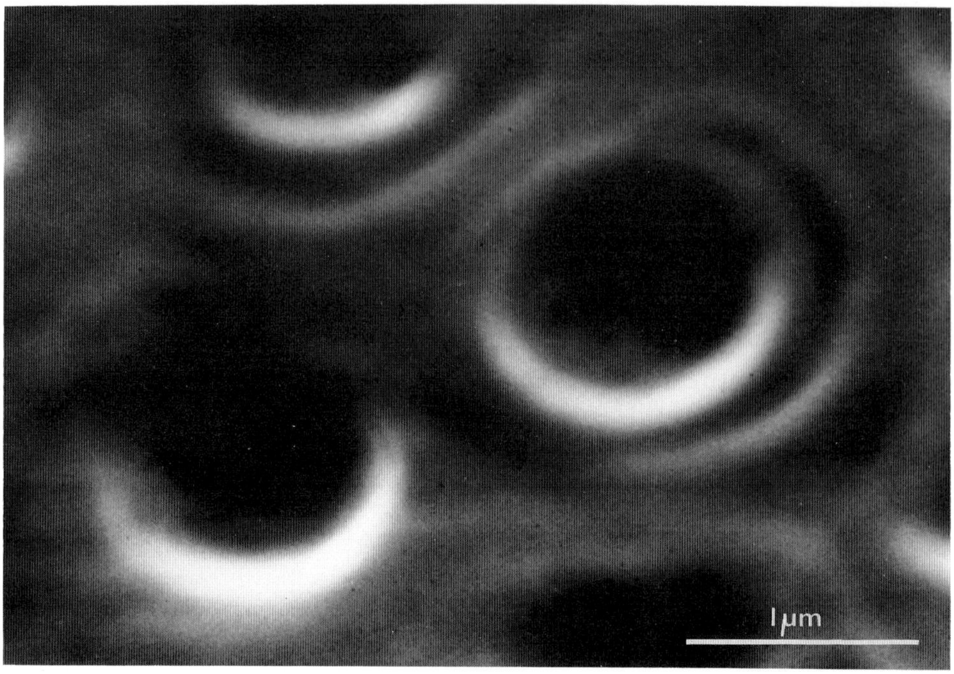

Figure 11.3. Contact x-ray micrograph of diatom obtained on VULCAN.

Improving the accessibility of laser x-ray microscopy

The laser systems used for these first x-ray microscopy experiments had been part of a multi-million pound national laser facility, and were not readily accessible on a day-to-day basis for use by university biology or medical departments. However, since the full capabilities of the Rutherford Laser systems had not been exploited in terms of the maximum intensities available (>10^{15}W/cm^2), the question arose as to what was the smallest laser pulse energy that could usefully be employed to produce single shot contact x-ray microscopy images.

A series of scaling experiments were therefore carried out using the VULCAN Nd laser system to determine empirically a lower limit on the pulse energy (Eason *et al.*. 1986; Rosser *et al.*, 1986). Using carbon targets, a laser irradiance of ~ 2-3 x 10^{13}W/cm^2 produces a

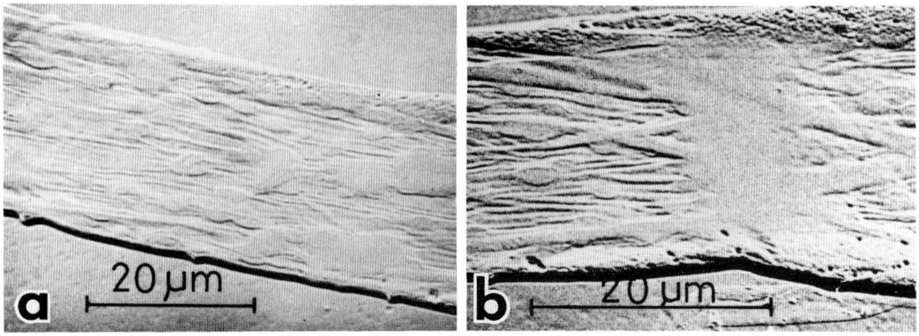

Figure 11.4. Images of epidermal hairs from the corolla of a foxglove.

plasma with a temperature of ~ 100eV, which efficiently emits x-rays in the "water window" spectrum from the denser regions of the laser-plasma core (Rumsby, 1985; Baldwin and Rosser, 1987). Therefore, the first parameter that requires scaling with the laser pulse energy (E) is the area of the focal spot (A) so as to maintain a constant laser irradiance (E/A W/cm^2) on the target. This picture is somewhat oversimplified, as cooling effects due to transportation across the plasma boundary become greater the smaller the size of the plasma plume. Hence, an empirical test of the laser plasma flux was employed to determine the usefulness of this scaling.

The second parameter that requires scaling is the x-ray flux incident on the sample, which is proportional (again disregarding edge effects) to E/l^2 (l being the source-to-sample separation). Hence both the laser spot size and the source-to-specimen distance must be decreased as the laser pulse energy is lowered. With these criteria in mind, the experimental conditions shown in Table 11.1 were chosen to determine the lower limit on the laser energy which could usefully be employed for contact x-ray microscopy.

The results show that by keeping the laser irradiance on target and the flux incident on the sample constant (within a factor of two), use-

Table 11.1: Results of the series of five 1ns 0.53μm scaling shots using the VULCAN laser.

Green laser energy E_L(J)	Spot diameter d(μm)	Irradiance	Source to sample separation	Relative Flux incident at sample	Type and time of development
34.2	600	1.2 x 10^{13}	60	0.57	1:1 MIBK, 1min
34.0	450	2.1 x 10^{13}	45	1.0	1:1 MIBK. 1min
14.4	300	2.0 x 10^{13}	30	0.95	1:1 MIBK. 1min
2.0	100	2.5 x 10^{13}	10	1.2	1:1 MIBK, 1min
0.5	70	1.3 x 10^{13}	7	0.6	ECA, 5min

* irradiance = $E_L/\pi d^2\tau$ (Wcm^{-2}), τ = pulselangth = 1nS.
** Source to sample separation l(mm)
*** Relative flux incident at sample = E_L/l^2

ful exposures yielding visible images for a 1 minute development time in 1:1 methyl isobutyl ketone/isopropyl alcohol were achieved for pulse energies $\geq$2J. However, for the 0.5J shot, 5 minutes development in ethyl cellusolve acetate was required (with subsequent degradation of the image), and this was deemed to be insufficient exposure for useful contact x-ray microscopy work.

It therefore appeared that small laboratory size lasers capable of producing several Joules of 0.53μm radiation in a single pulse would be sufficient to carry out contact x-ray microscopy of living biological samples. To test this, an experiment was subsequently carried out at the University of British Columbia (UBC), Canada, where a small Nd laser (3J in 2ns at 0.53μm) was used to produce x-rays from carbon and gold targets (Rosser *et al.*, 1985b). Useful images of the photoresist test grids on the Si_3N_4 windows were obtained with a 1 minute development time in 1:1 MIBK/IPA when 10 and 15mm source-to-sample separation distances were used respectively for the two targets. However, at such small distances, destruction of the Si_3N_4 windows occurred due to debris impact on many of the shots.

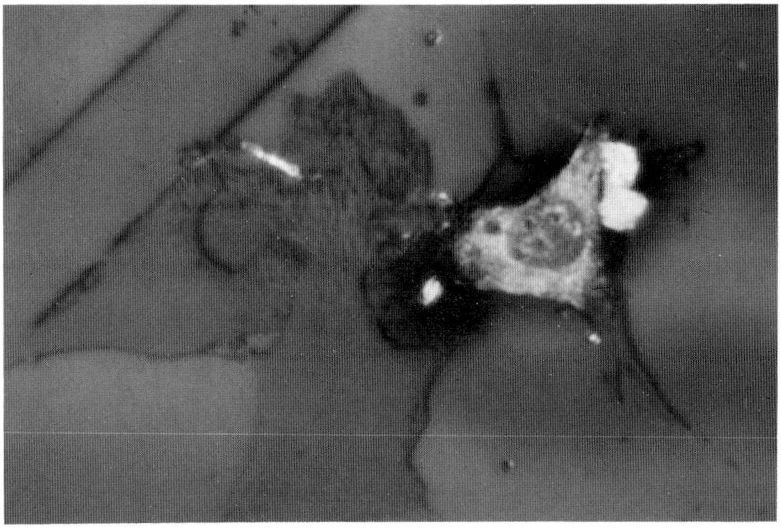

Figure 11.5

Further experiments have just recently been carried out at UBC using relay optics to image the x-rays onto the sample to avoid line-of-sight impact from target debris. A toroidal mirror system with a major radius of 2m and a minor radius of 6.25mm was interposed between the source and the specimen chamber. Despite losses in the efficiency of the mirror, the concentration of the x-ray flux on the specimen chamber produced adequate exposure of the photoresist using a single shot. Examples of some of the images are shown in Figure 11.5. The Si_3N_4 windows survived intact in all cases, indicating that the damage previously observed was due to debris impact and was not due to x-ray absorption. The advantage of this technique is firstly that there is no risk of damage to the resist surface either by direct impact or by post-exposure cleansing, and secondly that the vacuum windows can be left permanently in place. This should greatly facilitate live and wet cell preparation.

The Future

Further experiments have been carried out in other laboratories proving the viability of single shot x-ray contact microscopy using laboratory size laser systems. The Nd laser system in the Australian National University Laser Physics Centre for example has been used to produce images of blood cells with 2-3J of green light focussed onto lead targets. The portability of this x-ray microscopy system now opens up the possibility of bio-medical research in any major University department. This will greatly facilitate development in the field of contact x-ray microscopy which up until recently has been limited by the availability of sources. Now, however, it will be possible to concentrate on improving specimen preparation and handling as well as image processing in order to fully realize the potential of contact x-ray microscopy.

12. Soft X-ray Contact Imaging at CSRF

K.H. Tan[1], P. C. Cheng[2] and D. M. Shinozaki[3]

[1]Canadian Synchrotron Radiation Facility, Physical Science Laboratory, University of Wisconsin, Stoughton, WI 53589 USA

[2]IBM T. J. Watson Research Center, Yorktown Heights, NY 10598 USA

[3]Faculty of Engineering, University of Western Ontario, London, Ontario, N6A 5B9, Canada

Introduction

The advantages of using soft X-rays for the study of ultrastructure in biological materials have been described a number of times (Spiller *et al.*, 1979; Feder *et al.*, 1977) with some emphasis on the potentially high resolutions possible (Sayre *et al.*, 1978) for specimens thick relative to those used for electron microscopy. Because of the large differences in absorption coefficients for elements which are biologically important such as C, N, O, *etc.*, the inherent contrast is good for biological materials, and images can be obtained without the staining so often used in electron microscopy. In addition, the absorption coefficient of water is an order of magnitude smaller than that of protein in the soft x-ray range between the oxygen and carbon edges, pointing out the potential of imaging wet or perhaps even living biological material.

The image may be considered as consisting of a number of discrete picture elements. In order to resolve a small pixel, a minimum number of photons per pixel must be detected in the resist. In terms of the contact method therefore, high resolution soft x-ray imaging requires a high intensity at the specimen/resist interface. Synchrotron sources certainly provide this high photon flux. In addition, the use of a

X-ray Microscopy
Ed. by P.C. Cheng and G.J. Jan

monochromator in front of the specimen allows one to vary the wavelength of the incident radiation. If images are recorded above and below an elemental absorption edge from one specimen, the difference image will provide a map of the spatial distribution of the element in the microstructure. Thus tunable high intensity soft x-ray sources such as synchrotrons are potentially ideal for quantitative soft x-ray contact imaging of biological materials.

The Canadian Synchrotron Radiation Facility

The spectral output from an 800MeV synchrotron radiation source is close to ideal for contact imaging, with a usable flux in the photon energy range from 20 to 1000eV when used with the monochromator at CSRF. The ring can be operated at 1GeV from time to time if higher flux at higher photon energies is needed. In general, since the beamline has been operational, the electron current circulating in the ring immediately after injection is of the order of 120mA with lifetimes of the order of hours. It has been found that this provides an usable beam for imaging, and a wide variety of experiments in imaging have been possible.

A feature of the imaging beamline which is different from other contact imaging beamlines such as U15 at Brookhaven (operated by Kirz and co-workers primarily as a scanning x-ray microscope on the 740MeV VUV ring) is that the contact imaging at CSRF is accomplished in a windowless beamline. That is, there are no windows separating the specimen/resist area from the monochromator and ring itself. The low energy photons are therefore not filtered out to any great extent, and experiments on resists and specimens can be performed with monochromatic x- rays as soft as 20eV. This is of some importance from the viewpoint of radiation effects in polymeric materials (both synthetic and biological) since there are many unique chemical effects known in the UV region and many of the experiments planned for the CSRF imaging facility exploit this fact.

Since polymer resists and biological materials are not compatible with the ultra-high vacuum of the electron storage ring and the monochromator regions (1×10^{-10} torr), the experimental chamber is separated from the beamline proper by a differential pumping station capable of supporting a pressure difference of 1×10^{-6} and 1×10^{-10} torr. This means that the pressure in the beamline and electron storage ring will not significantly change as long as the pressure in the exposure chamber is in the range of 1×10^{-6} torr or better. In practice, the chamber for the imaging is maintained at pressures of 1×10^{-7} torr or better by means of a large cryopump acting directly on the specimen chamber. The large pumping rate capacity also is designed to allow other experiments which may result in some measurable outgassing during exposures.

The Monochromator and Beamline

The Canadian Synchrotron Radiation Facility was designed for the 1GeV electron storage ring at the University of Wisconsin, but has been used on the 240MeV storage ring (Tantalus), on which some early contact imaging was accomplished using a variety of borrowed specimen chambers (Dr. C. Olsen). The beamline was designed for gas phase photoelectron spectroscopy experiments (Dr. G. M. Bancroft and co-workers), and hence included the differential pumping station between the last optical element of the beamline and the experimental station (which could be either the imaging cross or the photoelectron apparatus). The differential pumping station consists of one 120 liters/sec ion pump located next to the refocussing mirror chamber which is itself pumped by a 120 liter/sec ion pump. Next to the ion pump is a 4200 liter/sec (pumping water) cryo pump, which is used to trap water and a variety of reactive gases very effectively.

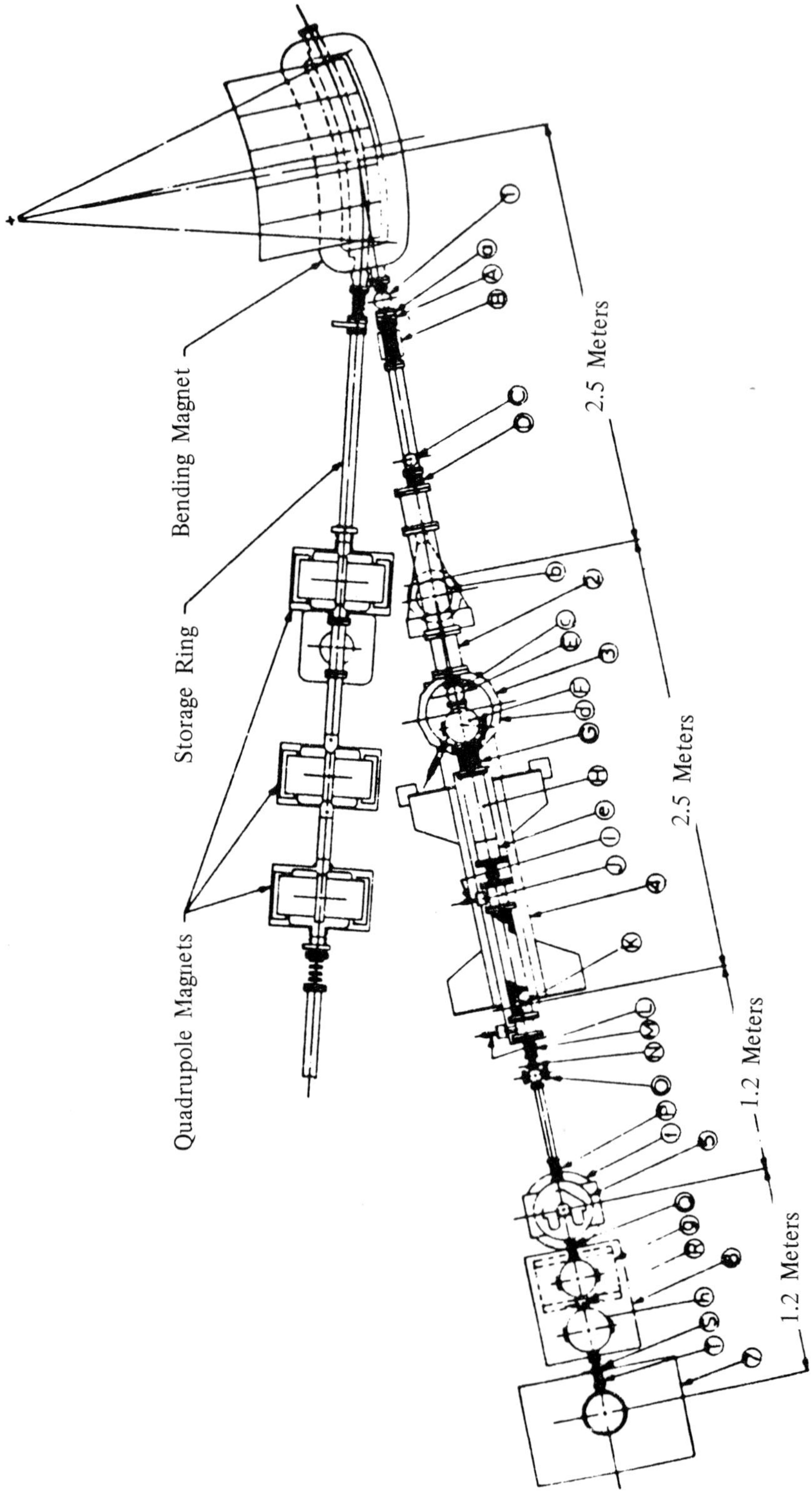

Figure 12.1. Canadian Synchrotron Radiation Beamline (CSRF).

A 1 1/2" pneumatic gate valve separates the cryo pump and the ion pump chambers. This is necessary so the cryo pump can be isolated and regenerated periodically. The pneumatic gate valve is interlocked to a nude ion gauge through the process control of the Granville Philips ion gauge controller. There are switches on the control box to open and close the valve as well as a relay latch to prevent unwanted cycling of the valve. For gas phase photo-emission experiments, there is an extra light pipe (10mm x 1mm x 80mm) physically to reduce the conductance between the experimental chamber and the beamline. With this installed, the experimental chamber was normally running at a dynamic gas pressure of 1 x 10^{-5} torr while the beamline was at 10^{-10} torr for the gas phase studies.

The complete beamline is shown in Figure 12.1, consisting of 7 sections : (1) the "front end" nearest the ring, (2) the Mo horizontal focussing mirror, (3) the monochromator pumping chamber, (4) the monochromator itself ("Grasshopper"), (5) the refocussing mirror, (6) the differential pumping station, and (7) the experimental chamber (interchangeable from imaging to photo- emission studies).

The "front end" consists of a water cooled shutter with a fixed aperture, a gate valve, a heavy metal radiation shutter and a 20 liter/sec ion pump. The first mirror (Mo) is a 1 meter long float glass,coated with 500 to 1000Å of gold and elastically bent to the required curvature. This mirror collects about 14mrad of the horizontal beam at a 2° grazing angle, and focuses the horizontal beam onto the exit slit of the monochromator. The heart of the beamline is the 2° grazing incidence Grasshopper monochromator, a high resolution monochromator usable in the photon energy range between 20 and 1000eV. The refocussing mirror is a cylindrical mirror 24cm x 2.5cm and with a ground in curvature of radius 4.2cm.

The horizontal focussing is determined by the grazing incidence angle "i" (2°), the ground in curvature "r", and the effective source point S (1.2m) and is given by the equation:

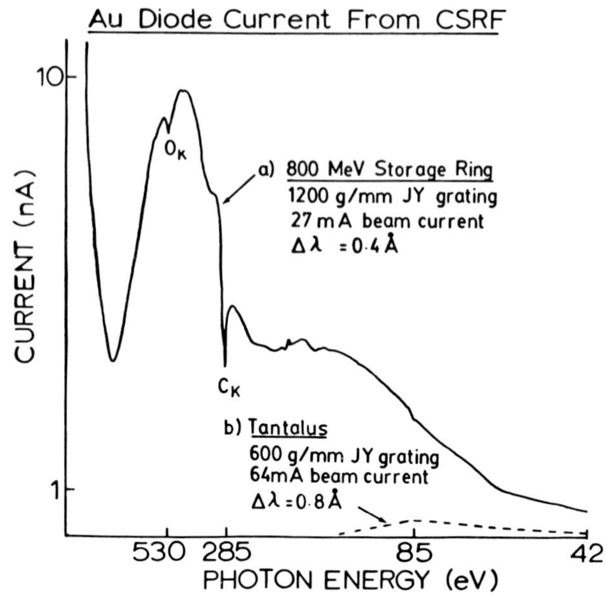

Figure 12.2. Gold diode current from CSRF beamline.

$$f = \frac{s \bullet r}{[2 \bullet s \bullet \sin(i - r)]} \tag{12.1}$$

The vertical focus depends on i (2°) and the adjustable bending radius "R" according to the equation:

$$f = \frac{s \bullet R \bullet \sin i}{(2s - R \sin i)} \tag{12.2}$$

The desired radius "R" is obtained by applying equal moments on each end of the mirror.

The two stage differential pumping station (6) described earlier provides sufficient differential pumping between the exposure station and the optical elements of the refocussing mirror chamber and the monochromator. The pressure in the entire beamline is better than 2x 10^{-10}torr with the beam off and is better than $2x10^{-9}$ torr with the beam on. The relative photon flux is measured as a function of photon energy from the total electron yield from a gold diode at the exit slit of the monochromator. The simple measuring circuitry consists of a 90V battery acting as a power supply and a standard electrometer to measure the photoelectric current. Figure 12.2 illustrates this measured spectrum plotted as gold diode current versus photon energy, for a bandwidth of 0.4Å using a 1200 lines/mm grating.

For x-ray imaging, the experimental chamber consists of a 6 way cross as shown in Figure 312.. The chamber can be isolated from the beamline with a pneumatic gate valve, and from the cryo pump acting directly on the chamber by a 8" gate valve. The chamber is let up to a dry nitrogen atmosphere and the specimen inserted. The prepumping is accomplished by a 110 liter/sec turbo pump which is then valved off when the cryo pump gate valve is opened. With moderately dry, small specimens used in our imaging experiments (consisting of a biological specimen on a 3mm diameter TEM grid and a layer of PMMA resist on glass or silicon), a typical turn around time from venting to opening the experimental chamber to the beamline is of the order of 1 hour. The specimen chamber is at a pressure of $3x10^{-8}$ torr or lower at this time. Chamber pressures of $1x10^{-5}$ torr are reached on the turbo within a few minutes of the start of pump-down. The turn around time is experimentally significant since imaging times are typically of the order of minutes or less, and pump-down times greater than 1 hour would be inconvenient. Hence if 5 to 10 specimens are mounted on the holder, the imaging process is quite reasonable in terms of total time used per exposure. For holders with a larger number of specimen positions, the 1 hour turn-around time is even less inconvenient. Thus the windowless experiment is not too much more time consuming than the experiments done in windowed beamlines.

The specimens are mounted on a linear feedthrough, or more recently on a VG X-Y-Z rotation feedthrough. An electron multiplier detector was used to monitor the intensities of the light transmitted

Figure 12.3. Experimental chamber

through the sample for certain kinds of absorption experiments on resists or biological materials. Small areas of the specimen holder were covered with P31 phosphor to facilitate the alignment of the sample with the center of the photon beam. The horizontal extent of the beam at the specimen is about 5mm while the vertical dimension is determined by the width of the slits in the monochromator.

Typical Image Experiments

Since the transfer of the beamline from Tantalus to the 1GeV storage ring in early 1986, most of the contact imaging has used PMMA resists spun on glass or silicon substrates. The biological specimens were about 10μm or less, and they were placed as close as possible to the resist to minimize the effects of diffraction and penumbral blurring on the image.

There are two ways to achieve the intimate contact between the biological specimen and the resist. The cells can be cultured directly on the surface of a thin formvar film, or culture cells on a Formvar coated EM grid which is placed in close contact with the photo resist. Although the first method provides the most intimate contact, the second technique is preferable because it is nondestructive and the specimen can be used repeatedly. Furthermore, if cells are cultured directly on the surface of the PMMA resist, removal of cells become very difficult or even impossible after x-ray exposure. This adhesive phenomenon is likely the result of of x- ray induced cross-linkage between the specimen and the resist.

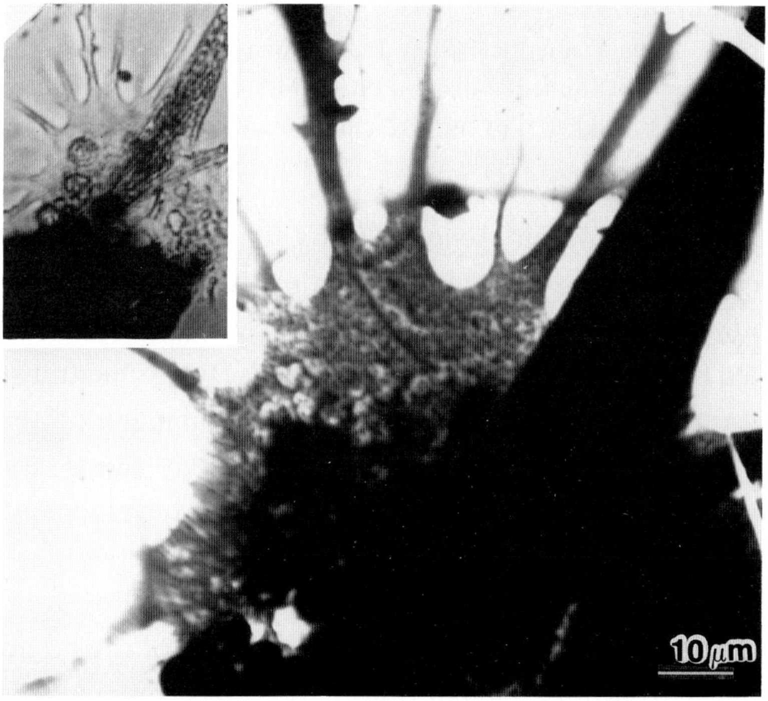

Figure 12.4. Image taken with polychromatic x-rays. TEM-magnified x-ray contact image of *Xenopus laevis* cultured cell. The contact image was formed on a free standing PMMA film (1.5μm) with white light synchrotron radiation (with a Al filter in the beam). Insert: phase contrast image of the original cell.

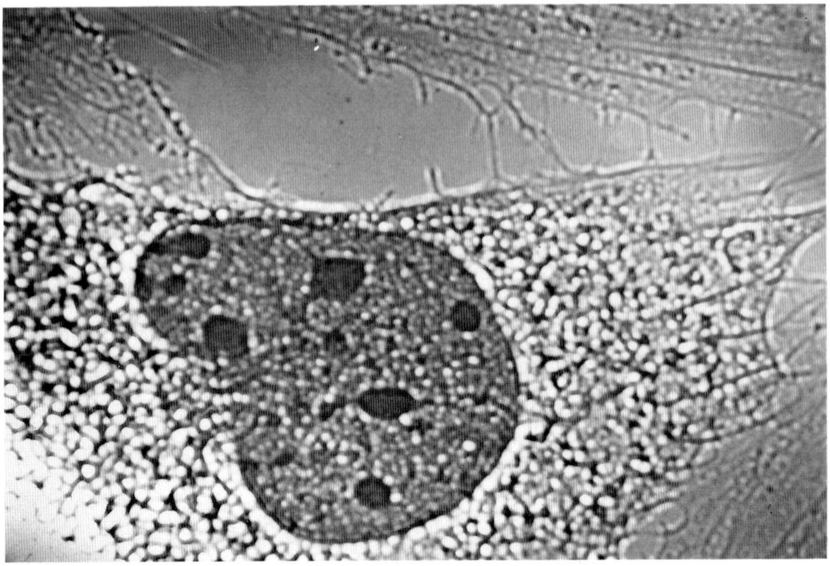

Figure 12.5. Image of a critical point dried human fibroblast taken with 77Å monochromatic radiation on PMMA resist. The contact image was magnified by using a transmission electron microscope.

Figure 12.4 shows an *Xenopus laevis* fibroblast cell imaged with Al-filtered white light synchrotron radiation. The cell was cultured directly on the surface of the PMMA resist, fixed in 1% glutaraldehyde, dehydrated in ethanol and critical point dried. Due to the transmission of the Al filter, the majority of the radiation impinging on the cell is $<$170Å After exposure, the cell was removed by the treatment of 1%Chlorax (10% sodium hyperchloride). After development of the resist, the resist was floated off from the glass wafer with a 1% HF solution (Cheng *et al.*, 1985), and supported by a Formvar coated copper grid. The insert is the phase contrast image of the original cell.

Figure 12.5 shows a human fibroblast imaged by monochromatic radiation (77Å). Fresnel diffraction fringes is evident. The cells were fixed in 1% glutaraldehyde in phosphate buffer, pH7.2, dehydrated in EtOH, critical point dried in CO_2. No heavy metal stain was used.

Conclusions

The contact soft X-ray imaging facility at the 1 GeV ring at the University of Wisconsin has been described. A windowless beamline suitable for illumination with photons from the UV to the X-ray wavelengths is used. The convenience of use for imaging has been ensured by using a high capacity large cryo-pump to which can be isolated from the chamber during specimen changes. A variety of related experiments including accurate photo-absorption spectra and soft X-ray induced mass loss from materials can be measured over the entire accessible photon range. The wavelength resolution over most of the range is excellent.

Acknowledgments

This work has been supported by NSERC, MRC and NRC (Canada). The helpful cooperation of the Physical Sciences Laboratory at the University of Wisconsin and the Synchrotron Radiation Laboratory is gratefully acknowledged.

13. Brief Report on the Present Status of the SRRC

Edward Yen

Synchrotron Radiation Research Center, Taipei, Taiwan, Republic of China, 10764

Introduction

In the past decades, the Republic of China has enjoyed fast economic growth and increase in scientific research activities. To keep up the rate of growth, it is important to go from labor- intensive to technology-intensive economy. It becomes obvious that basic and applied research will play a key role in our future development. In order to upgrade the quality of research activities, it is important to go from scattered, unrelated individual researches to organized, cooperative research, and to have a major facility to serve as the focal point for future development.

With this in mind, suggestions of construction of a dedicated synchrotron radiation facility were made by local research workers and overseas Chinese scholars including Drs. Luke C. L. Yuan, C. S. Wu, Lee Teng, Yuan T. Lee, Bill W. T. Weng, Hank Hsieh, Y. H. Kao, the late Robert Poe and others around 1980- 1981. Furthermore, the National Science Council (NSC) appointed a five-member team (Y. C. Liu, P. K. Tseng, K. C. Cheng, C. N. Chang, and Edward Yen) to carry out a feasibility study for a synchrotron radiation facility in Taiwan, Republic of China on January 1, 1982. A report was submitted to NSC in December 1982 and reviewed by an Advisory Committee consisting of six distinguished overseas Chinese scientists.

Here were some of the reasons cited on the report for building a synchrotron radiation facility in Taiwan, Republic of China. It would provide outstanding research opportunities in physics, chemistry, bi-

X-ray Microscopy
Ed. by P.C. Cheng and G.J. Jan

ology, engineering, material science, medicine, and other fields, and also stimulate industrial research and development. It will encourage companies here in Taiwan to start R&D programs and probably attract other companies to locate themselves near such a synchrotron radiation source, because it has great utility in industrial research and development. It will provide excellent training for students, which in the long run is the most important thing, and attract overseas scientists to return to Taiwan. A synchrotron radiation facility will not become obsolete for many years. A well-designed storage ring will produce synchrotron radiation and its capability can be exploited for ten or many more years. With new beam lines, new insertion devices, wigglers and undulators, it can be kept at the forefront for many years to come. Finally, it is a cost-effective tool. Just one storage ring, even if it is not very big, has very large capacity to run many experiments simultaneously and conduct a very broad range of researches.

Upon the recommendation of the feasibility study team and the Advisory Committee, the government of the Republic of China approved the establishment of the Synchrotron Radiation Research Center under the Executive Yuan on July 9, 1983.

Project Objective

The objective of this project is to establish the Synchrotron Radiation Research Center (SRRC) in order to promote basic and applied researches in science and technology. In Phase I, the goals are the construction and commissioning of a dedicated synchrotron light facility and training of users and organizing of research teams.

The facility will include a 1GeV - 1.3GeV electron storage ring and its injection accelerator, three beam lines with experimental stations, and supporting facilities including offices, in-house laboratories, machine shops and user preparation laboratories.

Organization

To underscore the importance of this project, the SRRC is to be established as an independent agency directly under the Executive Yuan.

The Board of Directors which was formed in September 1983 consists of five distinguished overseas scientists, Drs. Luke C. L. Yuan (BNL), C. S. Wu (Columbia Univ.), Lee C. Teng (FNAL), Samuel C. C. Ting (MIT), Yuan T. Lee (UC Berkeley), and five other prominent leaders from the internal scientific community, Drs. T. Y. Wu (Academia Sinica), C. H. Yen (Atomic Energy Council), K. T. Li (Minister Without Portfolio), Y. S. Tsiang (Science Advisory Board, National Security Council), and L. A. Chen (National Science Council). Dr. Luke C. L. Yuan is the Chairman.

A Technical Review Committee of international experts reviews and advises the technical aspects of the project. Dr. Herman Winick (SSRL) is the Chairman. Other members are Drs. Matthew A. Allen (SLAC), John Blewett (BNL), Albert Hofmann (SSRL/SLAC), Robert A. Jameson (Los Alamos National Laboratory), Gottfried Mulhaupt (BESSY), Arie van Steenbergen (NSLS) and C. S. Wu (Columbia University).

The SRRC Preparatory and Construction Office has the responsibility of implementing the facility construction and users' training program. Dr. L. A. Chen (Chairman of the National Science Council) is the Director. Mr. K. C. Liu (Director of the Institute of Nuclear Energy Research) and Professor Edward Yen (National Tsing Hua University) are the Deputy Directors. Under the office are the Technical Component Division, the Engineering Division, the Users' Training Division and the Administrative Division. An organization chart is presented in Figure 13.1.

At present, we have forty-six full-time and twenty-seven part-time staff (most of them have joint appointment with universities and re-

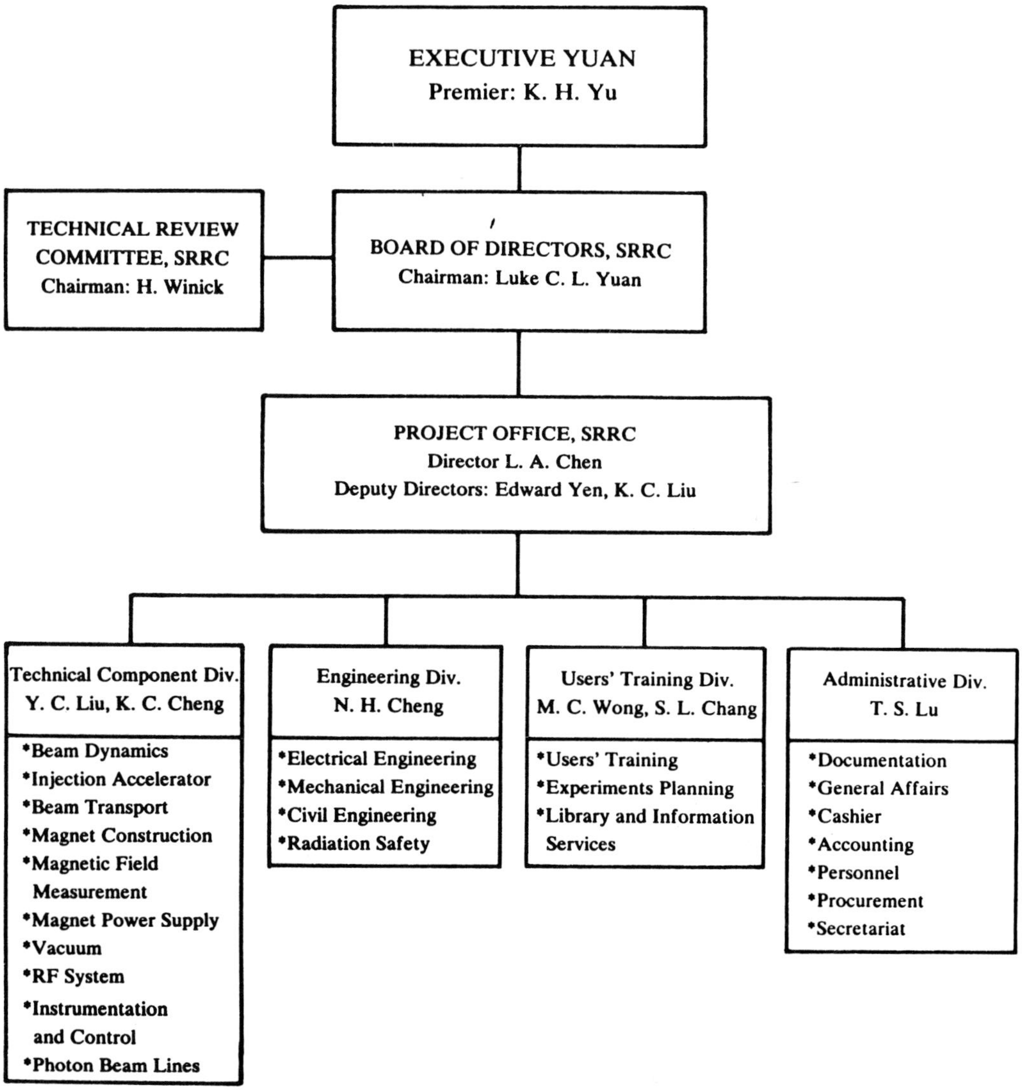

Figure 13.1 Organization chart.

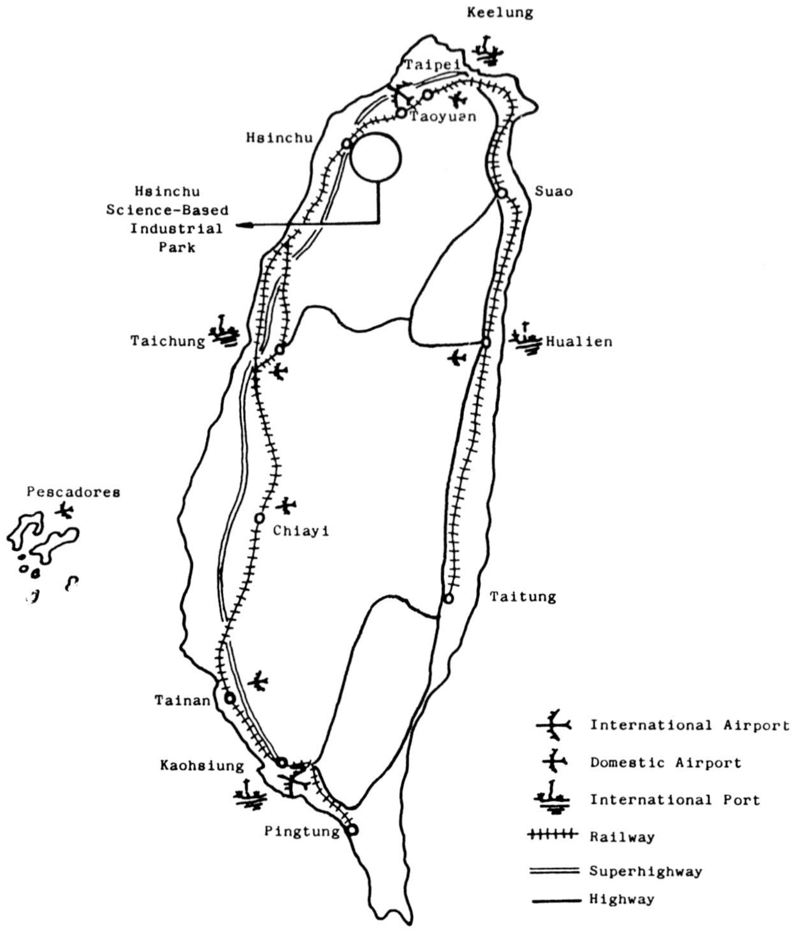

Figure 13.2. Map of Taiwan, the island province of the Republic of China, and location of Hsin-chu Science-based Industrial Park.

search institutes). With the progress of the project, we are expecting that this number will grow. Furthermore, we are also anticipating that the local research institutes and industries will provide technical and industrial support needed.

Budget

A preliminary budget of US$30,000,000 for the project has been approved. The government fully realizes the importance and significance of the project; therefore, not only is the funding source of the SRRC project independent of all other ongoing R&D programs, the funding policy is also rather flexible.

An additional funding of approximately US$5,000,000 may be needed if the recommendation of the Technical Review Committee of full- energy injection is adopted. Preliminary indication is that the government would probably concur with such a request.

Site

The site of the synchrotron radiation facility will be located at the northwest corner of the Hsinchu Science-Based Industrial Park. The locality will certainly enhance possible collaboration with the industrial R&D. The location is also near two major universities (National Tsing Hua University and National Chiao Tung University) and the Industrial and Technological Research Institutes (ITRI), a leading industrial research institute in Taiwan supported by the Ministry of Economic Affairs. A map of province of Taiwan and the location of the Hsinchu Science-Based Industrial Park is shown in Figure 13.2. The location of the SRRC site in the Industrial Park is indicated in Figure 13.3. Figure 13.4 is a preliminary layout of the SRRC site. The ground breaking ceremony was held on August 3, 1986. Preparations for construction are underway.

Major Parameters and Design Goals

After studying the major parameters of various existing synchrotron radiation facilities in the world and considering the need of the local users community as well as the possibilities for future development such as insertion devices, we set the design goals of the storage ring to be:

* the design energy should be 1.3 GeV
* the maximum energy should be 1.5 GeV or higher

* the circumference is about 120 to 140 m
* the ring should be optimized for high brilliance
* the beam emittance should be about 1.0^{-8} at 1 GeV
* the ring should have six long (6 m) insertions
* the total beam life time should be about 10 hr
* the RF frequency is 500 MHz to utilize the DESY cavities

Present Status of the Technical Groups

The major activities of the technical groups have been building up local capabilities of constructing the machine with outside consultation and advice, training machine scientists and designing the storage ring.

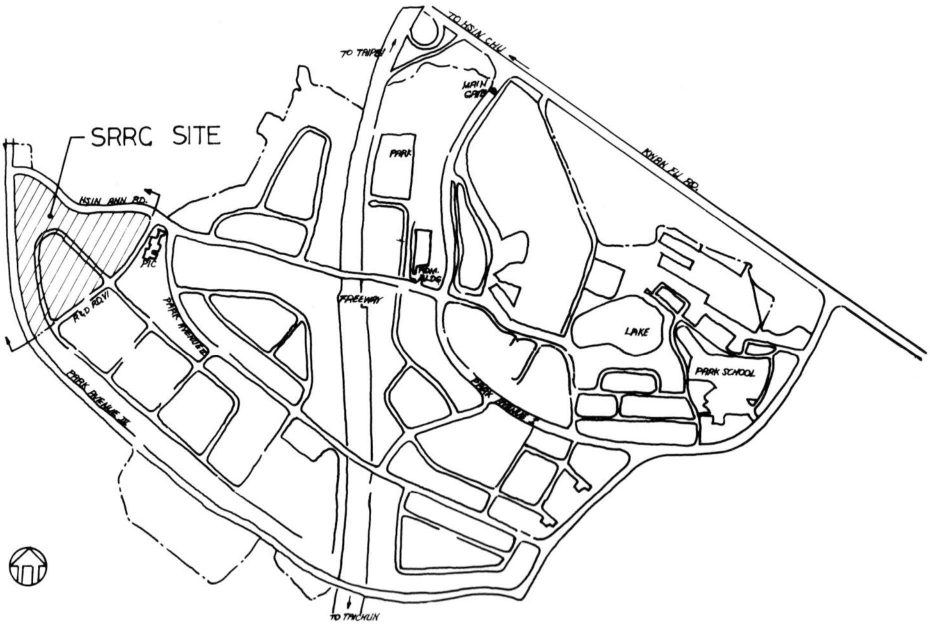

Figure 13.3. The location of the SRRC site in the Hsin-chu Science-based Industrial Park.

Summaries of the present status of the Beam Dynamics, Vacuum, Magnets, Injection and Beam Transport, Radio Frequency System, Control, and Photon Beam Lines Groups are as follows:

Beam Dynamics

Since the SRRC had no previous experience in accelerator physics, our Beam Dynamics Group started by studying different types of lattices which had been proposed. The lattices studied included Chasman-Green, FODO, TBA and FBA. From April to July (1986), our Beam Dynamics people worked three months at SSRL under the

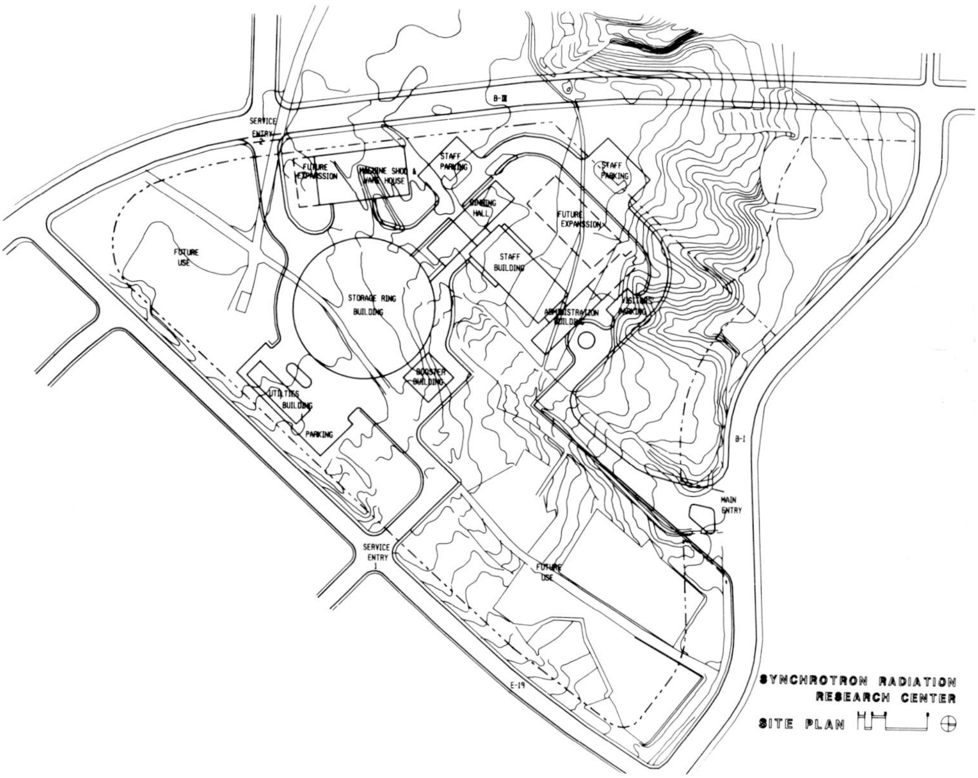

Figure 13.4. A preliminary layout of the SRRC site.

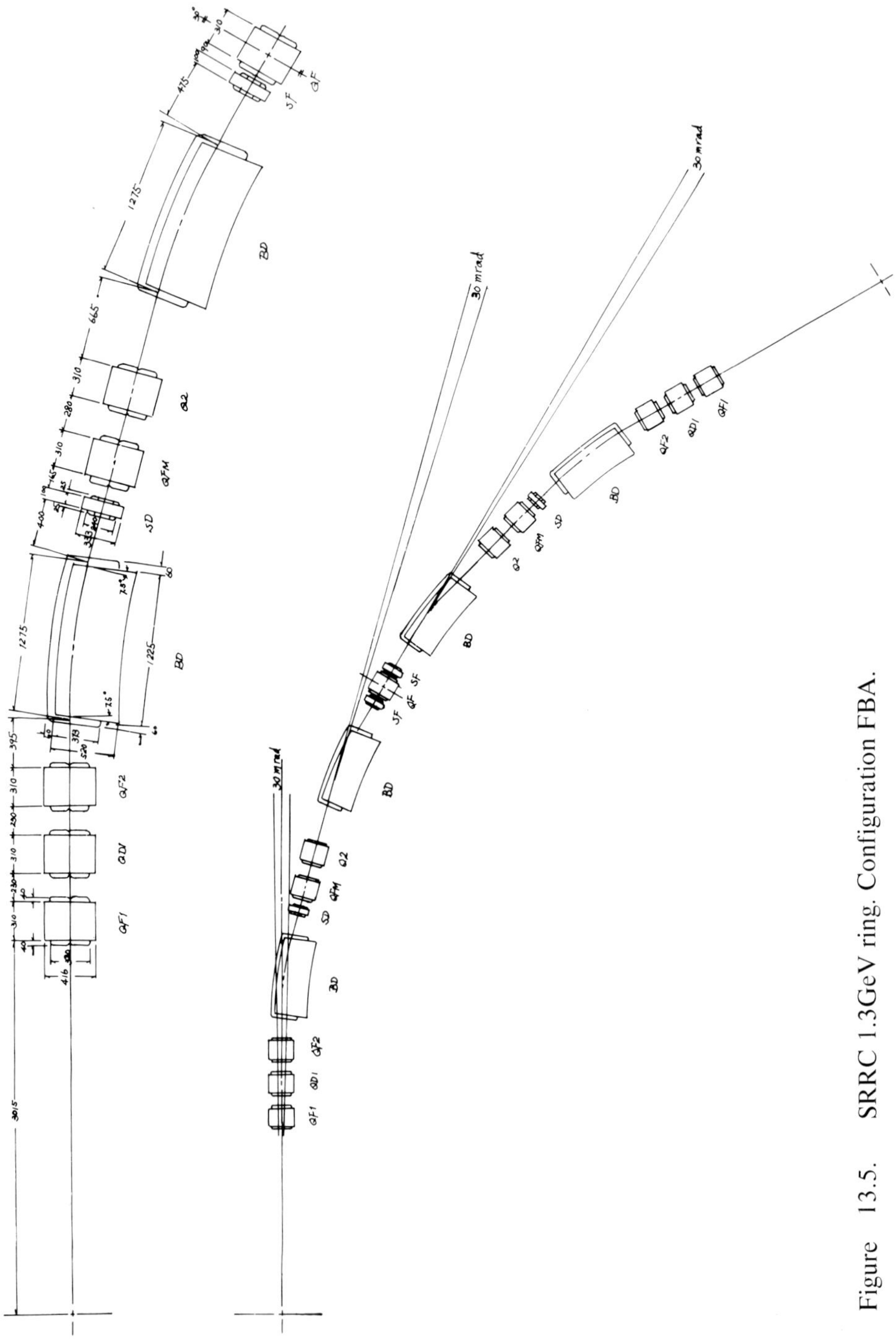

Figure 13.5. SRRC 1.3GeV ring. Configuration FBA.

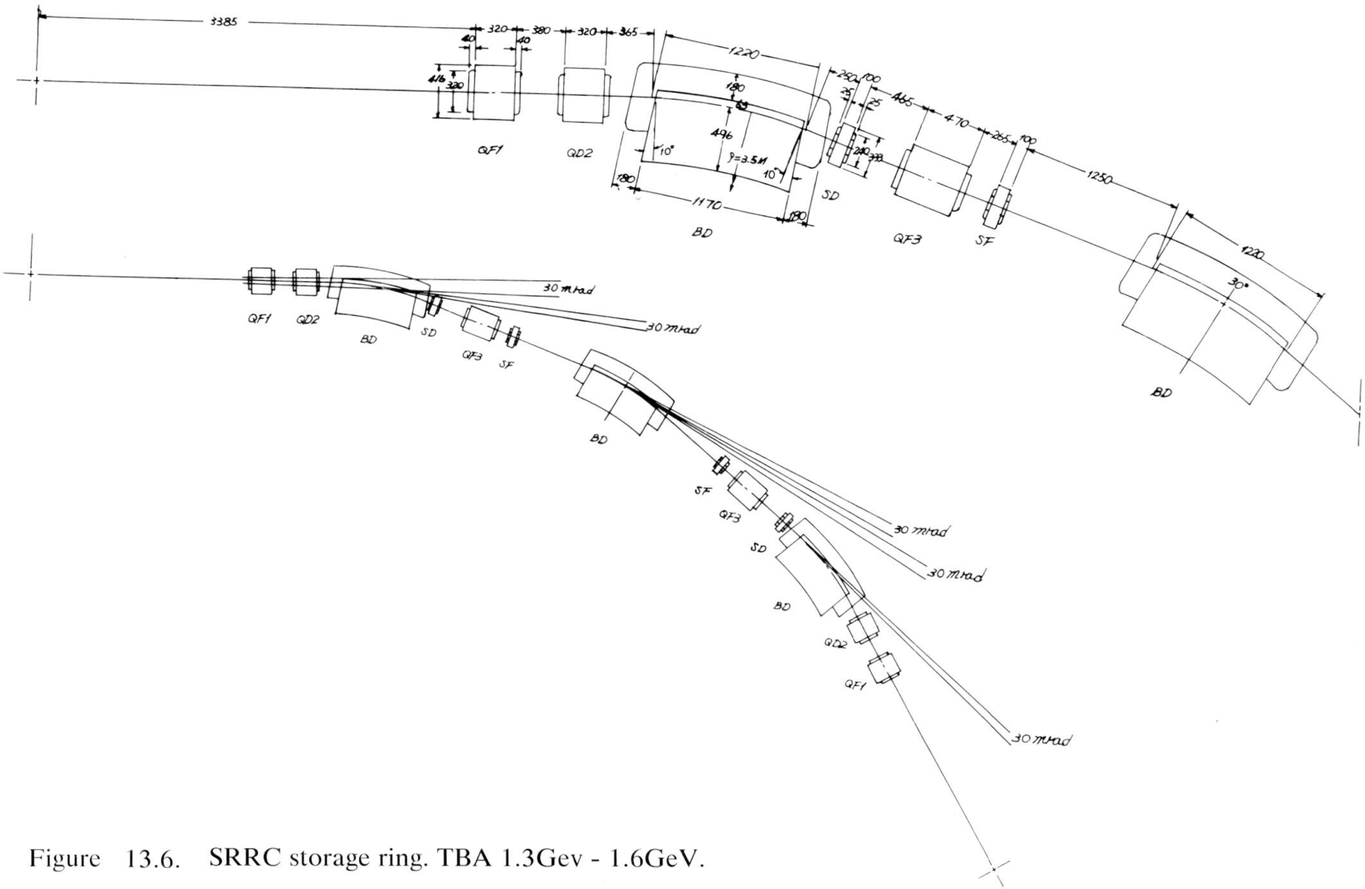

Figure 13.6. SRRC storage ring. TBA 1.3Gev - 1.6GeV.

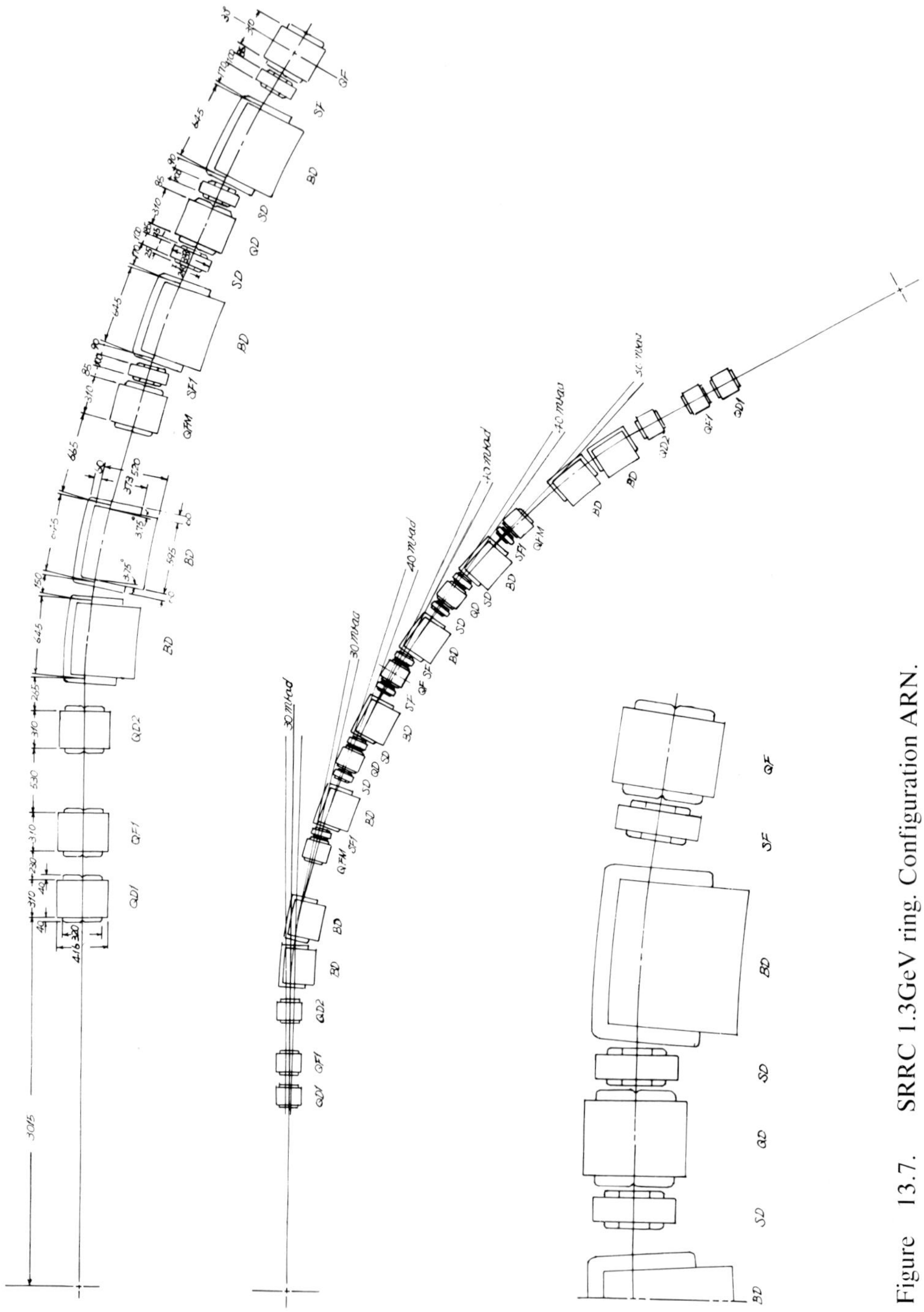

Figure 13.7. SRRC 1.3GeV ring. Configuration ARN.

guidance of Professor Helmut Wiedemann. They made a preliminary design for the storage ring. Some of the preliminary results are shown in Figures 13.5 - 13.8 and Table 13.1. However, further study has to be done before the decision on lattice will be made. Due to the importance of the lattice decision toward further progress, it is anticipated that by the end of 1986 the lattice for the storage ring will be frozen.

Vacuum

The Ultra-High Vacuum Group has been working on a 32 liter stainless steel test chamber. Using 150 l/s ion pump and 800 l/s titanium sublimation pump, an ultimate pressure of 4 x 10 torr has

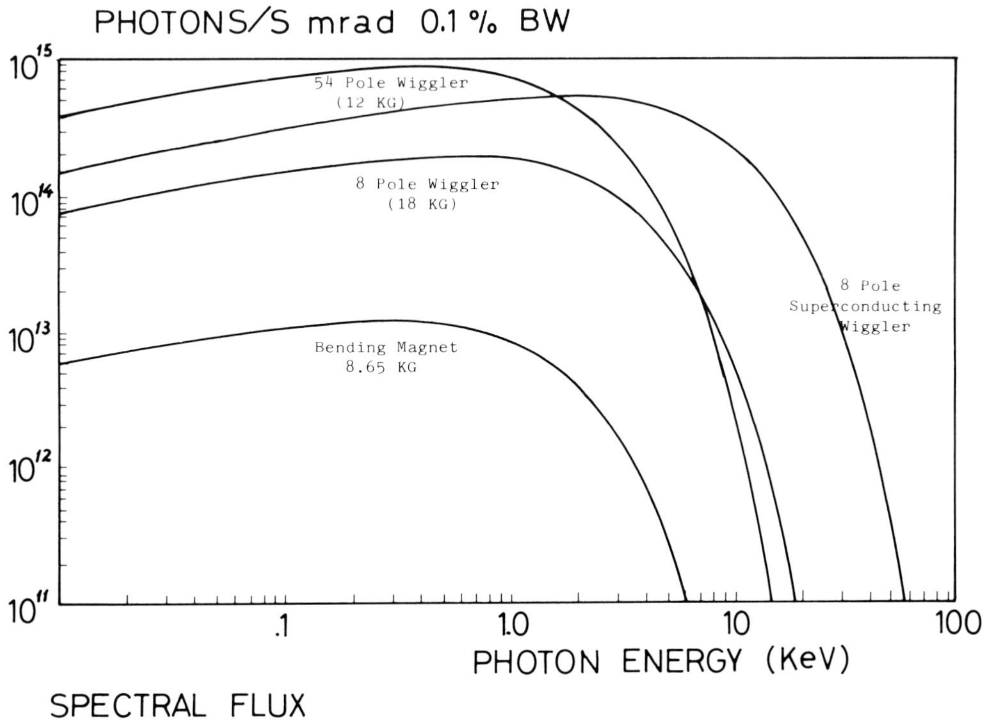

Figure 13.8. Photon flux spectrum at 1.3GeV.

Table 13.1: Comparison of FBA and FODO configurations.

	FBA	FODO
General ring parameters:		
Circumference (m)	124.80	124.80
Nominal energy (GeV)	1.30	1.30
Beam emittance (1.0^{-9} rad m)	12.80	13.80
Maximum energy (GeV)	1.50	1.50
Nominal beam current (mA)	100.00	100.00
Max. single bunch current (mA) (One RF system)	1.65	2.24
Number of insertions	6	6
Free length of insertions (m)	6.00	6.00
Bending magnets:		
Magnetic length (m)	1.275	0.645
Bending radius (m)	4.870	4.927
Magnet gap (mm)	50.00	50.00
Vertical aperture inside of vacuum chamber (mm)	40.00	40.00
Field @1.3GeV (kGauss)	8.900	8.975
Maximum field (kGauss)	14.00	14.00
Gradient (kGauss/cm)	0.286	0.00
Sagitta (mm)	41.66	10.55
Construction	laminated /curved /sector	laminated /straight /rectangular
Number of magnets	24	48
Photon beam lines		
Insertion beam lines	6	6
Bending magnet beam kines	12	30
Critical energy from bending magnet @1.3GeV (keV)	0.99	0.98
useful intensity up to (keV)	5.00	5.00
Bending magnet source size (typical, mm) in x	0.130	0.154
in y	0.019	0.022
Max. undulator photon energy (keV) 10mm gap, 20mm	0.66	0.66
period length in x (mm)	0.320	0.390
in y (mm)	0.027	0.017
Undulator pinhole brilliance (Ph/(sec • mrad^{2}mm^{2}0.1 BW)	6.0^{18}	7.8^{18}
Undulator condition	3.9^{-5}	3.9^{-5}
Beam divergence (Opt. coupling)	3.0^{-5}	3.3^{-5}

been reached. Measurement for an SUS 304 chamber gives an outgassing rate of 3 x 10 torr l/s cm.

Measurement for an A6063-EX aluminum chamber shows an even lower outgassing rate of 1 x 10 torr l/s cm. Measurement of pumping speed for various gases in various types of pumps has been established. Presently, works have been concentrated on the aluminum UHV system so that comparisons with the SUS UHV system can be made. The study of aluminum welding techniques is underway and preparations are being made to test the glow discharge and to construct a prototype straight section.

Magnets

The magnet design for two candidate lattices has been finished. Prototype magnets will be built for testing. Lamination steel and coil winding materials are on order for the prototype magnet. A locally-made DC magnet power supply has been constructed, having designed rating 60V x 500A. At present, the stability is less than 10 at 20V x 200A.

A complete set-up for magnetic field measurements of the various magnets needed for the synchrotron is underway and will be completed shortly. Preliminary analysis for carrying out magnetic field mapping and calculations is in progress.

Injection and Beam Transport

Due to lack of experience and manpower, the injection accelerator will be purchased from the industry. It will be a 1 - 1.3GeV machine so that the injection will be at full energy of the storage ring.

Radio Frequency System

An SRRC RF Group has been formed. Negotiations are in progress to purchase 6 RF cavities (500 MHz) with accessories from DESY in Germany for possible utilization in the SRRC synchrotron and for gaining practical experience by the RF Group.

Control

The control system structure for the storage ring has been designed to have three levels consisting of VAX-11/750, PDP-11/24 and Motorola VME microprocessors. The microprocessor-based intelligent local controller (ILC) has been constructed locally both in hardware and software. The data-link between ILC's is selected to be MIL-STD-1553B, which has also been constructed locally.

Basic plans for control have been initiated studies and preparations are in progress. The Control Group has completed the following tasks: (a) the MIL-STD-1553B data communications network, (b) hardware and software for part of the local control system of the synchrotron, (c) a 200A x 20V power supply for magnets with a regulation precision of 10^{-4}, and (d) automatic measuring and monitoring systems for the vacuum system.

Photon Beam Lines

The Photon Beam Lines Group has assembled an acoustic delay line (ADL) test facility to measure the relationship between delay time and the ADL parameters and compare with the theoretical calculations. Computer calculations were also done for the thermal gradient distribution in the optical elements under the illumination of synchrotron light. The results will be used for the estimation of thermal stress of optical elements.

Users' Training

The major task of the Users' Training Division is to train and recruit at least sixty users in the next five to six years. There are three main sources of users, local researchers interested in doing synchrotron radiation research, local graduate students recruited to go abroad for further training and experienced overseas users attracted to return to Taiwan, the Republic of China.

Past activities included organizing the Users' Office and formulating the users' training program with the participation of local scientists, planning of research laboratories for the preparation of users performing experiments on the storage ring, selecting young prospects and sending them abroad for training, collecting and distributing information, establishing a research library and organizing and sponsoring symposia and workshops on synchrotron radiation application.

Three important symposia were held in the past, the Symposium on Synchrotron Radiation Application on July 16 - 17, 1982 with speakers, Drs. Herman Winick (SSRL), P. Eisenberger (EXXON), J. Janin (Orsay), Y. T. Lee (University of California, Berkeley), T. N. Chang (University of Southern California), the Symposium on Synchrotron Radiation on August 20 -25, 1984 with speakers, Drs. David Lynch (Iowa State Univ.), Y.H. Kao (SUNY at Stony Brook), R. Klaffky (BNL), M. Ando (KEK), L. Teng (FNAL), H. Winick (SSRL), S. Y. Tong (University of Wisconsin at Milwaukee), K. Liang (EXXON), T. C. Chiang (University of Illinois), C. S. Wu (Columbia Univ.), E. Rubenstein (Stanford), and the symposium on Atomic and Molecular Science with synchrotron Radiation jointly sponsored with the Academia Sinica on March 21 - 25, 1985 with speakers, Drs. D. Shirley (LBL), K. M. Chen (Chung Shan University), D. L. Judge (USC), Y. Wang (National Taiwan University), D. L. Ederer (NBS), F. Wuilleumier (LURE), K. C. Cheng (SRRC), H. P. Kelly (University of Virginia), H. K. Tseng (Chung Yang University) and T. N. Chang (USC).

Future activities will be to send ten to fifteen students abroad for training every year, to support professors and researchers to go abroad to do synchrotron radiation research, to establish in-house laboratories (X-ray, VUV, Optical and Future Development, *etc.*) by 1987, to establish research capability before 1989 - 1990, to plan research projects and experimental program using the storage ring, and to co-operate with the construction groups to design and build the beam lines.

Outlook

With the ground breaking, we expect the construction work to start in earnest soon and the goals for Phase I to be achieved in five to six years. In the meantime, we hope to have ten to twelve groups of users ready to utilize the facilities and that in the future this synchrotron facility will be the core of the first national laboratories of the Republic of China on Taiwan.

Note added in proof: FBA configuration was chosen by the SRRC in March, 1987.

Ground breaking ceremony of the Taiwan Synchrotron Light Source (SRRC) at Hsin-chu Industrial Park on August 2, 1986.

14. Diffraction-Imaging Possibilities with Soft X-rays

David Sayre

IBM T. J. Watson Research Center, Yorktown Heights, NY 10598, USA

Introduction

Diffraction imaging is a form of microscopy which does not quite exist as yet with soft x-rays, but on which work is proceeding. It has, however, a long history with x-rays in the 0.1nm wavelength region (x-ray diffraction analysis), with electrons (electron diffraction), and in visible-light holography. Its interest lies in its being potentially the highest-resolution form of x-ray microscopy (see the previous article by the author in this volume), and in its special ability to obtain information regarding 3-dimensional structure in a single exposure. The present paper outlines the general principles of the method and gives a brief summary of its current status.

Recording the Signal

Diffraction-imaging microscopy constructs an image of the specimen from a recording of the coherently scattered, or diffraction, signal from the specimen. Figure 14.1 illustrates some of the different methods of recording the signal. In (a) the signal is recorded in the far field and outside the direct beam transmitted by the specimen. At the left is a screen which defines the beam; the diameter of the beam is usually made just larger than that of the specimen. In actual situations the screen is replaced by a collimator designed to produce minimum extraneous radiation accompanying the beam. In the absence of a specimen, most of the detector (here shown schematically as spherical)

X-ray Microscopy
Ed. by P.C. Cheng and G.J. Jan

receives almost no signal. When a specimen is inserted in the beam, the detector receives the scattered signal. Figure 14.2 shows a biological cell (diatom) similar to one used in an experiment of this kind, with the area covered by the beam indicated, and Figure 14.3 shows the signal recorded.

In Figure 14.1(b) the signal is recorded in the intermediate field and within the direct beam. The experimental arrangement is identical to that for contact x-ray microscopy, except that the detector is moved a short distance away from the specimen. Figure 14.4 shows a typical recording of this kind. The interference effects caused by the coherently scattered signal combining with the intensity-modulated transmitted signal are clearly seen.

In Figure 14.1(c) the signal is again recorded in the far field, but is mixed with some of the incident radiation to form a Gabor hologram. Figure 14.5 shows a recording of this type.

Figures. 14.3 - 14.5 show that the soft x-ray diffraction signal from specimens which are microscopic in size can be recorded, but it should be understood that to date only a few such recordings have been made. More important than this, the recordings to date have not extended to large diffraction angles (angle between incident and scattered photons.) Since the resolution in imaging from the diffracted signal depends upon the maximum diffraction angle 2θ as $\lambda/2\sin\theta$, the importance of recording to large diffraction angles is evident. At the maximum possible value for 2θ, $180°$, the resolution is $\lambda/2$ (*i.e.* 1nm for x-rays of wavelength 2nm), but for $2\theta = 5.7°$ the resolution worsens to 10λ.The basic premise of diffracted-field imaging (that by avoiding the imperfections of focussing optics it can image at higher resolution than scanning or imaging microscopy) is thus true only if it is successful in recording at large diffraction angles. To date the largest diffraction angles 2θ recorded with soft x-rays and microscopic specimens are approximately $4.5°$, $11.4°$, and $2.3°$ for the methods of Figure 14.1(a), 14.1(b), and 1(c), respectively. This situation should

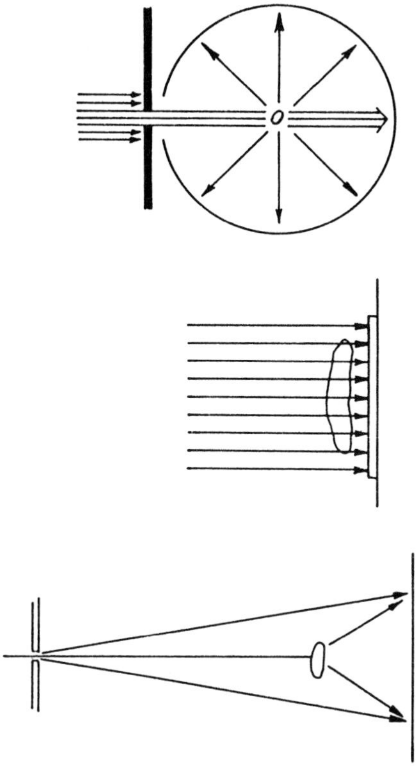

Figure 14.1. Three basic geometries for recording the diffraction signal: (a) in the far field; (b) in the intermediate field; (c) as a Gabor hologram.

improve with better sources and apparatus, but until that occurs diffraction imaging will be of limited practical interest.

Imaging from the Signal

We now turn from the recording of the diffraction signal to a consideration of the information contained in it. We assume that all the scattering is coherent and that all the photons incident on the specimen have a single wavelength and direction. Then the field surrounding the specimen is stationary and generally non-uniform, its non-uniformity reflecting the interference pattern of the N signals

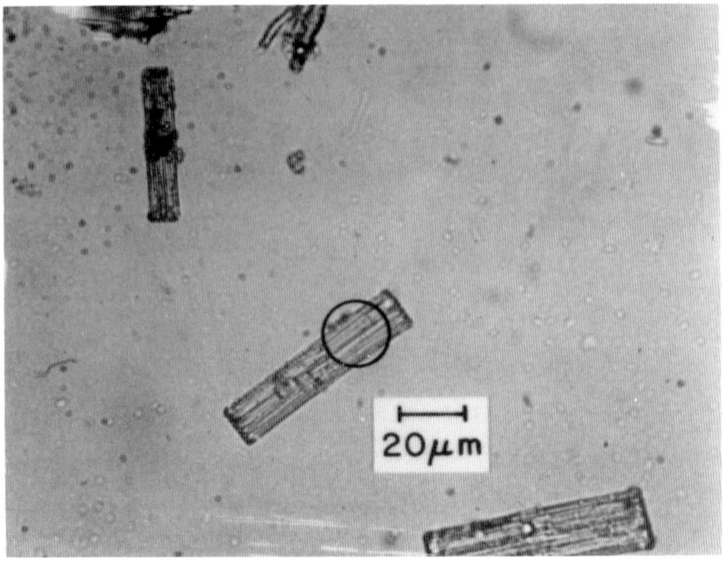

Figure 14.2. Optical micrograph of a diatom similar to that used to record the diffraction pattern shown in Figure 14.3. The circle shows the portion of the specimen illuminated by the direct beam. (From Yun *et al.*, 1987)

scattered from the N atoms present. Qualitatively it is easy to see that this pattern is rich in information concerning the positions and scattering strengths of the atoms present. Quantitatively we consider here only the simplest case, in which the effects of the specimen on the field are small (case of the first Born approximation). In this case it can be shown that the diffraction field gives the Fourier transform of the function describing the spatial distribution of scattering strength in the specimen; this will be equivalent to being given the Fourier transform of the image, if one is willing to accept the scattering-strength function as an image of the specimen. In slightly more detail, if the scattering-strength function is $d(\vec{x})$, it Fourier transform is $d(\vec{X})$, $\vec{x}$ and $\vec{X}$ being the position vectors in specimen space and transform space respectively, then the measurement of the diffraction signal in the experiment of Figure 14.1(a) yields the value of $\hat{d}(\vec{X})$ on the 'Ewald sphere' shown in Figure 14.6. Here the diffraction signal is thought of as having magnitude and phase; *i.e.* it and $\hat{d}$ are both complex. Assuming now that even a single experiment of the type of Figure 14.1(a) has been done, the information on the Ewald sphere regarding $\hat{d}(\vec{X})$ can be Fourier inverted to give a medium-quality 3-dimensional image of $d(\vec{x})$. Alternatively, if the experiment of Figure 14.1(a) is repeated with many different orientations of the specimen, the value of $\hat{d}(\vec{X})$

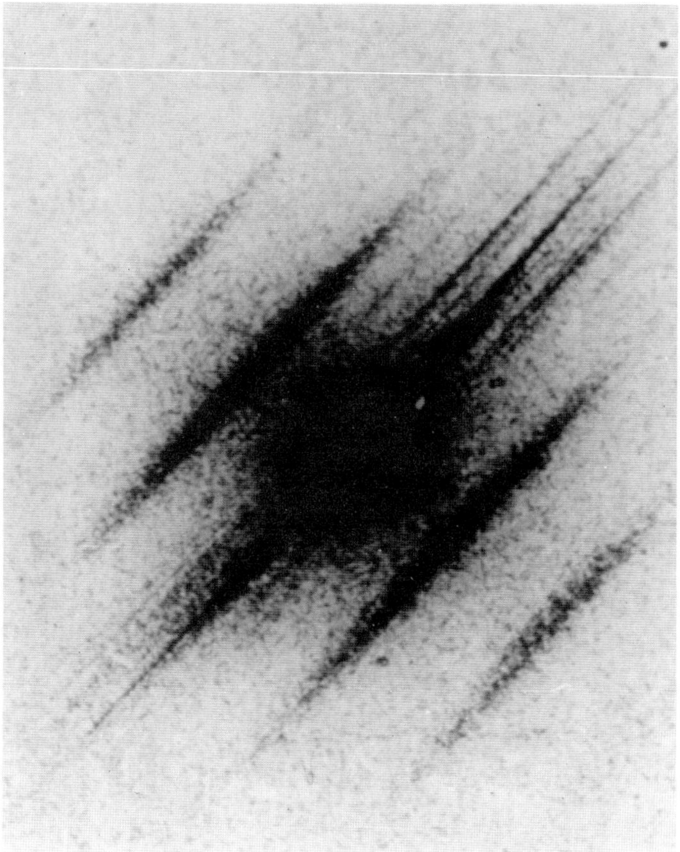

Figure 14.3. Recording of the far-field diffraction signal from a portion of a diatom using 3.2nm wavelength x-rays. The geometry was that of Figure 14.1(a); see also Figure 14.2. The source of x-rays was the U15 beamline at the National Synchrotron Light Source at Brookhaven, and the detector was a Kodak type 101 silver halide plate. (From Yun *et al.*, 1987)

will be determined everywhere within the sphere $|\vec{X}| < 2/\lambda$, and when this information regarding $\hat{d}(\vec{X})$ is Fourier inverted, a high-quality and fully 3-dimensional image of $d(\vec{x})$ to resolution $\lambda/2$ is obtained.

Although we have discussed here the information in the far-field portion of the diffraction field, the situation is basically the same throughout the field: as long as the field is measured on a surface which fully surrounds the specimen it essentially yields a measurement of $\hat{d}(\vec{X})$ on the Ewald sphere (Devaney 1985). The choice of a particular region for measuring the field is thus governed more by experimental and processing considerations than by the nature of the information produced.

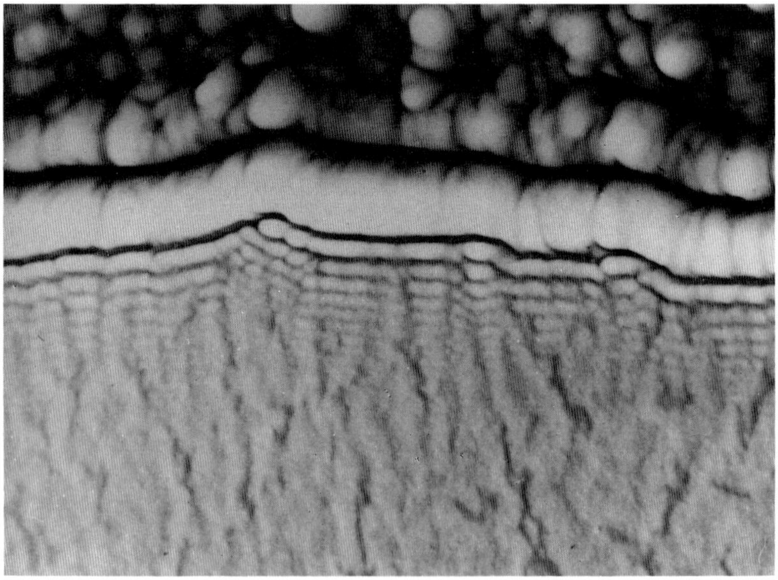

Figure 14.4. Recording of the intermediate-field diffraction signal from a section of rabbit psoas muscle using 3.6nm wavelength radiation. The geometry was that of Figure 14.1(b). The source of x-rays was the U15 beamline at the NSLS at Brookhaven, and the detector was polymethylmethacrylate (PMMA) x-ray resist. (Courtesy R. Feder).

We are now in a position to summarize the diffraction-imaging process. The diffraction field produced by the specimen is measured to as large a diffraction angle as possible, and if possible with the specimen in many orientations relative to the incident beam; the information is converted into information on $\hat{d}(\vec{X})$; and $\hat{d}(\vec{X})$ is Fourier inverted to give the image $d(\vec{x})$. As we shall see, this process has not yet been carried out in a fully satisfactory manner with soft x-rays. To give an idea of the result, however, we show Figure 14.7, which is the result of carrying out the process in x-ray diffraction analysis.

The Need to Phase the Signal

In our account thus far there has been one major omission, which we must now discuss. We have spoken of the detector as measuring

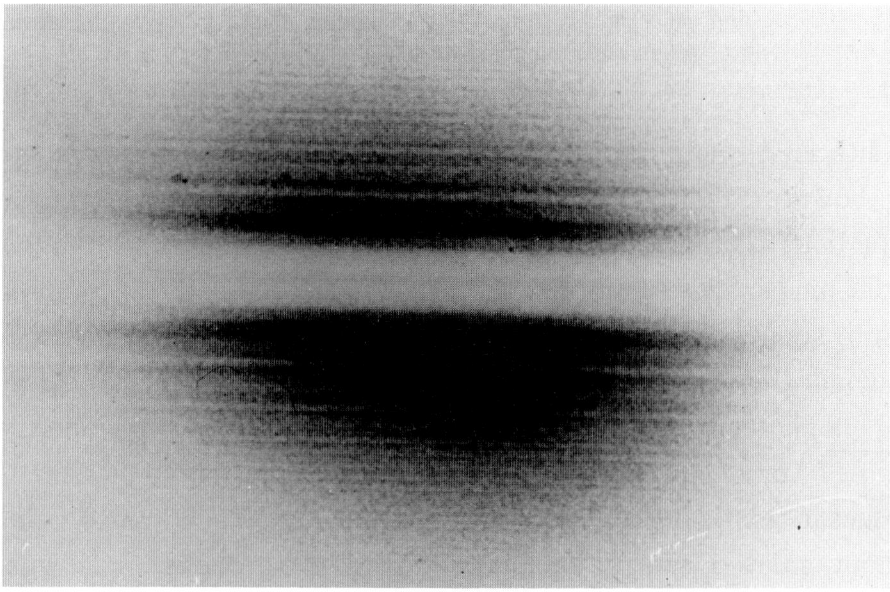

Figure 14.5. Gabor hologram of a thin (12.5μm) wire using 3.1nm wavelength radiation. The geometry was that of Figure 14.1(c). The source of x-rays was the U15 beamline at the NSLS at Brookhaven, and the recording medium was Kodak type 131-01 film. (From Howells *et al.*, 1986)

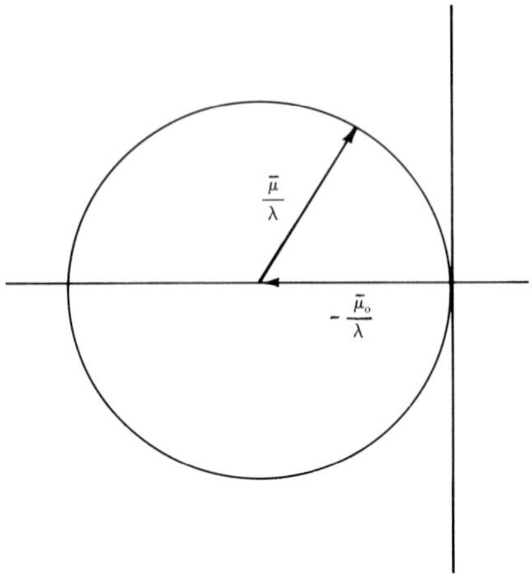

Figure 14.6. The Ewald sphere. The sphere is located in Fourier space, *i.e.* the space containing $\hat{d}(\vec{X})$ (see text). Its center is at $-\vec{u}_0/\lambda$, where $\vec{u}_0$ is a unit vector in the direction of the incident beam in Figure 14.1(a), and its radius is $1/\lambda$. The values of $\hat{d}(\vec{X})$ on this sphere, and the values of the diffraction signal at the detector of Figure 14.1(a), are the same.

the diffraction field in both magnitude and phase. At x-ray frequencies, however, there are no detectors capable of measuring the phase of the field. The recording of the diffraction signal, which is the basis of the image construction, is thus incomplete, and the information must be supplied before imaging can occur. Fortunately techniques for overcoming the problem are known. In x-ray diffraction analysis, the field magnitude may be recorded from the original specimen and then recorded again from a slightly modified form of the specimen; from the changes in the field magnitude the field phase can usually be obtained without great difficulty. This so-called isomorphous replacement technique was the method used in forming the image of Figure 14.7. In a variant of this method, called the anomalous dispersion technique, the specimen is left unchanged but the wavelength of the radiation is changed; again changes are induced in the field magnitude from which the field phase can usually be calculated. Again in the holographic

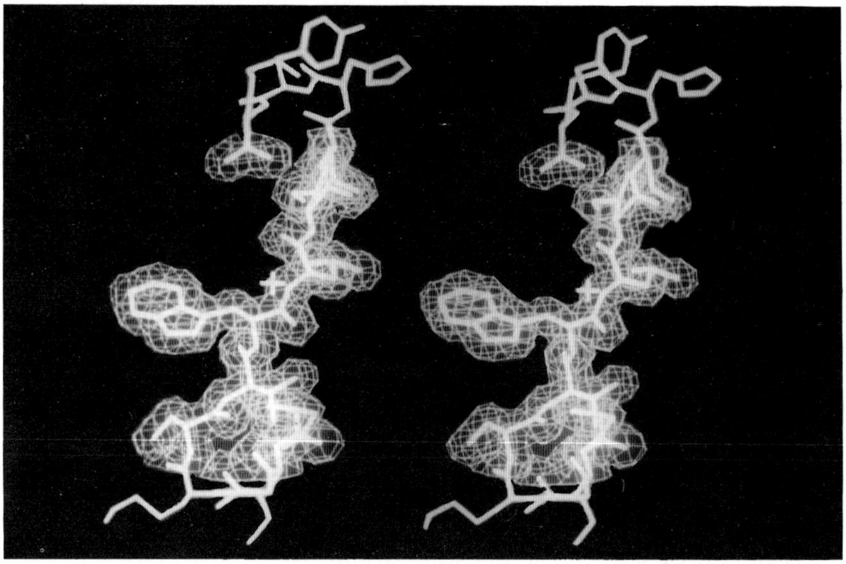

Figure 14.7. Example of diffraction-imaging microscopy. The image is fairly typical of the work done in x-ray diffraction analysis today. The specimen was a small crystal of cytochrome B5, a respiratory enzyme protein. The computer display is here limited to a small part of the image, in the vicinity of residue 23 (tryptophane) of the protein. The 3-dimensional image is shown here as a pair for stereo viewing. The image is shown by the 'chicken wire' lines; the solid lines show an atomic model of the structure based on the image. Because the image here is based on diffraction of 0.15nm x-rays, the resolution is considerably higher than can be expected with soft x-rays. (Courtesy C.D. Barry).

technique, which exists mainly in the visible light region, the magnitude measurement is made on the sum of the diffraction field and a known reference field. There are also other methods in use for phasing the diffraction field from recordings of its magnitude, but the three which have been mentioned are the ones which are most likely to be transportable to the soft x-ray case. To date one of these, holography, has produced soft x-ray images (Aoki and Kikuta, 1974; 1986; Howells *et al.*, 1986), but not as yet at interesting resolutions.

Recently Sayre (1986) has suggested a fourth method, based on the notion of measuring the field magnitude at several different positions in the intermediate field. Using the relation

$$\hat{f}(\vec{R},z_2) = \hat{f}(\vec{R},z_1)e^{i2\pi\Delta z\sqrt{\lambda^{-2} - |\vec{R}|^2}} \tag{14.1}$$

(where z_1 and z_2 are two distances from the specimen, $\hat{f}(\vec{R},z_1)$ and $\hat{f}(\vec{R},z_2)$ are the Fourier transforms of the diffraction field $f(\vec{r},z_1)$ and $f(\vec{r},z_2)$ as measured at those distances, $\Delta z = z_2 - z_1$, and λ is the wavelength), he has been able to calculate the phase of $f(\vec{r},z)$ from knowledge of $|f(\vec{r},z)|$ at several values of z. Relation (1), which expresses the structure of any stationary monochromatic scalar radiation field in free space, is known as the propagator relationship (Goodman, 1968). The probable field recording technique for this method would resemble that of Figure 14.1(b), except that a detector stack would replace the single detector shown there. Preliminary attempts at recording using up to 3 detectors in a stack have recently been made by Howells and Jacobsen (1986). One advantage of measuring in this geometry is that it is relatively easy to record to large diffraction angles, which as noted earlier is a necessity for imaging at high resolution. It must be noted, however, that this technique of phasing requires very high spatial accuracy in the recording, which may be difficult to achieve.

Summary and Conclusions

Diffraction-imaging microscopy with soft x-rays potentially offers imaging resolutions as high as 1 to 3nm, plus high-quality 3-dimensional imaging with multiple exposures or medium-quality 3-dimensional imaging with a single exposure. The principles of the technique have been described, and examples showing that the recording of the necessary diffraction data may be feasible have been shown. The necessary next steps appear to be methods for obtaining better diffraction data and methods for supplying phases for the data.

15. X-ray Microholography, Exciting Possibility or Impossible Dream?

Eberhard Spiller

IBM T. J. Watson Research Center, Box 218, Yorktown Heights, NY, 10598, USA

Introduction

Bright, spatially coherent soft x-ray sources are now becoming available and high-resolution imaging of biological specimens by x-ray holography has been discussed as an exciting application for the new sources (Robinson, 1985; Underwood and Attwood, 1984; Solem and Baldwin, 1982; Solem and Chapline, 1984; Matthews *et al.*, 1985). However, even with fully coherent sources, there are some severe limitations for 3-D imaging which make it questionable if x-ray holography will ever be of any practical use for the imaging of microscopic objects. The purpose of this note is to point out these difficulties and speculate about possibilities to alleviate the problem.

Let's assume an object in the shape of a cube. The number N_s of spatial modes emerging from one face of the cube is equal to the number of resolvable spots of size δ_x, δ_y on the surface of this face (Laue, 1914; Gabor, 1961).

$$N_s = (4\,S/\lambda)\,\sin^2\frac{\theta}{2} \qquad (15.1)$$

$$\delta_x = \delta_y = \lambda/2\,\sin\frac{\theta}{2} \qquad (15.2)$$

where S is the surface area of the face of the cube and $\frac{\theta}{2}$ is the half angle of the cone of light emerging from the cube.

X-ray Microscopy
Ed. by P.C. Cheng and G.J. Jan

For a 2-dimensional object we have a match between the number of pixels in the object and the number of spatial modes in the radiation field emerging from the object. However, the radiation field has not enough modes to carry the additional information which is required, if we want to image N_z planes inside the cube with the same resolution. Matching of the information for a general 3-D object requires that we either accept a poorer resolution $\delta_{3d} = \delta_x \sqrt{N_Z}$ or take N_z different exposures, for example, by rotating the object or the illumination.

In the reconstruction of one plane of a 3-dimensional object from a hologram or when viewing one plane of the object with coherent light, the information about the other planes is encoded as a background fringe pattern. For a general object with many object points a statistical treatment of the fringe (speckle) pattern becomes appropriate and the intensity I of the speckle pattern can be described by the probability

$$p(I) = \frac{1}{< I >} e^{-\frac{I}{<I>}} \tag{15.3}$$

with a contrast

$$C = \frac{\sqrt{< (I - < I >)^2 >}}{< I >} = 1 \tag{15.4}$$

where $< I >$ is the average intensity. The average size of the speckle, i.e. the correlation distance in the speckle pattern is equal to the diffraction limited pixel size as defined in eq. (15.2) (Martienssen and Spiller, 1967; Martienssen, 1966; Dainty, 1984; Goodman, 1985) .

Therefore, a single image point is reconstructed on a background of a speckle (noise) pattern with contrast one and cannot be distinguished from it. The large fluctuations in the speckle background can be reduced by averaging the intensity over larger areas. For that case, the contrast of the speckle pattern given in Eq. (15.4) is reduced to

$$C_{AV} = 1/\sqrt{N} \tag{15.5}$$

where N is the number of 2-dimensional resolution elements which are used for the averaging.

As an example, let's assume we want to reduce the contrast of the background noise to 1%; this requires a value of N = 10,000 or an average over an area 100 $\delta_x \cdot 100\ \delta_y$. Therefore, if we look at feature sizes which are 100 times larger than the resolution limit, the effective contrast of the speckle is reduced to an acceptable value. Nearly all holographic work in the visible light uses this type of speckle reduction with reduced spatial resolution. An equivalent method of averaging is obtained in the holographic reconstruction by dividing the hologram into N subholograms, thus obtaining N different reconstructions whose intensities can be added to reduce the speckle contrast to a value given by Eq. 15.5. (Martienssen, 1966). In general, the smaller subholograms have correspondingly smaller resolution; however, one can adjust the shape of the subhologram for specific objects and purposes: for example a subhologram in the shape of a cross or a star produces a loss of contrast in the image but no loss of resolution (Dainty, 1984).

Three-dimensional reconstructions of an object can be obtained without loss of contrast and resolution if a large number of full size holograms is made, with the illuminating beam or the specimen being rotated between the different hologram recordings. The intensities of the reconstructions from the different holograms are added to average out the speckle background fluctuations (Martienssen and Spiller, 1967).

In the visible region holography is not used today for 3-D imaging for microscopic objects. The preferred method is to construct the 3-D image from different incoherent images, which are obtained by focussing through the object and/or rotating the object. The main advantage of incoherent imaging is the elimination of speckle, because speckle is already averaged out in each individual image. For imaging with soft x-rays the incoherent methods are also preferable. A scanning x-ray microscope with diffraction limited spot is the equivalent to a completely incoherent imaging system and represents the preferred method to obtain the large number of 2-D images required for a full 3-D reconstruction. Feder and Sayre (1980) have produced stereo-images of biological objects with contact microscopy by taking two micrographs at two different angles, and it appears as a natural extension of this method to take more pictures and combine them to a 3-D reconstruction by computer. From an optics point of view the best possible method for a 3-D reconstruction (with the best signal/noise ratio and lowest radiation damage) consists of slicing the specimen and recording the 2-D image of each slice separately.

For an un-sliced specimen the number of individual pictures that can be taken is limited by the radiation damage to the specimen. Most living specimens are killed by the dose required for a single, high-resolution picture. Solem and Baldwin (1982) pointed out, that this problem can be avoided if a hologram is produced with a very short pulse from an x-ray laser. In that case the specimen might be completely destroyed by the pulse; however, for a sufficiently short pulse the hologram can be recorded before there was enough time for the object to change. In this scenario, a single hologram or a single 2-D picture is the only information that can be recorded and the reconstruction of a 3-D image becomes practically impossible.

The possible techniques for the reconstruction of 3-D images from a 2-D recording can be divided into two groups, either reducing the information content in the object to the amount that can be stored in a plane or by using *a priori* information which has been obtained by some other method to supply the missing information in the reconstruction. The implementation of the first method is relatively straightforward. Accepting a lower resolution is always possible and permits one to reduce the speckle by incoherent addition of subareas of either the hologram or the image (Eq. 15.5).

An object that is mostly empty or contains some regular pattern (for example periodic substructures) contains less information and one should be able to reconstruct such an object with high spatial resolution. In the "empty object" (isolated object points separated by large space) the diffraction patterns from different object points are separated far enough, such that one can recognize and understand the different components in the reconstruction and remove them. An example for this technique with visible light is the recording of aerosol particles in a 3-dimensional volume (Thompson *et al.*, 1967). Work in x-ray holography reported up to now uses "empty" objects and low resolution (Howells, 1984; Reuter and Mahr, 1976; Aoki and Kikuta, 1972; Baez, 1952). Specimen that consist of periodic areas can be treated in a similar way because they become "empty" objects after a Fourier transform.

The incorporation of *a priori* knowledge can be easy in specific cases and represents a big challenge for future work in the general case. Figure 15.1 gives an example of an easy case. We can remove the speckle in the hologram reconstruction in the test chart of Figure 15.1 by using the knowledge that the test pattern consists of groups of parallel lines of different orientation and spacing. We can get a good estimate of the linewidth and orientation of each group by averaging over the total area covered by this group (possible using a Fourier transform) and can than use this information to redraw a speckle-free test pattern.

The general case is much more difficult. Let's assume we have made a single hologram of a not "empty" biological object and want to reconstruct this specimen in three dimensions with the best possible resolution. The specimen has been destroyed during the exposure of the hologram, therefore, no additional information on this specific object is available. However, we have available a big library of holograms, contact radiographs, electron microscope pictures, *etc.* of the some type of biological specimens prepared in a similar way.

The intuitive feeling is that one might be able to feed all this information into a computational method for a reconstruction, such that the most probable speckle-free reconstruction of the specimen con-

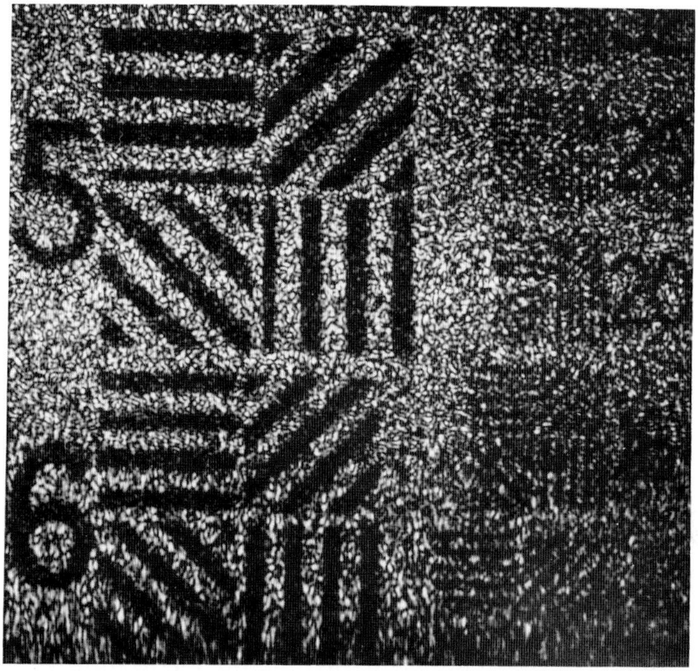

Figure 15.1. Holographic reconstruction of a test card illuminated by a diffusing screen (Martienssen and Spiller, 1967): The object can be considered as a 3-dimensional object consisting of 2 planes, that of the diffuser and that of the test card. The object is not representative of a real biological object in the x-ray region, because phase shifts are smaller for x rays. However, even a plane with random intensity modulation only produces a random phase distribution after propagation over a distance larger then the depth of focus and produces a similar speckle statistics as the diffuser in our model.

sistent with all the information in the hologram and in our library of previous work is obtained. While work continues and our library of images grows one might hope that the process might converge to a full 3-D reconstruction with diffraction limited resolution.

There are many open questions to this approach. How should the library information be quantified and used? Does a general algorithm exist which is valid for many types of objects or does each specimen require its own procedure? For a full 3-D object with details down to the diffraction limit, the information contained in any single plane is only a small fraction of the total information content, and it is very difficult to be optimistic about the possibility for a solution to the problem.

We conclude that imaging of 3-dimensional objects with x-ray holography will still present challenging problems even after fully coherent, bright x-ray sources become available. Holography pays for the depth resolution with a loss in spatial resolution or signal/noise ratio in each of the image planes. Contact x-ray radiographs of biological specimens show rich detail down to the resolution limit, and it is questionable if interesting biological specimens exist, where the extra depth resolution is worth the information loss in each plane. One can speculate that incorporation of *a priori* knowledge during the image reconstruction can solve the discussed problems; however, this approach appears not to be very promising for general 3-D objects with large information content.

Acknowledgment

This note was originally presented at the First Symposium on the Applications of Laboratory X-ray Lasers, Pacific Grove, CA, Feb. 1985. The author acknowledges helpful discussions with L. Beese and D. Sayre about the incorporation of *a priori* knowledge into the reconstruction algorithm.

16. Proposal for a Phase Contrast X-ray Microscope

G. Schmahl and D. Rudolph

Forschungsgruppe Röntgenmikroskopie, Universität Göttingen, Geismarlandstr. 11, D-3400, Göttingen, Fed. Rep. of Germany

Introduction

In x-ray microscopy experiments performed up to now the contrast is dominated by photoelectric absorption. The resulting radiation dose, though much less than in electron microscopy, limits fundamentally the resolution which can be obtained when investigating biological materials in the natural (intact wet and unstained) or living state, Sayre *et al.* (1977). Investigations to decouple resolution and radiation dose leads to a proposal for a phase contrast x-ray microscope. This work is based on experimental results on phase zone plates, Hilkenbach and Thieme (1986), and on the atomic scattering factors published by Henke (1981).

Phase Contrast X-ray Microscope

In Figure 16.1 the x-ray optical set up of an x-ray phase contrast microscope is shown. The object is coherently illuminated. A thin plate called the phase plate is placed in the back focal plane of the micro zone plate. This phase plate retards or advances the phase of the central order with respect to the diffraction spectra by one-quarter of a period.

We consider an object with the thickness t, consisting of different components with the refractive indices n_i according to Figure 16.2. With $n = 1 - \delta - i\,\beta$, the amplitude transmission is given by

X-ray Microscopy
Ed. by P. C. Cheng and G. J. Jan

$$T_i(x,y) = e^{\frac{-2\pi}{\lambda}\beta_i(x,y)t} e^{i\frac{2\pi}{\lambda}(1-\delta_i(x,y))t} \tag{16.1}$$

If we consider a pure phase object and are only interested in the differences of the phase shift caused by the different components, we can write

$$T_p(x,y) = e^{i\frac{2\pi}{\lambda}\delta(x,y)t} \tag{16.2}$$

With

$$\frac{2\pi}{\lambda}\delta(x,y)t = \phi(x,y) << 1 \tag{16.3}$$

we can write

$$T_p(x,y) = 1 + i\phi(x,y) \tag{16.4}$$

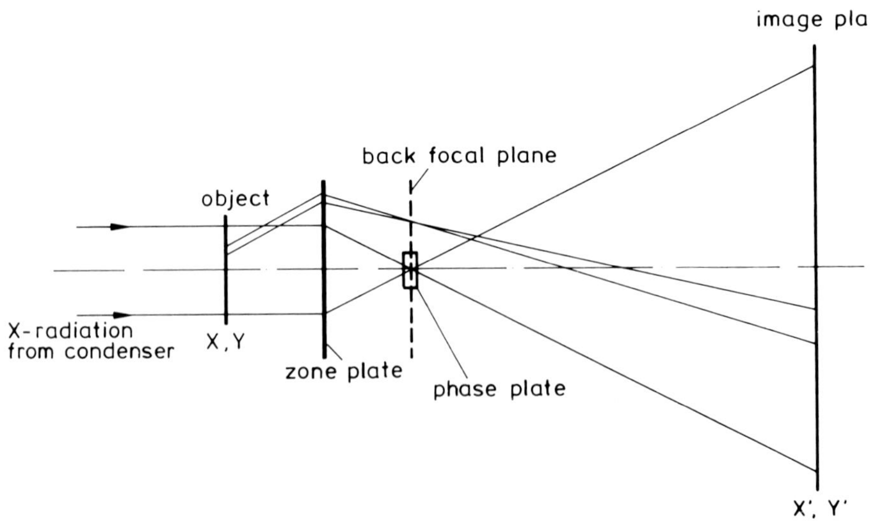

Figure 16.1. Schematic of a phase contrast x-ray microscope

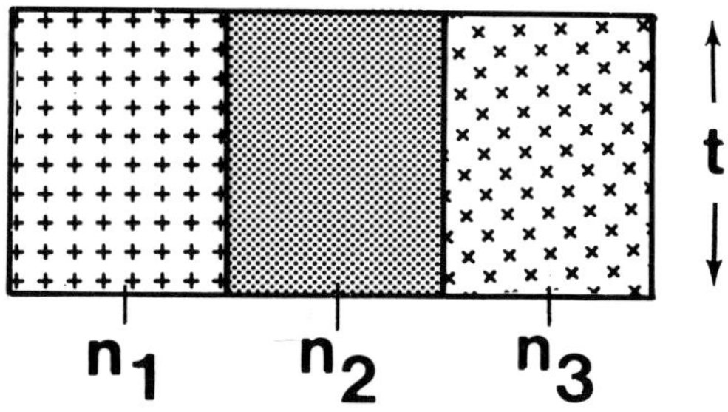

Figure 16.2. Object consisting of different components with refractive indices n_i

The phase plate in Figure 16.1 will cause a light distribution in the image plane which represents a fictitious amplitude object

$$\hat{T}(x,y) = \pm i + i\phi(x,y) \qquad (16.5)$$

The intensity in the image plane is therefore proportional to

$$I(x,y) = |\hat{T}(x,y)|^2 = 1 \pm 2\phi(x,y) \qquad (16.6)$$

with x' = V x, y'= V y, V = x-ray magnification. In this case, it is assumed that the transmission of the phase plate is 100%.

We consider now two different object structures, both of thickness t, and the resulting intensities

$$I_1 = 1 \pm 2\phi_1(x,y)$$

and

$$I_2 = 1 \pm 2\phi_2(x,y) \qquad (16.7)$$

In the following we restrict to the upper sign in (16.7) which described the case where the phase of the central order is retarded with respect to the diffraction spectra.

The contrast in the image plane $K = \dfrac{I_1 - I_2}{I_1 + I_2}$ is given by

$$K_p = \frac{\phi_1(x,y) - \phi_2(x,y)}{1 + \phi_1(x,y) + \phi_2(x,y)} \tag{16.8}$$

If we observe the same structures without the phase plate, *i.e.* in amplitude contrast caused by photoelectric absorption, only the first factor in eq. (16.1) counts resulting in the transmission function

$$T_A(x,y) = e^{-\frac{2\pi}{\lambda}\beta(x,y)t} \tag{16.9}$$

With the linear absorption coefficient $\mu_l = \dfrac{4\pi}{\lambda}\beta$ and weak absorption, *i.e.* $\mu_l t << 1$, one obtains

$$T_A(x,y) = 1 - \frac{1}{2}\mu_l(x,y)t \tag{16.10}$$

Considering again two different object structures the resulting intensities in the image plane are proportional to

$$\begin{aligned} I_1(x,y) &= 1 - \mu_{l,1}(x,y)t \\ I_2(x,y) &= 1 - \mu_{l,2}(x,y)t \end{aligned} \tag{16.11}$$

The contrast K_A is, therefore, given by

$$K_A = \frac{(\mu_{l,2}(x,y) - \mu_{l,1}(x,y))t}{2 - (\mu_{l,1}(x,y) + \mu_{l,2}(x,y))t} \tag{16.12}$$

It should be mentioned that the illumination of the object, shown as coherent illumination in Figure 16.1, in practical cases will be partial coherent.

Phase Contrast Versus Amplitude Contrast - Numerical Examples

To calculate the phase and amplitude contrast according to (16.8) and (16.12) the δ- and β- values have to be derived using the relations

$$\delta = \frac{r_o\lambda^2}{2\pi} n\bar{f}_1 \qquad \beta = \frac{r_o\lambda^2}{2\pi} n\bar{f}_2 \tag{16.13}$$

with

$$\bar{f}_1 = \sum_k n_k f_{1,k} \qquad \bar{f}_2 = \sum_k n_k f_{2,k} \tag{16.14}$$

n_k is the percentage number of atoms of type k in the compound, n is given by

$$n = \frac{6.022 \times 10^{23}}{M} N\rho[cm^{-3}] \tag{16.15}$$

with M = molecular weight, *i.e.* the weight in gram of one mol of the compound, N = number of atoms of the compound, ρ = density in gram cm^{-3} r_o is the classical electron radius, $\frac{r_o}{2\pi} = 4.485x10^{-14}cm$. The atomic scattering factors $f_{1,k}$, $f_{2,k}$ are listed in tables published by Henke *et. al.* (1977). A compound differs from a molecule by a constant factor for all atoms in the molecule. For example we consider a protein with $\rho = 1.35gcm^{-3}$ with the relative mass fractions m_H = 0.065, m_C = 0.530, m_N = 0.160, m_O = 0.230, m_S= 0.015 corresponding to a composition of the compound (empirical formula) $C_{94}H_{139}N_{24}O_{31}S$. In this case it is M = 2132.4 g mol^{-1}, N = 289, n = 1.10 x 10^{23} and *e.g.* n_H = 0.48.

In the following we consider a cube of protein with 50nm length with the above mentioned composition surrounded by water. Table 16.1 shows the phase contrast K_P and amplitude contrast K_A for the three wavelengths λ = 0.62nm, λ = 2.48nm and λ = 4.5nm.

λ [nm]	K_P	K_A
0.62	1%	0.003%
2.48	4.9%	3.5%
4.50	2.8%	0.02%

Table 16.1. Phase contrast K_P and amplitude contrast K_A for a 50nm protein structure surrounded by water

The table shows that phase contrast is larger than the amplitude contrast. More detailed investigations show that this is true for the whole wavelength range interesting for soft x-ray microscopy and for other organic material in wet and dry state.

One conclusion is that the wavelength region suited for high resolution x-ray microscopy can be extended to shorter wavelengths as considered up to now. In another paper of this volume an x-ray image of a part of a human fibroblast, critical point dried, is shown, Meyer-Ilse *et al.* (1986). The picture has been made with $\lambda = 4.5$nm, showing a part of the nucleus in amplitude contrast. Calculations show, that 50nm dry protein structures have an amplitude contrast of 1.4% which corresponds closely to the value shown in Table 16.1 for protein in water for the wavelength $\lambda = 0.62$nm in phase contrast.

The phase contrast values of Table 16.1 have been calculated under the assumption that the phase plate has an amplitude transmission of 1. In practice, all phase plates for soft x-radiation have an amplitude transmission of less than one. This leads to a better adaption of the intensities of the central order and the diffraction spectra and, therefore, to an enhanced phase contrast. The phase contrast in this case is in good approximation given by eq. (16.8), divided by the amplitude transmission A of the phase plate. Table 16.2 shows thickness, transmission and phase contrast enhancement for different materials and wavelengths.

	t[nm]				T				1/A			
λ[nm]	Cr	Ni	Au	Re	Cr	Ni	Au	Re	Cr	Ni	Au	Re
0.62	450	350	328	400	67%	54%	46%	2.6%	1.2	1.4	1.5	6.2
2.48	230	133	123	100	57%	51%	5.3%	9%	1.3	1.4	4.4	3.3
4.50	105	77	149	91	45%	29%	1.3%	2.8%	1.5	1.9	8.8	6.0

Table 16.2. Thickness, transmission and phase contrast enhancement for different materials and wavelength. t[nm]: thickness of the phase plate, T: transmission of the phase plate, 1/A: phase contrast enhancement

There are several advantages in using shorter wavelengths than used up to now:

1. The radiation dosage applied to the object is considerably reduced.
2. Thicker specimens than up to now can be investigated.
3. The numerical apertures of zone plates are smaller resulting in larger depths of focus and larger object distances.
4. The absorption in air and helium is much smaller.
5. The absorption of the supporting foils of zone plates, filters and specimen holders is rather low.
6. The detective quantum efficiency (DQE) of CCD cameras is higher.

Disadvantageous is that zone plates have to be built with higher aspect ratios and that the total length of the x-ray microscope will be increased.

We hope that the phase contrast method in x-ray microscopy will play an important role for future biological and medical investigations as it is the case in light and electron microscopy.

Acknowledgement

This proposal has been reported at the ESRF - EMBL Workshop on X-ray Microscopy, Heidelberg, December 12, 1986.

17. Soft X-ray Microscope with Free-standing Zone Plates

Sadao Aoki[1], Yasushi Kagoshima[1], Hironao Yamaji[1], Masami, Kakuchi[2], Toshiaki Tamamura[2], Misao Sekimoto[3], Akira Ozawa[3], Hideo Yoshihara[3], Masami Ando[4], Hideki Maezawa[4], and Kazuyuki Hyodo[4]

[1]Institute of Applied Physics, University of Tsukuba, Sakura, Ibaraki 305 Japan

[2]NTT Electrical Communications Laboratories, Nippon Telegraph and Telephone Corporation, Tokai, Ibaraki 319-11, Japan

[3]NTT Electrical Communications Laboratories, Nippon Telegraph and Telephone Corporation, Morinosato, Atsugi, Kanagawa, Japan

[4]Photon Factory, High Energy Physics Laboratory, Oho, Ibaraki, Japan

Introduction

High resolution soft x-ray microscopes have been demonstrated by using zone plate optics (Schmahl, 1983; Kirz and Rarback, 1985). Zone plates can be fabricated by the laser holographic lithography or the electron beam lithography. The latter method is generally used because a wide variety of zone plate specifications can be achieved. The collection efficiency of the zone plate depends on its structure. It is usual that the zone plate is supported by a thin substrate. The absorptivity of the substrate material cannot be negligible in the ultra soft x-ray region. The recent advance in fabricating an x-ray lithography mask make it possible to produce a free-standing zone

X-ray Microscopy
Ed. by P.C. Cheng and G.J. Jan

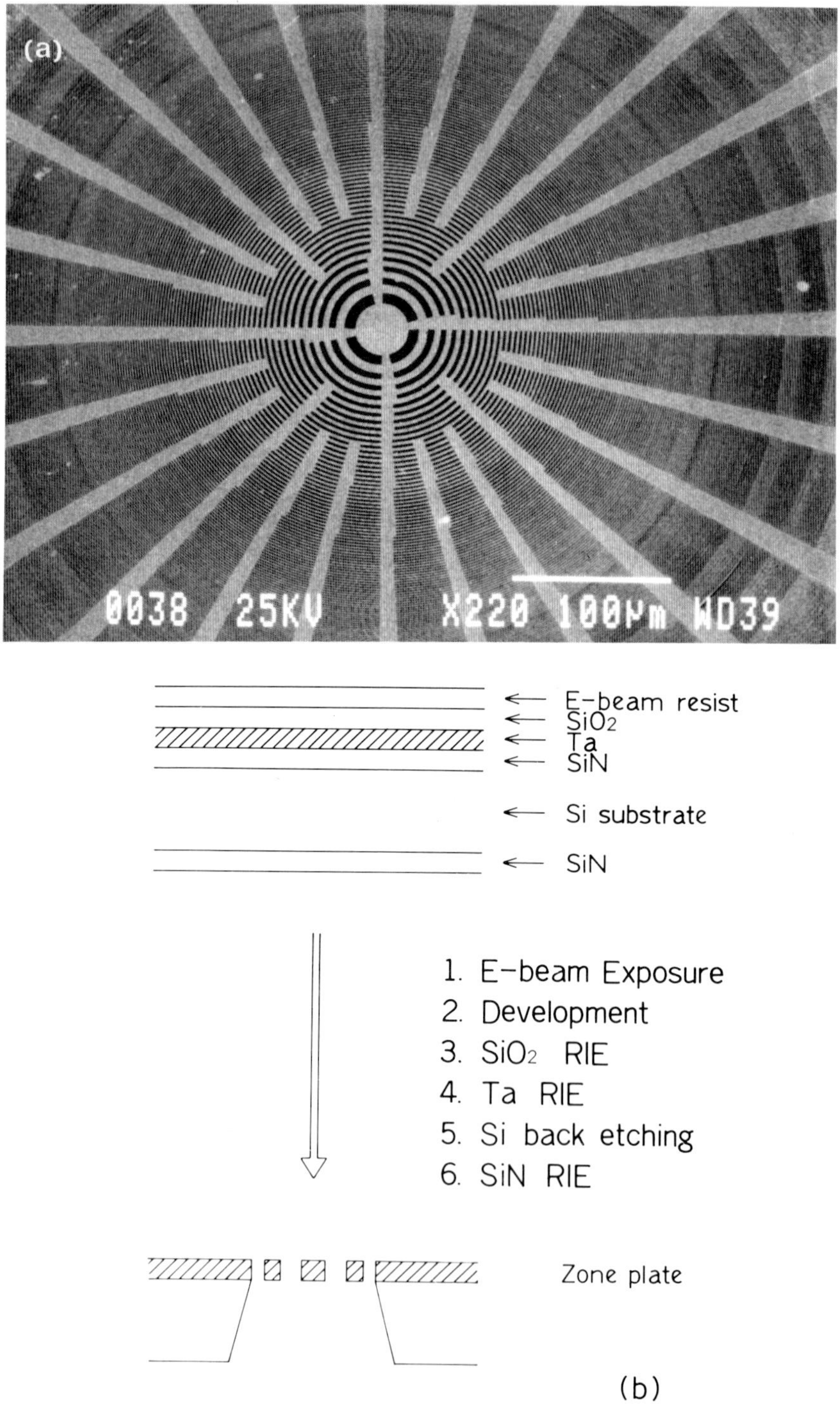

Figure 17.1. (a) SEM micrograph of a condenser zone plate. (b) Schematic diagram of the fabricating process of a zone plate.

plate. It can be used in a wide range of the spectrum from 10Å to 1000Å.

An imaging or scanning zone plate microscope needs high brilliance synchrotron radiation because of their low efficiency. Fortunately the advent of very high brilliance undulator radiation accelerate the development of the x-ray microscope. In the following sections some recent results are described.

Zone Plate

Zone plates have been fabricated by means of the electroplating of gold. The electroplating technique has several drawbacks to produce fine and large free-standing structures which are often collapsed by the remained stress. A newly developed x-ray lithography mask structure which is made of tantalum can be applicable to the zone plate. The remained stress of Ta can be successfully controlled by the sputtering conditions. The structure and the fabrication process of the Ta zone plate are shown in Figure 17.1. A relatively large condenser zone plate and a micro zone plate were fabricated. Parameters of these zone plates are shown in Table 17.1. It is very convenient to use tantalum as an opaque zone because of its high density. The transmission of tantalum at wavelength of 10Å and 20Å are 0.13 and 0.0007, respectively.

Experimental Results

Experiments were performed at station BL2B in Photon Factory (PF) Tsukuba, using undulator radiation from the straight section of the 2.5GeV storage ring as the x-ray source. The first harmonic of this radiation can be tuned from 13Å to 30Å by changing the magnetic field, and higher-order harmonics are also included in this case. The bandwidth of the first harmonic is about 1/20 (Maezawa *et al.*, 1986). A schematic diagram of the experimental arrangement is shown in

Table 17.1 Zone plate parameters

	Condenser zone plate	Micro zone plate
Focal length (mm)		
$\lambda = 18\text{Å}$	----	20.0
$\lambda = 25\text{Å}$	90	14.4
Radius of the innermost zone (μm)	15.0	6.0
Outer diameter (mm)	1	0.35
Thickness of Ta (μm)	0.5	0.5

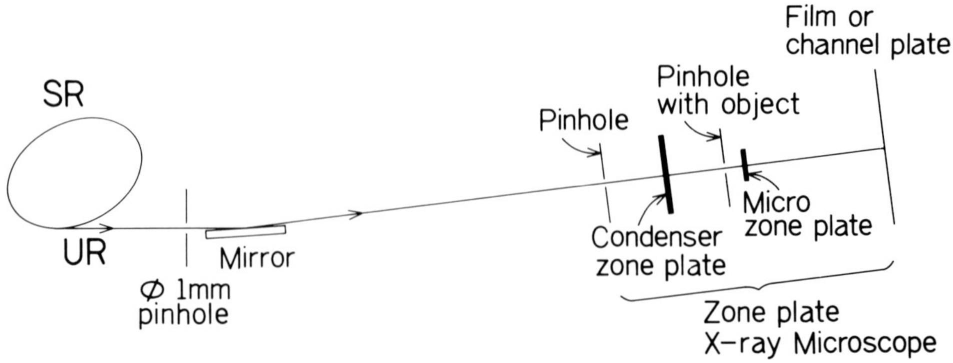

Figure 17.2. Schematic arrangement of the x-ray microscope at the Photon Factory storage ring.

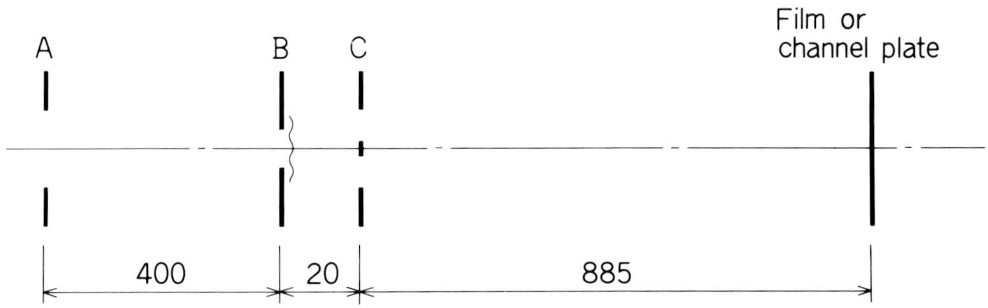

Figure 17.3. The optical arrangement of the x-ray microscope without the condenser zone plate.

Figure 17.2. The undulator radiation was collimated by a 1mm diameter pinhole and deflected by a flat mirror made of platinum-coated silicon carbide. As the grazing incidence angle is 2°, x-rays shorter than 4.7Å are cut off. A micro-channel plate was used to monitor the magnified image. The entrance of the x-ray microscope was set about 25m from the undulator radiation source.

The first preliminary experiment was made with a very simple optical system. Figure 17.3 shows the optical arrangement of the x-ray microscope with no condenser element or monochromator. Because undulator radiation is highly collimated and quasi-monochromatic, no illuminating optical system could be used. Only a 100μm diameter pinhole was used to eliminate the stray light. Figure 17.4 (a), (b), (c) show the magnified x-ray images of the copper mesh #1000 with three different wavelengths at 17Å, 18Å, and 19Å. The film used was the

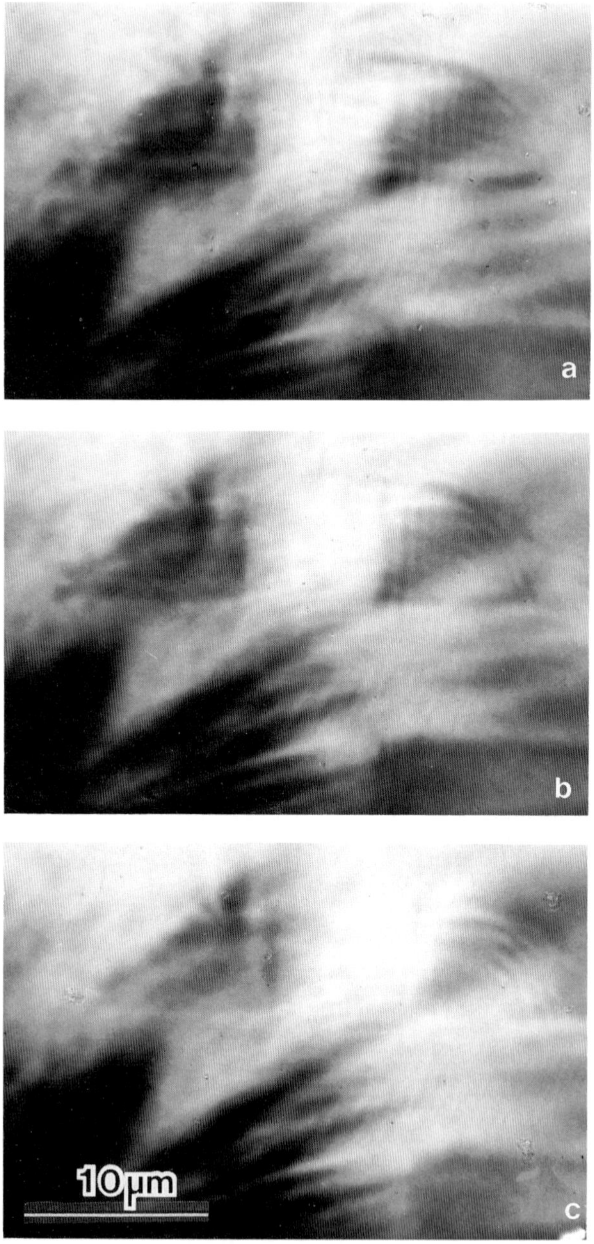

Figure 17.4. Image of a copper mesh #1000 with three wavelengths. (a) 17Å, (b) 18Å (well focused), (c) 19Å. The over-exposed region is due to zeroth order radiation.

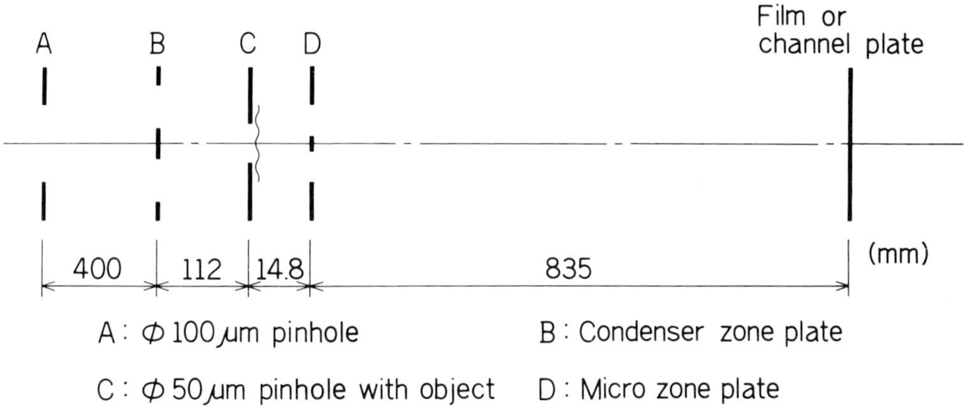

Figure 17.5. The optical arrangement of the x-ray microscope with the condenser zone plate.

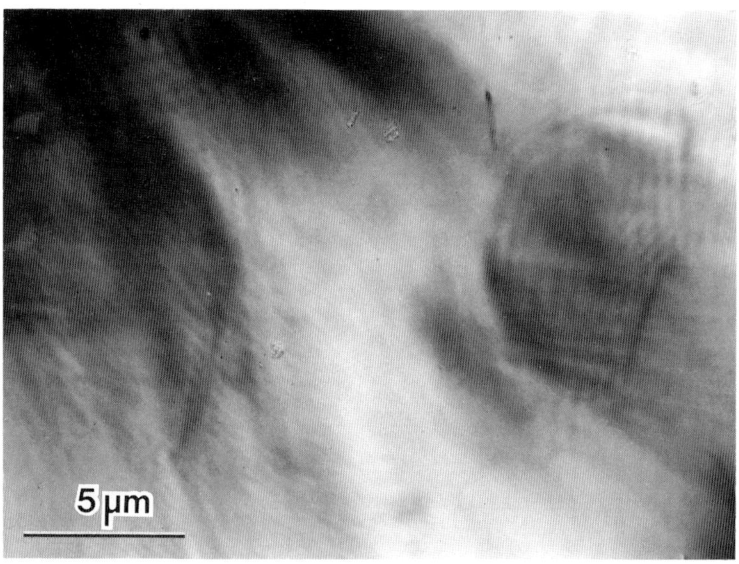

Figure 17.6. Image of a copper mesh #1000 with the wavelength of 25Å.

Agfa 10E56 holographic plate. The exposure time was 10sec at the electron current of 93mA. Magnified images were not focused at the center of the image field, but at the off-axis region to avoid overlapping with zeroth order radiation.

The second experiment was made with a popular optical system. Figure 17.5 shows the optical arrangement of the x-ray microscope with a condenser zone plate. The condenser zone plate was used to image the first pinhole on the object plane. The object was imaged by the same micro zone plate as was used in the first experiment. No monochromator was used because the condenser zone plate acted as a linear monochromator.

Figure 17.6 shows the magnified x-ray image of the copper mesh #1000 with the wavelength of 25Å. Round corners of the x-ray image were not due to the blur of the image but due to the contamination. The magnified image was also focused at the off-axis region. The exposure time was 15sec at the electron current of 222mA.

Conclusions

Two experiments described above showed that the free-standing zone plate could be applicable to soft x-ray microscopy. The most critical problem of our optical system was zeroth order radiation. In order to solve this problem both the condenser and the micro zone plate must be apodized.

18. Zone Plate Replication by Contact X-ray Lithography, and Its Application to Scanning X-ray Microscopy

C. J. Buckley[1], K. Ogawa[1], M. T. Browne[1], P. Charalambous[1], J. M. Kenney[1], R. Rosser[2], H. Ade[2], I. McNulty[2], R. Feder[3], P. C. Cheng[3]

[1]Physics Department, Kings College, London, UK

[2]Physics Department, State University of New York at Stony Brook, Upton, NY 11794 USA

[3]IBM T. J. Watson Research Centre, Yorktown Heights, NY 10598 USA

Introduction

The resolution obtainable with an x-ray microscope which uses a Fresnel zone plate as the focusing element is primarily dependent on the width of the finest ring of the zone plate and the accuracy of the zone plate pattern (Simpson, 1984). High accuracy zone plates with outermost zone widths of only a few hundred angstroms have been made by carbon contamination e-beam lithography at Kings College London (Buckley *et al.*, 1985). Whilst these zone plates have the finest zone widths yet produced, the fabrication of one zone plate takes ten hours and the carbon material from which the absorbing zones are formed is semitransparent to soft x-rays. This lack of opacity results in a reduction of focusing efficiency of about a factor of ten compared to the theoretical figure. The reduced efficiency increases the length of time required to form an image in a scanning x-ray microscope (Kenney *et al.*, this volume). The time taken to form an image in an x-ray microscope should be as short as possible to minimize specimen deterioration and maximize throughput. For this reason, the repli-

X-ray Microscopy
Ed. by P.C. Cheng and G.J. Jan

cation of carbon zone plates in a higher contrast material, to improve efficiency and minimize imaging times, is highly desirable.

This paper discusses the replication of carbon zone plates by x-ray lithography using synchrotron radiation. Results of some of the replication work carried out at National Synchrotron Light Source (N.S.L.S.), Brookhaven National Laboratory in April 1986 are presented.

X-ray Lithography and Zone Plate Replication

The replication scheme transfers the image of a carbon zone plate pattern on to an x-ray resist (PMMA) by exposing the resist to x-rays via the carbon zone plate. During exposure the carbon zone plate is held in close contact (less than 5μm separation) with the replicating substrate. The replicating substrate is a 100nm thick Silicon Nitride support window which is the bottom layer of a sandwich of 100nm of gold and 200nm of PMMA resist (top layer). After exposure, the resist is developed to leave islands of resist (in pattern of Fresnel zones) on a base of gold. This is then argon ion milled, which removes the uncovered gold to create a gold zone plate pattern on a silicon nitride support window.

X-ray Source and Choice of Wavelength

Of the range of soft x-ray sources available (Spiller and Feder, 1977) synchrotron radiation is the most suited for zone plate replication because it has excellent collimation and wavelength tunability. The choice of wavelength used to replicate carbon zone plates is a compromise between contrary requirements. The wavelength of the x-rays used for replication should lead to satisfactory absorption by the features of the zone plate mask, minimize diffraction effects, and not lead to deterioration of resolution of the replica due to the range of photo electrons excited in the resist.

To obtain satisfactory contrast between the transparent and absorbing areas of the carbon mask, the replicating wavelength is confined to the region close to the carbon absorption edge at 4.4nm. However at this wavelength diffraction can be a significant factor in degrading the resolution obtained in the replica, as there is a finite separation between resist and mask. Shorter wavelengths can be used to reduce this effect at the expense of mask contrast. Also, the range of photo electrons liberated in the resist increases as the wavelength decreases which can also degrade the replica resolution. Thus the wavelength must thus be chosen to minimize the degradation of the quality of the replicated pattern for a given carbon thickness and finest zone width.

The wavelength used to produce the replicas presented here was 3.5nm. At this wavelength a 100nm thick free-standing carbon zone plate mask absorbs about 50% of the incident radiation. The contact distance between zone plate mask and resist was in the range 1 to 5 microns. Diffraction effects produced by zone widths of 75nm are not negligible, at this wavelength and separation. However the degrading effect that diffraction has on the replica can be reduced by manipulating the exposure dose (see next section). The range of the photo electrons at a wavelength of 3.5nm is less than 10nm (Rishton *et al.*, 1982) and the effect these have on replica resolution is insignificant at the zone widths considered here. The influence that these and other factors have on zone plate replication is considered at length in a forthcoming publication (Buckley, 1987).

Exposure

The amount of exposure or dose received by an x-ray resist will determine the rate at which resist is dissolved away by solvent during development. The dissolution rate of resist is a complicated function of absorbed dose (Spiller and Feder, 1977; Thompson *et al.*, 1983).

At absorbed doses in excess of 100 Joules per cubic centimeter the dissolution rate of PMMA resist has a large variation for only small differences in dose. Thus it is possible to produce a workable aspect ratio in the resist replica of a mask which has low contrast. For example, It is possible to obtain a 1:1 aspect ratio in PMMA when exposed via a mask which has features that absorb only 50% of the incident radiation. This phenomenon also helps to reduce replica feature blurring produced by diffraction. The beneficial effects of replicating with high doses have been discussed in a previous publication (Buckley, 1986).

Development and Ion Milling

The exposure dose used in the replication experiments was chosen to satisfy two goals; (1) To give a high dissolution rate differential between areas exposed via the absorbing and non-absorbing areas of the carbon zone plate mask, and (2) to give a controllable development time. In practice an average development time of 60 seconds (equivalent to 200nm/min) in IPA gave sufficient control over development depth, and the exposure was optimized for this.

After development, the resist pattern acts as a protective mask during an argon ion milling process. The ion milling removes the gold uncovered by the development and provides a gold zone plate pattern supported by a silicon nitride membrane. The replica is then ready for use as a high resolution focusing element in an x-ray microscope.

Results

The replication scheme was successful for zone plates with finest outer zones of 100nm and 75nm (200nm and 1500nm periods). Figures 18.1 and 18.2 show transmission electron micrographs of a 50 micron diameter, 100nm finest zone gold replica, and a 45 micrometer, 75nm gold replica respectively. The 100nm finest zone replica shown

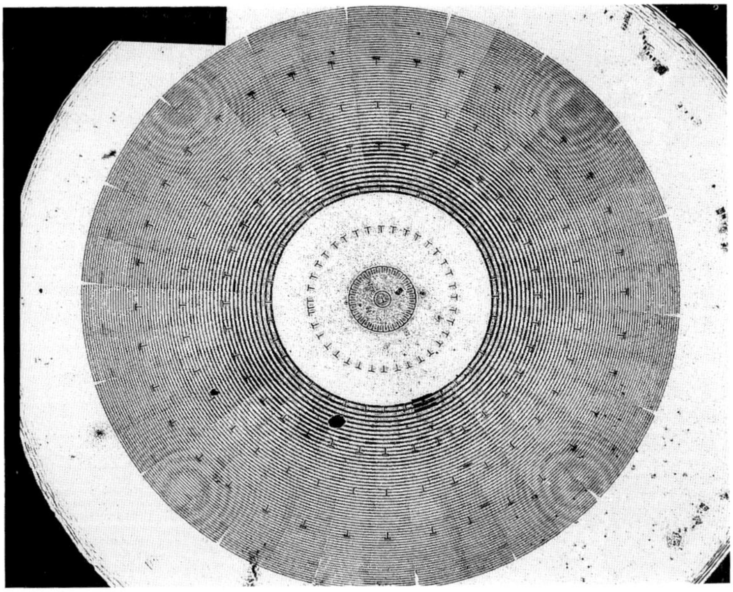

Figure 18.1. Transmission electron micrograph of a gold replica zone plate. The diameter of the zone plate is 50 microns, and the finest zone width is 100nm.

in Figure 1 has been used as the focusing element in the Stony Brook STXM at U15, VUV ring, National Synchrotron Light Source, Brookhaven National Laboratories. Initial tests showed that the zone plate was capable of a spatial resolution approaching the diffraction limit, with a focusing efficiency of approximately 4% - an improvement of about a factor of ten over the carbon original (Ade *et al.*, 1986).

Conclusions and Future Work

The 100nm finest zone replicas tested so far have shown that the replicating technique can produce gold zone plates which have considerably higher efficiencies than the carbon originals. Both 100nm

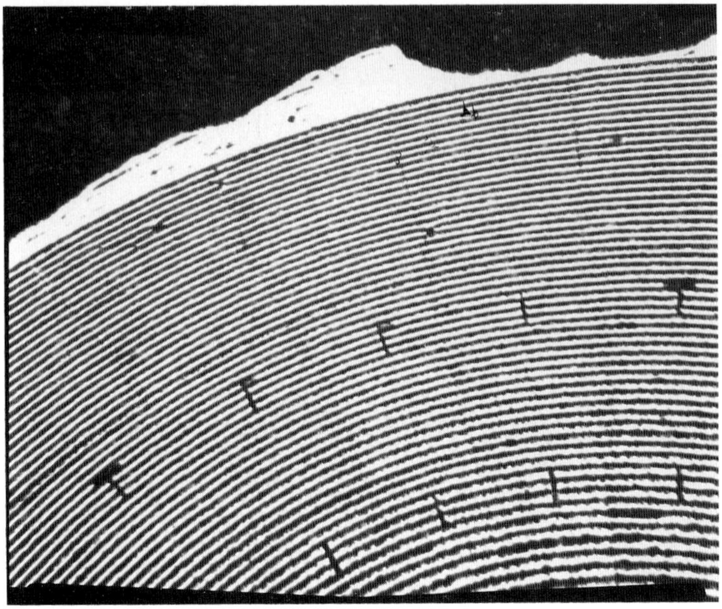

Figure 18.2. Transmission electron micrograph of part of a gold replica zone plate. The diameter of the zone plate is 45 microns and the finest zone width is 75nm.

and 75nm gold replicas will be used on the Kings College STXM which will go into operation in August 1986, and the performance will be reported in a future publication.

The finest zone widths which can be replicated by this method has been limited to 75nm. The limitation on resolution at this scale is primarily due to feature blurring due to diffraction. At the wavelength of 3.5nm, the separation between the carbon mask and the replicating medium which is required to replicate features of less than 75nm is less than 5 microns. The profile of the support apertures on which the carbon originals are formed have prevented the separation distance from being less than a few microns, thus limiting the replica resolution. However, it is hoped that zone plates fabricated by the contamination method will be formed on thin, flat silicon nitride windows in the future. This will allow for separation distances of one micron and less and should facilitate the replication of carbon zone plates into gold at line widths below 75nm.

Acknowledgements

The Authors would like to acknowledge the direct and indirect support of the following persons: R. E. Acosta and the staff of the instrument machine shop at IBM T. J. Watson Research Centre, Yorktown Heights. D. M. Shinozaki of the Faculty of Engineering, University of Western Ontario, London, Ontario, Canada. C. Jacobsen, J. Kirz, H. Rarback and the technical staff of the N.S.L.S. and SUNY, Department of Physics. R. E. Burge, A. G. Michette, P. J. Duke (S.R.S. Daresbury Laboratory) and W. J. Luckhurst of Kings College, London. The research at Kings College and Daresbury Laboratory is funded in part by S.E.R.C. of UK and at Stony Brook/BNL by the N.S.F. and the D.O.E. of USA.

19. A 10keV X-ray Microprobe with Grazing Incidence Mirrors

Sadao Aoki[1], Yohichi Gohshi[2] and Atsuo Iida[3]

[1]Institute of Applied Physics, University of Tsukuba, Sakura, Ibaraki, 305, Japan

[2]Department of Industrial Chemistry, Faculty of Engineering, University of Toyko, Hongo Bunkyo-ku, Toyko, 113, Japan

[3]Photon Factory, National Laboratory for High-Energy Physics, Oho, Ibaraki, 305, Japan

Introduction

An x-ray microprobe has been expected to be a promising analytical tool for spectroscopic and structural studies of the elemental constitution of matters with spatial resolution (Sparks, 1980). Detectable signals are fluorescent x-rays, photoelectrons, EXAFS and so on. Although these signals are examined rigorously in the spectroscopic study, only a few experiments on micro-analysis have been reported (Horowitz and Howell, 1972).

Recent advances in x-ray sources and x-ray optical elements made it possible to design various types of x-ray microprobes. Bent crystals are favorable for high energy x-rays above 10keV. Grazing incidence mirrors are appropriate for low energy x-rays below 10keV. Zone plates are useful for soft x-rays below 1keV. Multilayer mirrors will be widely used for focusing x-rays.

Very accurate fabricating techniques are required to obtain a small x-ray spot less than a few microns. X-ray mirrors and zone plates an-

X-ray Microscopy
Ed. by P.C. Cheng and G.J. Jan

swer this purpose because these x-ray optical elements are now being rapidly developed in the study of x-ray microscopy.

In this paper we show that the combination of grazing incidence tandem mirrors can be applied to produce a small x-ray spot in the energy region up to 10keV. Some preliminary results obtained with synchrotron radiation are presented.

Design considerations

It is known that the refractive index in the x-ray region can be written by complex number n:

$$n = 1 - \delta - i\beta \tag{19.1}$$

x-rays can be totally reflected at the grazing incidence angle smaller than the critical angle (θ_c) which is given by

$$\begin{aligned} \theta_c &= \sqrt{2\delta} \\ &= 1.6\text{x}10^{-3}\lambda\sqrt{\rho} \end{aligned} \tag{19.2}$$

where λ is the wavelength in Å and ρ is the density of the reflection surface in g/cm^3. For a grazing incidence angle θ, the refelctivity R is described by the Fresnel equation, which is given by (Compton and Allison, 1935)

$$R = \frac{(\theta - a)^2 + b^2}{(\theta + a)^2 + b^2} \tag{19.3}$$

where

$$a^2 = \frac{(\sqrt{(\theta^2 - 2\delta)^2 + 4\beta^2} + \theta^2 - 2\delta)}{2}$$

$$b^2 = \frac{(\sqrt{(\theta^2 - 2\delta)^2 + 4\beta^2} - \theta^2 + 2\delta)}{2}$$

Mirror materials must be carefully selected to obtain a smooth surface and a high reflectivity. In general high Z materials are efficient for relatively high energy x-rays except their absorption edges. We consider a platinum-coated mirror. Figure 19.1 shows the calculated reflectivities for typical x-ray energies.

Some relayed mirror optics must be used to obtain a few microns x-ray spot from synchrotron radiation sources because the large demagnification optics is not feasible with a single mirror. Although an ellipsoid of revolution gives a perfect image of the axial source point, the image of the off-axis point degrades rapidly due to coma. In order to reduce coma the optical system of Wolter's type I (Wolter, 1952) is considered for focusing x-rays. Wolter's optical systems have been successfully used in x-ray astronomy and laser fusion diagnostics. For focusing synchrotron radiation we use two combined optical systems, the first of them is a pair of long mirrors which is designed to collect x-rays as many as possible, the second is a relatively short ones designed to make a small x-ray spot. A schematic diagram of the beamline layout is shown in Figure 19.2. A 90° surface of revolution is used for the condenser mirror. A complete 360° surface of revolution will be used for micro-focusing mirror. Practical limits to fabricating these surfaces are the length of the mirror and the radius of rotation. For a 10keV microprobe a grazing angle must be less than 8mrad with a platinum-coated mirror. Very long mirrors must be used to collect an appreciable amount of x-rays with this grazing angle. Parameters of the condenser mirrors and micro-focusing mirrors are shown in Figure 19.3. The first mirrors demagnify the source 1/20 diameter and then the second mirrors demagnify the first image 1/13.

Expected intensities from Photon Factory storage ring at 2.5GeV with these focusing mirrors can be estimated by using geometrical optics. The spectral intensity of radiation at 10keV is 1×10^{11} photons/s mA • mrad • 1% bW (Sasaki, 1980). The FWHM value of the vertical dimension of the source is calculated to be about 0.8mm and in the

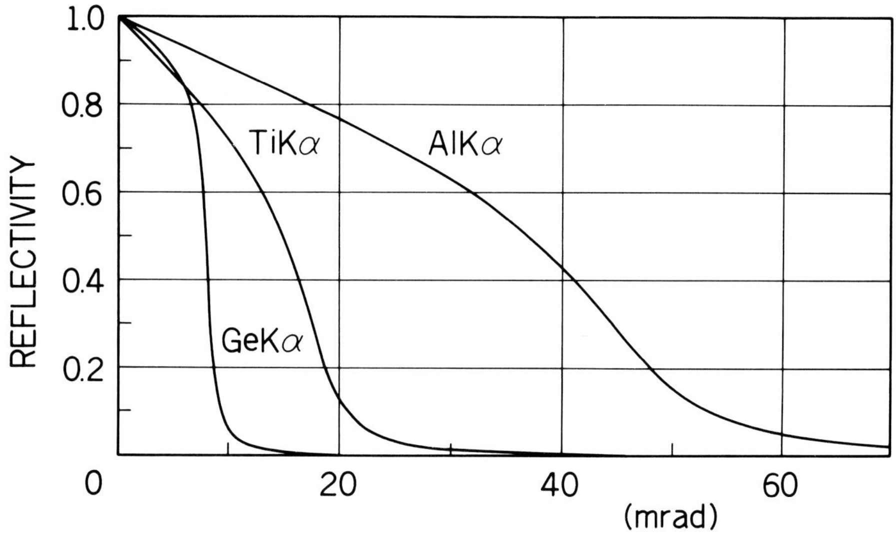

Figure 19.1. Calculated reflectivities of a platinum-coated mirror with three different x-ray energies. Reflectivities are plotted as a function of a grazing incidence angle.

horizontal dimension 10mm at station BL4A. The first focused spot size will be $500x40\mu m^2$. Double $250\mu m$ - thick beryllium windows are set in the beam line. Taking account of the electron current (200mA), the angular acceptance (0.2mrad), the bandwidth ($\Delta E/E = 0.5$), the absorption of the windows and the reflectivity of the mirror (76% at 7mrad), the intensity of the first focal spot is estimated to be $5.5x10^8$ photons/s • μm^2. The convergent angle of the x-ray beam at the first focus is 4mrad. The front end of the second mirrors has a 10mm-diameter annular aperture which subtends 4mrad at the first focus. About 10% of the divergent x-rays from the first focal plane are collected by the second mirrors. The final spot size will be 3x40 μm^2. The intensity of the spot is estimated to be $5.4x10^9$photons/s bullet μm^2.

Experimental Results

The condenser optical system was examined with synchrotron radiation. The long condenser mirrors were produced by conventional grinding, polishing and lapping. The base material of the mirror was fused silica. Each conicoidal surface was polished by using the shaped cast iron mandrel. The pitch polisher with cerium oxide was used for final lapping. Figure 19.4 shows the ellipsoidal and the hyperboloidal mirror coated with 1000Å thick platinum. The condenser mirrors were tested at station BL4A in Photon Factory (PF). The continuous, slightly divergent radiation from the storage ring was filtered by beryllium foils and focused by the mirrors. A demagnified x-ray image of the synchrotron radiation source was focused on the photographic plate. The grazing incidence angle for the first mirror surface varied from 6.9mrad to 7.1mrad and that for for the second mirror surface varied from 5.0mrad to 9.6mrad. X-rays higher than 13.3keV were cut off at the angle of 6.9mrad. The additional filter used was 1.92mm thick aluminum which transmitted a quarter of the incident 10keV x-rays. The width and the height of the entrance aperture of the mirrors were 5mm and 4mm, respectively. Figure 19.5 shows the demagnified

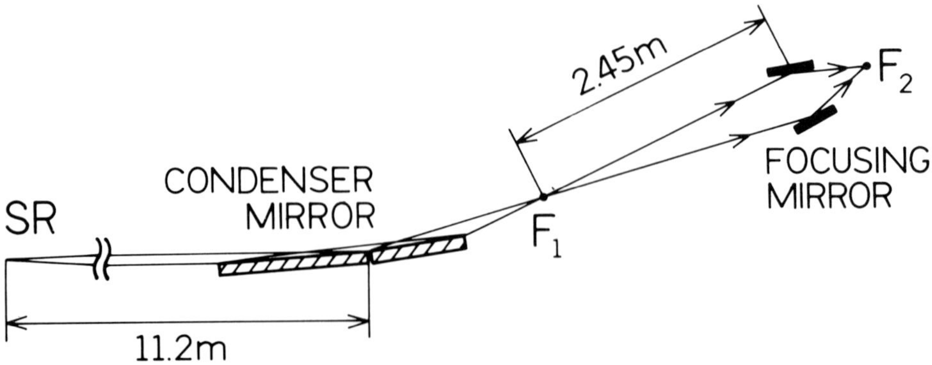

Figure 19.2. Beamline layout for the microprobe using Wolter's type I optical systems.

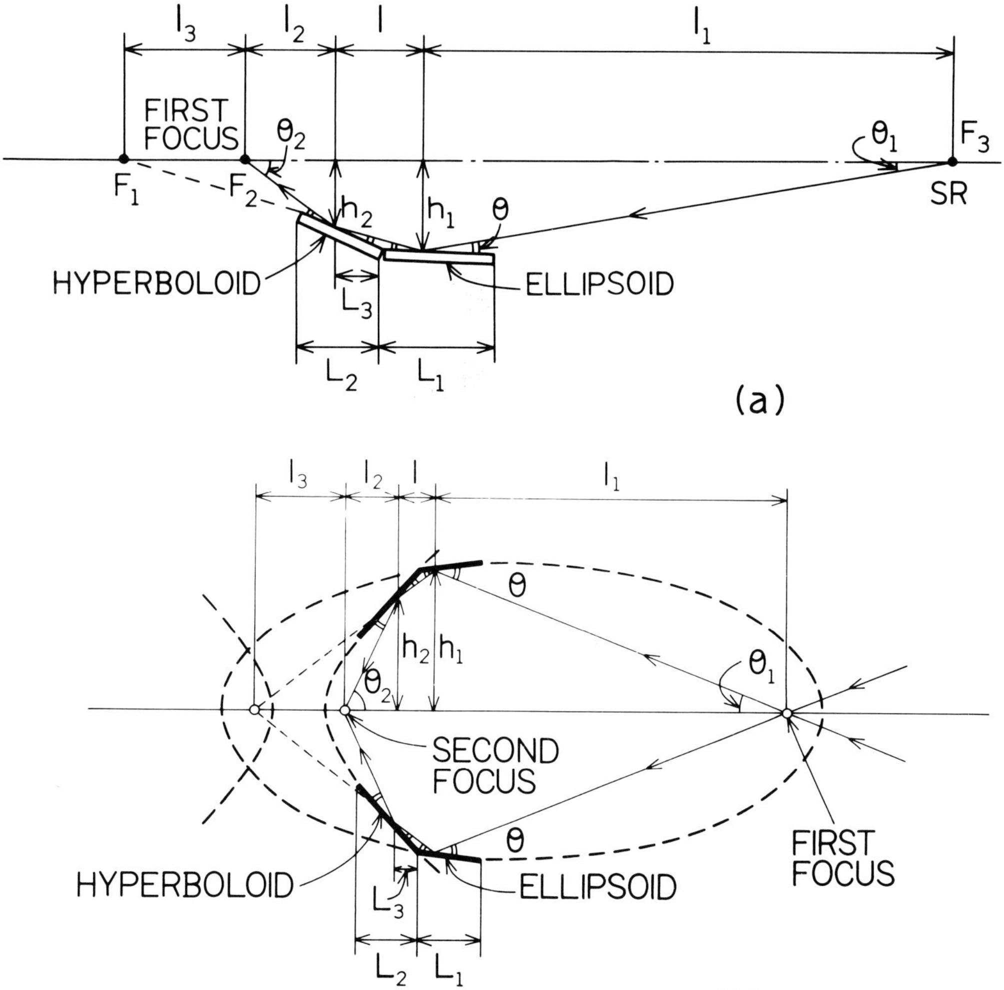

Figure 19.3. Numerical data of the condenser mirrors (a) and the micro-focusing mirrors (b). Parameters are

(a): $\theta = 7$mrad, $\theta_1 = 1.33$mrad, $\theta_2 = 26.67$mrad $h_1 = 14.663$mm, $h_2 = 9.596$mm, $l = 400$mm $l_1 = 11000$mm, $l_2 = 360$mm, $l_3 = 398$mm $L_1 = 494$mm, $L_2 = 324$mm, $L_3 = 200$mm.

(b): $\theta = 7$mrad, $\theta_1 = 2$mrad, $\theta_2 = 26$mrad $h_1 = 4.9$mm, $h_2 = 4.54$mm, $l = 30$mm, $l_1 = 2450$mm, $l_2 = 174.6$mm, $l_3 = 203.7$mm, $L_1 = 35$mm, $L_2 = 35$mm, $L_3 = 15$mm.

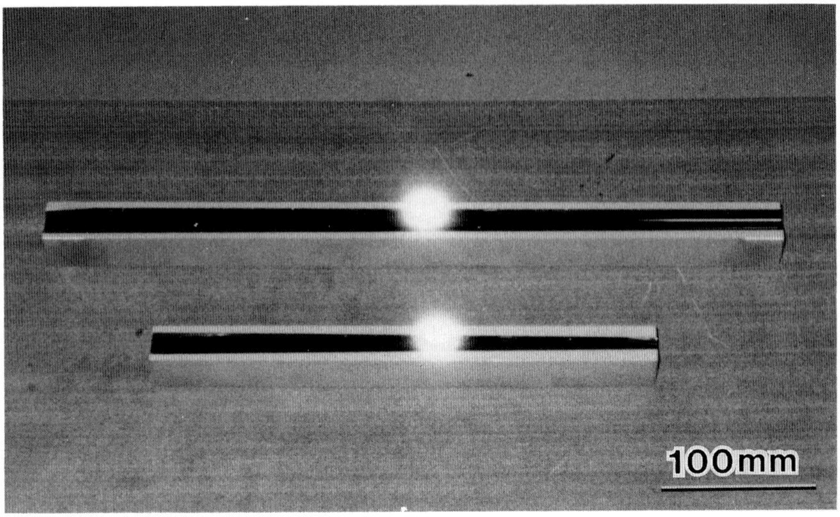

Figure 19.4. The ellipsoidal mirror (long) and the hyperboloidal mirror (short).

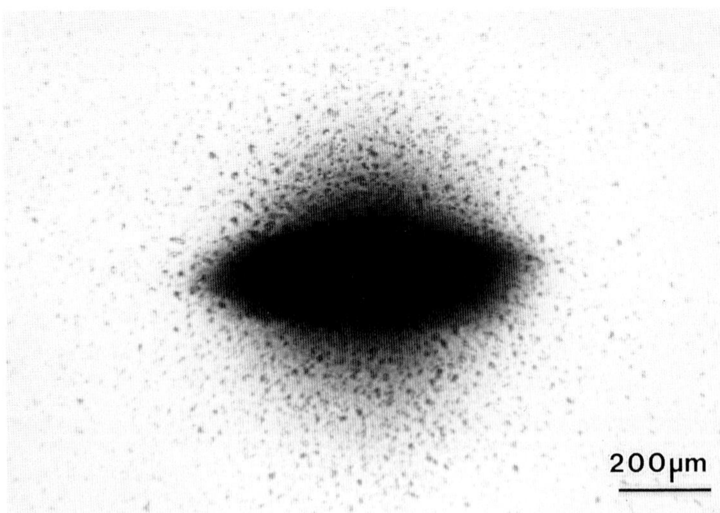

Figure 19.5. 20:1 demagnified x-ray image of synchrotron radiation source.

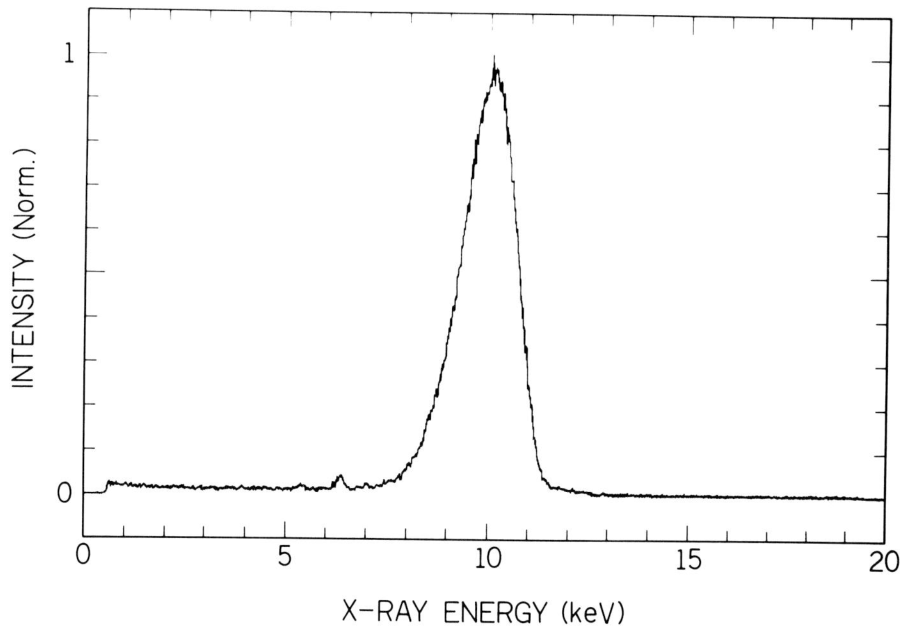

Figure 19.6. Reflected spectrum of the condenser mirrors.

x-ray image of the synchrotron radiation source at the electron current of 165mA. The film used was the Fuji industrial x-ray film 80. Although the horizontal image size was near the expected value, the vertical one is enlarged by a factor of about 5. The reason why the difference is so large is under investigation. The reflected spectrum from the mirrors was examined by the scattered spectrum from an acrylic plate. Figure 19.6 shows the reflected spectrum. A 0.66mm thick aluminum filter was used in this case. The sharp energy cutoff of the mirror and the filter made a wide bandpass monochromator.

Conclusion

It has been demonstrated that the long grazing incidence Wolter's type I mirrors were useful for focusing synchrotron radiation up to 10keV x-rays. A small x-ray spot less than 5μm will be produced by the combination of the condenser mirrors and the micro-focusing mirrors.

Acknowledgement

We wish to acknowledge the help of H. Yamaji, K. Sakurai, and S. Hayakawa. One of us (YG) was supported by Grant-in-Aid for Scientific Research from the Ministry of Education, Science and Culture. This work was supported in part by the Nissan Science Foundation.

20. Feasibility Study for the Observation of Biological Materials in VUV Wavelength Regions. Using Zone Plates Fabricated by Electron and Ion Beam Lithographies

Yoshinori Nagai[1], Yasushi Nakajima[2], Yutaka Watanabe[3], Shigetaro Ogura[3], Koshi Uyeda[4], Yoshio Shimanuki[5] and Hiroshi Kihara[6]

[1]Laboratory of Molecular Biology, School of Veterinary Medicine, Azabu University, Fuchinobe, Sagamihara 229, Japan.

[2]Department of Physics, School of Science and Engineering, Waseda University, Shinjuku-ku, Toyko 160, Japan.

[3]Electronic Material Division, Research Department 34, Canon Research Center, Canon Inc., 5-1 Morinosato-Wakamiya, Atsugi, Kanagawa 243-01, Japan.

[4]Technical Development Division, Akashi Seisakusho Ltd., 2-5020 Hironodai Zama, Kanagawa 228, Japan.

[5]Department of Oral Anatomy, School of Dental Medicine, Tsurumi University, Tsurumi 2-1-3, Yokohama 230, Japan.

[6]Department of Physics, Jichi Medical School, 3311-1 Yakushiji, Minamikawachicho, Tochigi 329-04, Japan.

Introduction

In biological system, the important things is to find out what really we should know. There are so many phenomena spreading in wide scale in space (10^{-11} - 10^{2}m) and in time (10^{-13} - 10^{7}sec) in biological world. A purpose of biological science is to explain such various biological phenomena (or more appropriately biological function) in terms of more established concepts in physical science such as thermodynamics and/or conformation. It should be stressed that the observation of biological substances is not the goal of biology, but

X-ray Microscopy
Ed. by P. C. Cheng and G. J. Jan

merely the start of the understanding of the complicated biological systems in various space and time scale as desired. However it will be very much helpful in the fundamental understanding of biological function. This is the reason why we require methods to enlarge or reduce sizes in space and in time.

In biology, many interesting subjects are spread into a hierarchy as shown in Figure 20.1. Biologically characteristic phenomena appear sometimes in atomic level and sometimes in higher order. Even if we have a powerful tool to look into molecular structures in atomic level, it would be almost impossible to understand biological function in terms of structure without observing gross structure corresponding to the biological function directly. It is, therefore, very important to develop various probes for observating structure in each hierachical level.

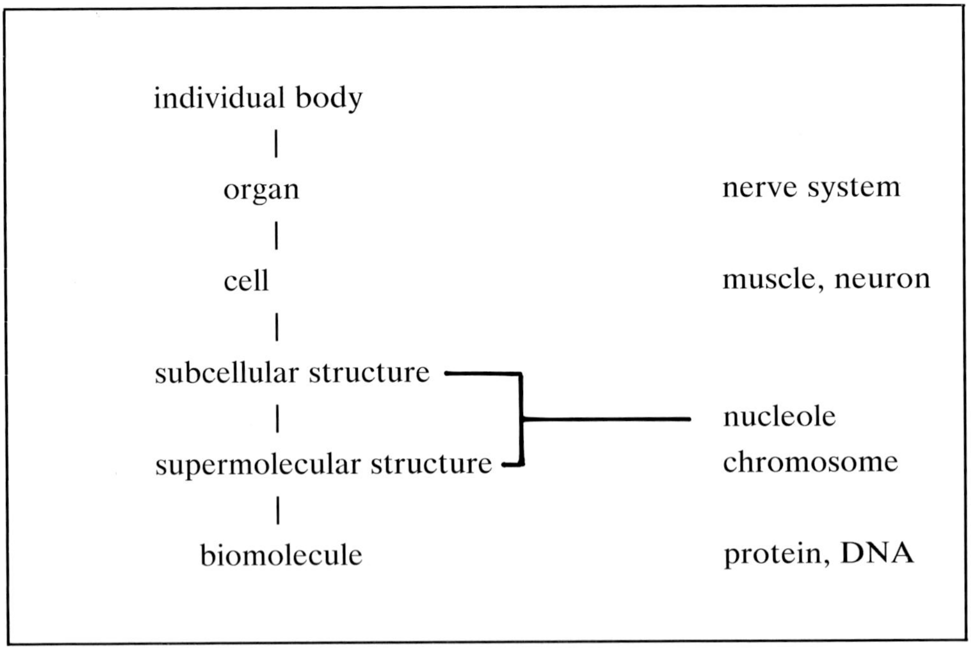

Figure 20.1. **Biological Hierarchy.**

In Figure 20.2, examples of biological substances distributed in spacial scale are shown. For materials smaller than $1\mu m$, very limited methods are applicable such as electron microscope, x-ray scattering and x-ray crystallography. X-ray microscopy, if developed appropriately, will be counted as a powerful tool for the microscopic observation as well. For instance, "sliding theory" in muscle contraction, illustrated in Figure 20.3a, has not been well established because of the lack of direct structural evidence how a myosin slides on an actin filament. Time-resolved x-ray diffraction could not help the situation, probably because myosin molecules do not move in phase. Another example is based on an attractive hypothesis by Oosawa (1985), who suggests that F_1 part of proton- ATPase rotates against membrane-bound F_0 part (Figure 20.3b). Third example shown in Figure 20.3c suggests that disk part of flagella may rotate against its counterpart, which is the driving force of flagella movement (Oosawa and Masai, 1982). Such attractive targets are waiting for the development of x-ray microscopy to be investingated with.

Besides the importance of high resolution in space, resolution in time scale should be also stressed. It is not overstated that static measurement cannot elucidate true nature in biology. In Figure 20.4, various biological phenomena are plotted as a function of their characteristic times. Replication and folding are slow, whereas thermal movement and rotational diffusions are comparatively fast. Although biological motions are spread into wide time scale, it is easily noticed from the figure that physiologically important phenomena are converged only in milli-sceond or slower time ranges (Kihara, 1985; Albery and Knowles, 1976), which strongly encourages the construction of a tool for time-resolved study working in milli-second time region. Millisecond time might be too fast to be monitored by x-ray microscopy. Reactions in biological systems, however, can be slowed down by cooling. Douzou demonstrates that enzyme reactions are slowed down in the order of 2 or 3 by decreasing temperature down to -20° by the use of antifreezer such as ethylene glycol (Douzou, 1977). Thus, biological important phenomena can be monitored by time-resolved x-ray microscopy in seconds time region, which will be feasible.

In summary, x-ray microscopy will become a powerful tool in biology if resolution in space becomes better than 0.1μm and resolution

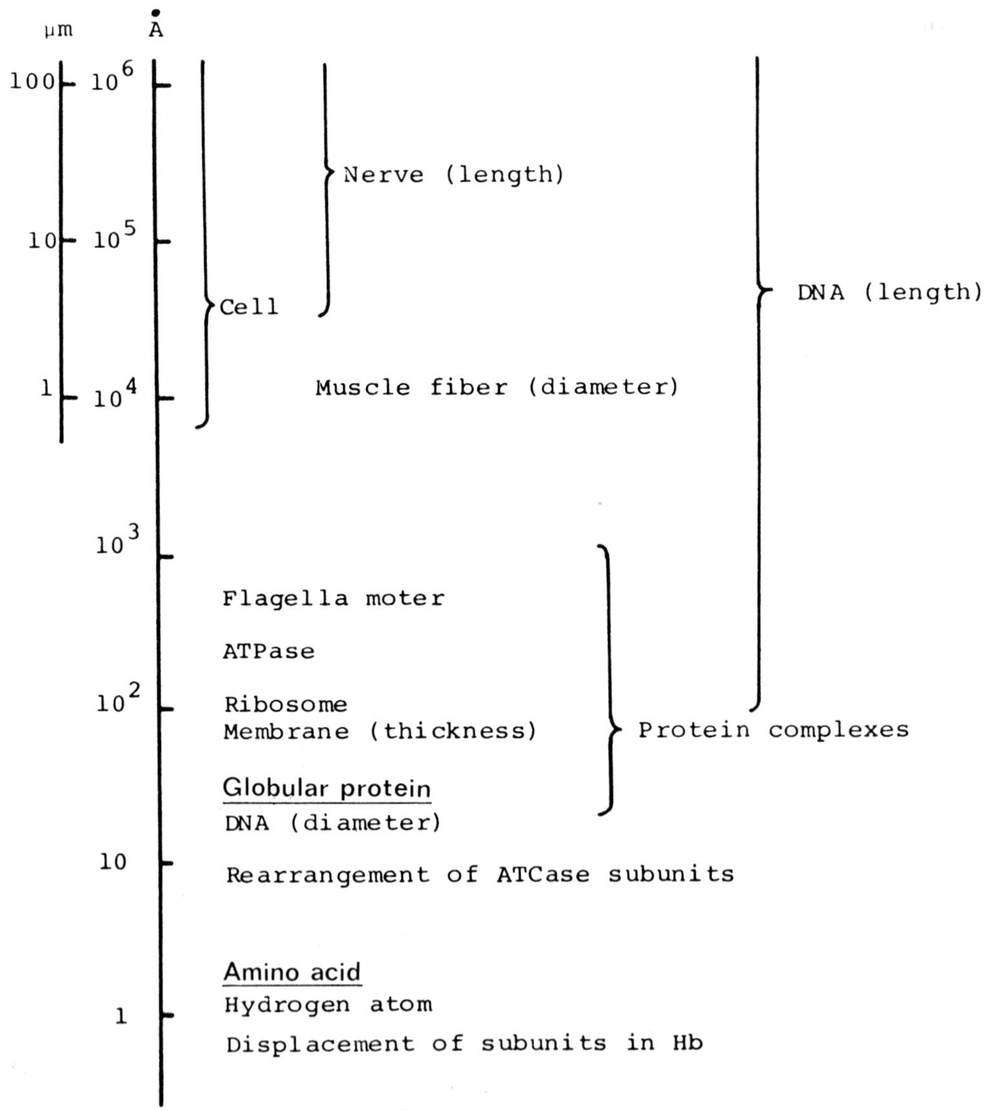

Figure 20.2. Characteristic lengths of biosystems in spatical domain.

in time is better than several seconds. Cooling is also useful to prevent the specimen from radiation damage.

Biological Materials and Their Mass Attenuation Coefficients

Biological materials can be classified as protein, nucleic acid, lipid, bone, saccharide, cytoplasma, *etc.* based on their chemical compositions. Most common atoms in biological materials are C, H, O, N, S, P, ..., and Na, Cl, K, ... specially in cytoplasma. It is, therefore, most common to develop x-ray microscopy in the wavelengths between 24Å to 45Å for the observation of biological materials in water, by ultilizing the difference in the absorption edge of C (or N) and O. In biology, however, less common atoms such as Ca, P, Mg, Fe, Cu and

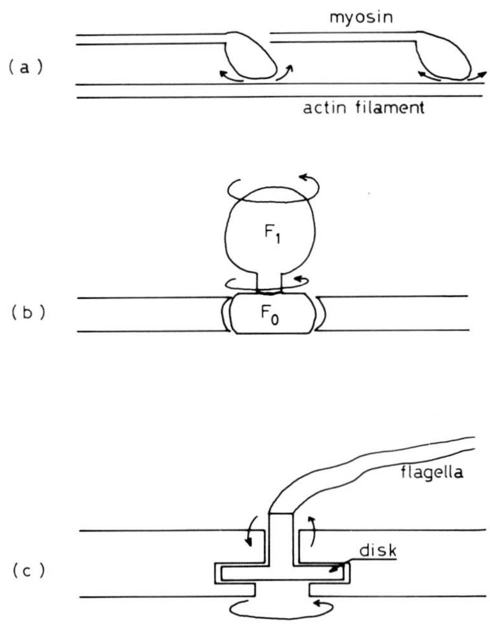

Figure 20.3. Schematic views of biomolecular mothions relevant to their functions. (a) Sliding theory: myosin sliding along actin filament in muscles. (b) Rotation of F_1 part of proton ATPase against its counterpart. F_0. (c) Flagella motor. Disk rotates against its counterpart, which is the driving force of cell movement.

Mn play also important roles as they are used in regulation (Ca, P, Mg) or in active center (Fe, Cu, Mn). Especially, if these atoms can be seen selectively, it will give us very useful information. Distribution of Ca in bones reproted by Kenney *et al.*, (1985) demonstrates the usefulness of the application of the absorption edge of the atom.

In the present paper, we have done some feasibility studies on the utilization of x-ray microscopy in VUV region from the following reasons:

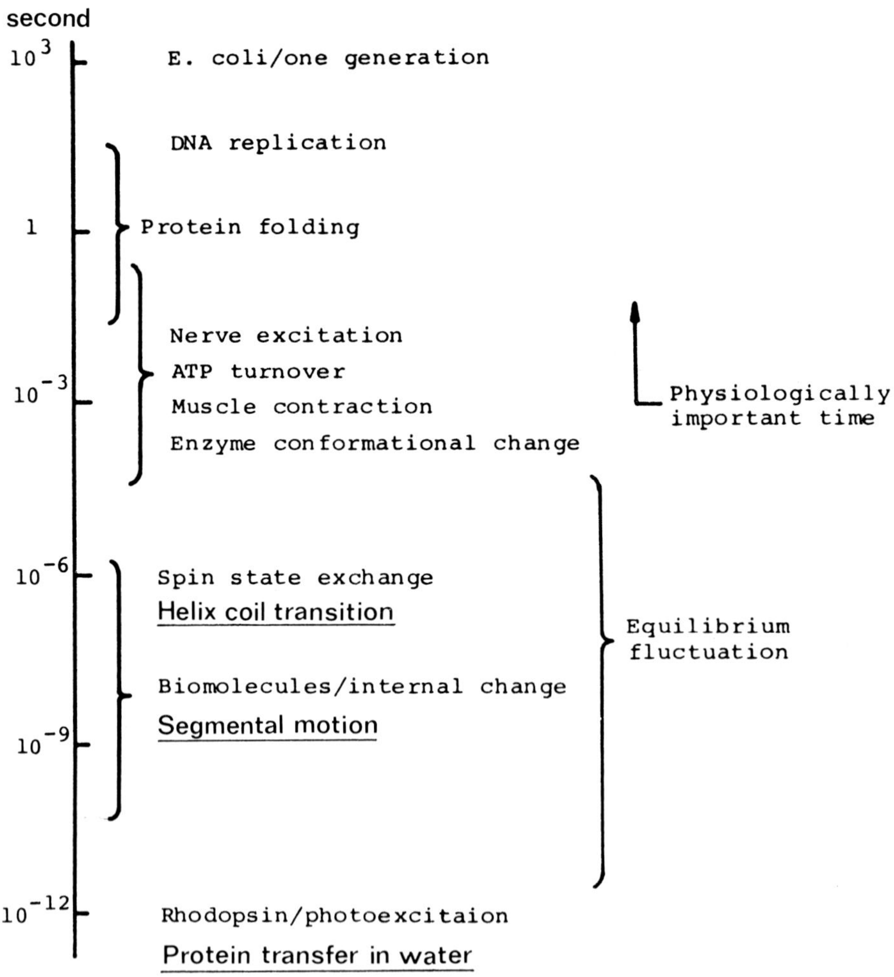

Figure 20.4. Characteristic times of biosystems in time domain.

(1) Extensive studies have been done in soft x-ray regions. However, less examinations have been done in longer wavelength region, relating to biological materials.

(2) It is usually believed that the measurement with the shorter wavelength gives better special resolution. However, as shown in Figure 20.5, it would be difficult to obtain high contrast if size of observed materials is much smaller than its thickness (see details, in caption of Figure 20.5). Hence it is not self-evident that the measurement at the shorter wavelength gives higher resolution in any cases. Best resolution will be obtained when wavelength of monitoring x-ray, size of observed materials and the thickness of specimen are compromised.

In Figure 20.6a and 20.6b, mass attenuation coefficients of atoms relevent to biological materials are plotted against the wavelength. In the wavelength regions concerned, absorption edges of O, N, C, Ca, K, Cl, S and P are noticed. Absorption edges of O, N, C and Ca have been already extensively discussed elswhere. Feasibility of L-absorption edges of P, S, K, and Cl for x-ray microscopy will be examined later in this paper.

In Figure 20.7, mass attenuation coefficients of several biological materials are plotted against wavelength. Their chemical compositions are shown in Table 20.1. In order to examine feasibility of x-ray imagings in these wavelength regions in more detail, ratio of linear absorption coefficients of biological materials to those of water or protein are plotted against wavelength, which was shown in Figure 20.8. From Figures 20.6-20.8, we can deduce following speculations.

(a) Usefulness of x-ray imaging between absorption edges of oxygen and carbon.

Figure 20.8 clearly demonstrates that good contrast was obtained for proteins, lipid (DPPC) and DNA in water between 24Å and 45Å.

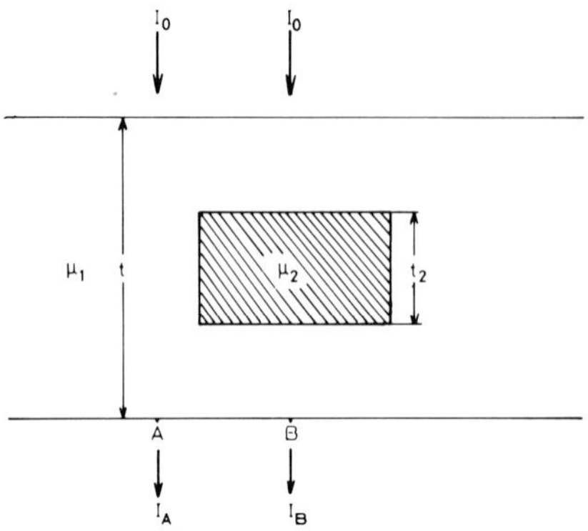

Figure 20.5. X-ray transmission through a material in the media. Let us denote thickness of the observing material as t_2 and that of surrounding media as t_1 (total thickness is equal to t), and absorption coefficients of each part as μ_1 and μ_2 respectively. Transmission I_A through the media at position A is

$$I_A = I_o e^{-\mu_1 t} \tag{F1}$$

and transmission I_B through the media and the material at position B is

$$\begin{aligned} I_B &= I_o e^{-\mu_1 t_1 - \mu_2 t_2} \\ &= I_o e^{-\mu_1 t_1 - (\mu_2 - \mu_1) t_2} \end{aligned} \tag{F2}$$

where I_o is incident intensity. Hence

$$\frac{I_B}{I_A} = e^{-(\mu_2 - \mu_1) t_2} \tag{F3}$$

Since μ_i is represented as

$$i = \sum_{j=3}^{4} A_{ij} \lambda^j \tag{F4}$$

except regions near absorption edges (Victoreen, 1943; 1948)., Equation F3 becomes

$$A = e^{-\sum_{j=3}^{4} (A_{2j} - A_{1j}) \lambda^j t_2} \tag{F5}$$

which gives contrast. Equation (F5) indicates that the contrast becomes larger when wavelength increases.

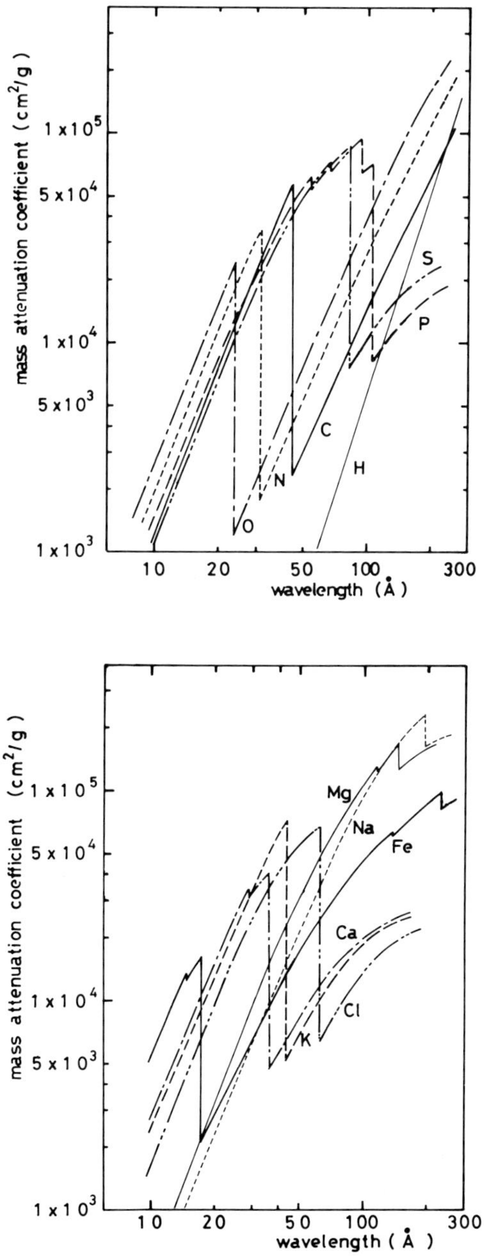

Figure 20.6. Mass attenuration coefficients of atoms common in biological materials against wavelength (10). (a) Atoms common in biological substances. (b) Atoms essential in regulation and activity in biological function.

In soft x-ray and VUV regions examined, no other wavelength regions show better contrasts. If one wants to observe biological substances in water, best choice is the imaging with the use of x-rays in these wavelength regions.

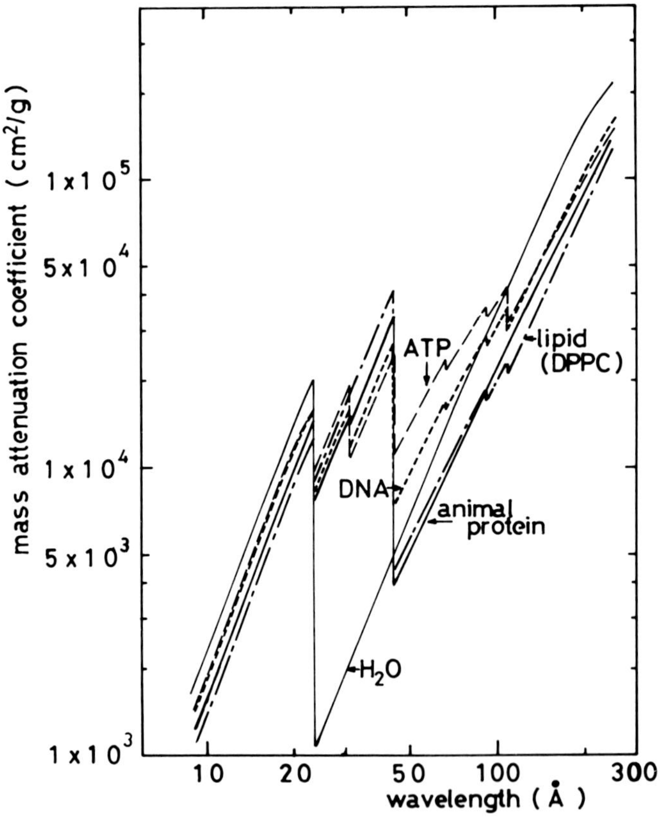

Figure 20.7. Mass attenuation coefficients of several biological materials against wavelength. They were calculated based on Henke and Elgin (1970) and Henke *et al.* (1982). DPPC: dipalmitoylphosphatidyl-choline, DNA: deoxyribonucleic acid, ATP: adenosine triphosphate.

(b) DNA in proteins

Situations discussed in the last paragraph will change if one wants to observe biological materials in other biological materials, *e.g.* DNA in proteins. Absorption coefficient of DNA does not differ so much from that of proteins between 24Å and 45Å. Insteads, absorption by DNA becomes larger than protein when wavelength is longer than 45Å (Ratio of μ_{DNA} to $\mu_{protein}$ changes from 1.03 at 39.8Å to 2.34 at

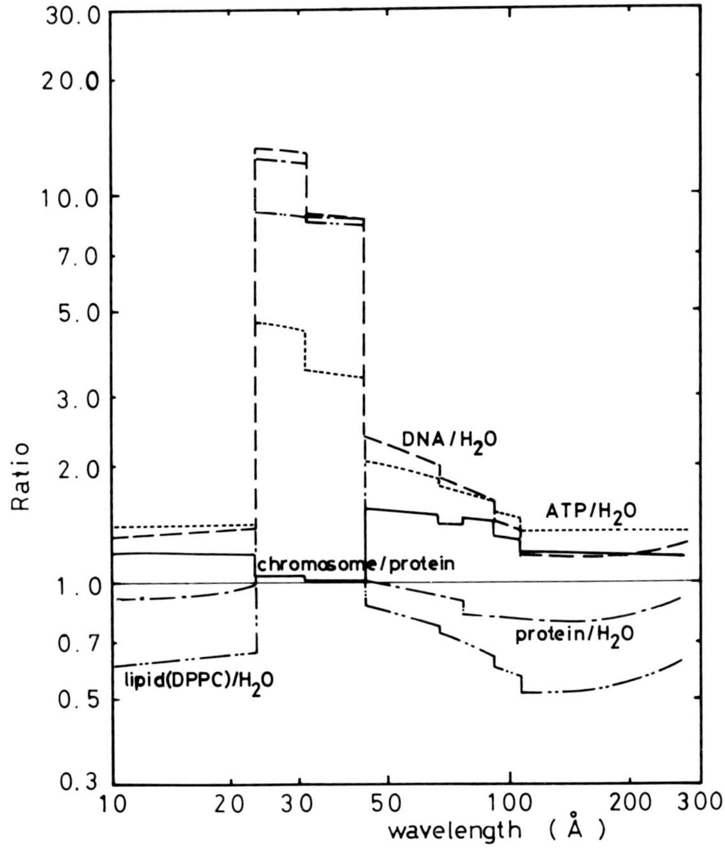

Figure 20.8. Ratio of absorption coefficients of some biological materials to that of water. The calculation of absorption coefficients of chromosome were shown in the main text. Density of DNA and protein are from Kirz and Rarback (1985).

Table 20.1. Chemical composition of some biomoelcules

	components (weight %)						
	C	H	O	N	S	P	Ca
Animal protein[×]	52.5	7	22.5	16.5	1.5	-	-
DNA	34.3	3.3	37.7	15.5	-	9.1	-
ATP	23.7	3.2	41.0	13.8	-	18.3	-
Lipid (DPPC)	65.4	11.1	17.4	1.9	-	4.2	-
$Ca_5(PO)_4OH$ (Hydroxylapatite)	–	0.2	41.4	–	–	18.5	39.9
$Ca_5(CO_3OH)_3OH$	8.0	0.9	46.4	–	–	–	44.7

[×]by B. L. Henke and R. L. Elgin (1970)

44.7Å). Therefore, best wavelength regions for observing DNA in proteins would be longer than 45Å. Ratio of 2.34 will be hopefully detectable by x-ray microscopy in near future.

DNA usually exists as chromosome in living cells. Therefore, more exactly, we have to set up our problem whether chromosomes are detectable in proteins. To make clear this point, absorption coefficient of chromosome, μ_C, is represented as

$$c = \frac{V_{DNA}\rho_{DNA}}{V_c}\left(\frac{\mu}{\rho}\right)_{DNA} + \frac{V_h\rho_h}{V_c}\left(\frac{\mu}{\rho}\right)_h \qquad (20.1)$$

where suffices c and h denote chromosome and histone respectively, and ρ and V means density and volume, respectively. With an assumption that $V_c = V_{DNA} + V_h$, Equation (1) becomes

$$\frac{\mu_c}{\mu_h} = \frac{1}{1 + \dfrac{V_{DNA}}{V_h}} \left\{ 1 + \frac{V_{DNA}\rho_{DNA}}{V_h\rho_h} \, \frac{(\frac{\mu}{\rho})_{DNA}}{(\frac{\mu}{\rho})_h} \right\} \tag{20.2}$$

The calculated result, *i.e.* the ratio of μ_c to μ_h is also plotted in Figure 20.8. The ratio is 1.55 at 44.7Å and 1.00 at 39.8Å. Ther result indicates that chromosomes would be distinguisable from proteins when wavelength of monitored x-ray is longer than 44Å, which encourages the preparation of x-ray microscopy working in VUV region.

(c) Protein in lipids

Many proteins are bound in lipids as membrane proteins. Therefore, it is an important question that proteins are detectable when they are buried in lipids. Lipids as well as proteins show large contrast against water between 24Å and 45Å. However, as is the case of DNA in proteins, proteins are not detectable in lipids between 24Å and 45Å (ratio of absorption coefficients are 1.04 at 39.3Å). Ratio becomes as much as 1.4 when wavelength becomes shorter than 27Å or longer than 93Å.

(d) ATP detection in living cell

As shown in Figure 20.8, best contrast of ATP against water will be expected at wavelwngth between 24Å and 45Å. However, if we want to monitor ATP distribution selectively in living cell, we should distinguish ATP from other materials such as proteins, DNA and lipids, which would be difficult without the use of L- absorption edge of phosphor. In Table 20.2, ratio of absorption coefficient of ATP to that of water are summarized. Differences of the ratio at two wavelengths (before and after $L_{II,III}$ absorption and $L_{II,III}M$ edges) are also shown. The result demonstrates that 1M of ATP is detectable in water with the use of $L_{II,III}M$ and/or $L_{II,III}$ absorption deges of phosphor. ATP concentration is ~10mM in physiological condition. However

Table 20.2. Ratio of absorption coefficients of ATP solution to that of water. Ratio ($\mu_{ATPsol.}/\mu_{H_2O}$)

ATP	wavelength (Å)			| difference |	
concentration	83.4	93.4	108.7	83.4/108.7	93.4/108.7
10mM	1.0066	1.0050	1.0035	0.0031	0.0015
100mM	1.0661	1.0498	1.0349	0.0312	0.0149
1M	1.6611	1.4982	1.3492	0.3119	0.1490
2M	2.3223	1.9964	1.6983	0.6240	0.2981
3M	2.9834	2.4946	2.0475	0.9359	0.4471
10M	7.6112	5.9810	4.4916	3.1196	1.4902

Ratio of absorption coefficients of ATP solution to that of water around $L_{II,III}$ absorption edge (94Å) and $L_{II,III}$ M edge (193.8Å) of phosphor (ref). The ratios were calculated accoding to equations described below: mass attenuation coefficient of ATP solution is calculated as

$$\left(\frac{\mu}{\rho}\right)_{sol} = \frac{W_{water}}{W_{sol}}\left(\frac{\mu}{\rho}\right)_{water} + \frac{W_{ATP}}{W_{sol}}\left(\frac{\mu}{\rho}\right)_{ATP} \quad \text{(T1)}$$

where $\left(\frac{\mu}{\rho}\right)_{sol}$, $\left(\frac{\mu}{\rho}\right)_{ATP}$ and $\left(\frac{\mu}{\rho}\right)_{water}$ are mass attenuation coefficients of ATP solution. ATP and water respectively, and W_{sol}, W_{ATP} and W_{water} are weights of ATP solution, ATP and water in volume V respectively. Absorption coefficient of ATP solution can be deduced as

$$\mu_{sol} = \frac{W_{water}}{V}\left(\frac{\mu}{\rho}\right)_{water} + \frac{W_{ATP}}{V}\left(\frac{\mu}{\rho}\right)_{ATP} \quad \text{(T2)}$$

with the assumption that V_{sol} is equal to V_{water}, ratio of mass attenuation of ATP solution to that of water is

$$\frac{\mu_{sol}}{\mu_{water}} = 1 + \frac{nW_{ATP}}{1000}\,\frac{\left(\frac{\mu}{\rho}\right)_{ATP}}{\left(\frac{\mu}{\rho}\right)_{water}} \quad \text{(T3)}$$

where n means molar number in 1 liter and M_{ATP} denotes molecular weight of ATP. Absorption coefficients in the table were calculated according to Equation T3.

we may find a biosystem in which ATP or some other phosphor compounds are concentrated as much as 1M.

Detection of phosphor compounds in biology besides ATP and DNA is important in biology, since it is used in regulation of enzyme action (phosphorylation and dephosphorylation) *etc.*. However, the concentration of phosphor in biosystem is much less than 1M.

(e) Apatite (McConnell, 1973)

Main component of teeth, bone and some kinds of "shells" are hydroxyapatite (Lazzari, 1976), $Ca_{10}(PO_4)_6(OH)_2$ (density = 3.156), of which mass attenuation coefficients were calculated and plotted in Figure 20.9. Good contrast of calcium distribution is expected with the use of absorption edge of calcium, as demonstrated by Kenney *et al.* (1985). Absorption edge of phosphor (Cheng, 1985) is not so useful as the edge of calcium.

In apatite of tooth enamel, usually 1.5 to 2.9% of carbonate compounds of apatite are included. Carbonate content increases from enamel surface to dentioenamel junction. Apatite, replaced the phosphate groups by carbonate is named carbonated hydroxyapatite. After the replacement, apatite crystal cannot grow up to the large size, and becomes friable, which will be a cause of carious tooth. Detection of microcrystal of carbonated hydroxyapatite (or dahllite), of which chemical formula is written as $(Ca,X)_{10}(P, C)_6(O, OH)_{26}$, will be then very useful to dentry. Hence X in the formula denotes H_3O, H_2 O, Na, Mg *etc.*. As a feasibility study, we have estimated mass attenuation coefficients of a hydrated calcium carbonate hydroxide, $Ca_{10}(CO_3OH)_6(OH)_2$, which are also shown in Figure 20.9. The figure indicates mass attenuation coefficient of the latter compound is significantly smaller than that of hydroxapatite at wavelengths longer than 45Å. X-ray imaging at wavelength longer than 45Å might, then, become a powerful tool in detection of local structure of decayed teeth. The method is probably also applicable to microstructure of bones, shells and fossiles. Especially, the method may be useful in elucidating the aged effect of fossiles.

(f) Observation of S, K, Na, Mg, Fe and Cl distribution

Sodium, potassium and chloride ions are distributed in cytoplasma. Then with the use of absorption edge of Cl, cytoplasma in the cell is observable, i.e. negative imaging of tissues.

If local distributions of K and Na are detectable, x-ray microscopy will contribute to nervous excitation study, as active potential is created by the K and Na ions flow through excitable membrane. It is

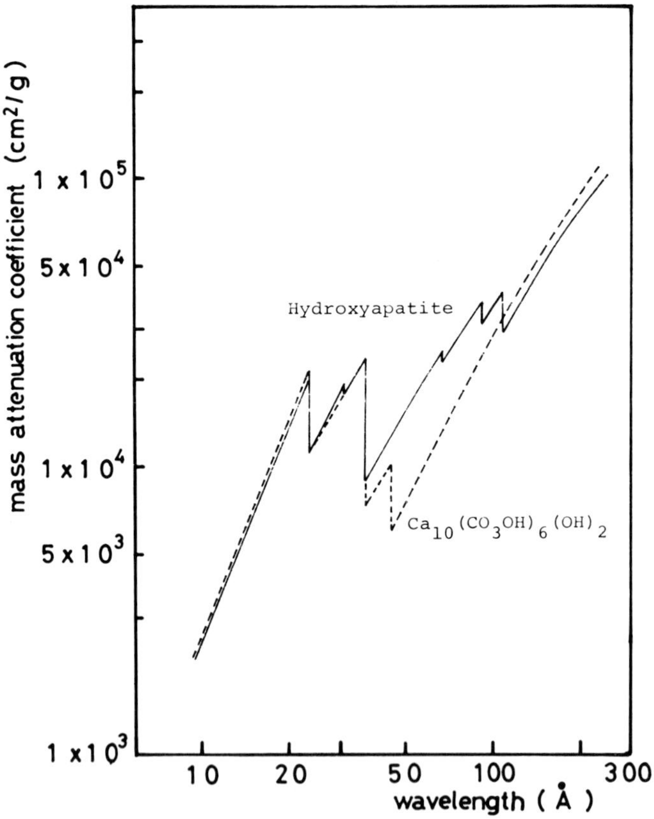

Figure 20.9. Mass attenuation coefficients of hydroxyapatite ($Ca_{10}(PO_4)_6(OH)_2$) and hydrated calcium carbonate hydroxyoxide ($Ca_{10}(CO_3OH)_6(OH)_2$) against wavelength.

worthwhile to test the applicability of the absorption edge of K to detect the local distribution of potassium ion in cells.

Magnessium, sulfur and iron are important materials in biology as they are used in key parts of biological activity such as active center of hemoproteins or disulfide bonds. If they are detectable, x-ray microscope will become very powerful. However we are suspicious to the usefulness of normal x-ray microscopy as the concentration of these atoms in cells are rather low. If micro-focussed EXAFS is developed, it will be more promising.

Zone Plate Fabrication by Electron and Ion Beam Lithographies

For the purpose of studying biological specimens in VUV regions, we have started fabricating zone plates as an enlarging element, using electron and ion beam lithographies. Characteristics of the zone plates and their fabrication procedure by electron beam lithography are shown in Table 20.3 and Figure 20.10, respectively. The pictures shown in Figure 20.11a to 20.11d, indicate a zone plate pattern on PMMA resist (corresponding to Figure 20.10(2)), reactive-ion-etching PIQ patterns in the vicinity of the periphery, near the center of the same zome plate (Figure 20.10 (4)), and a finally obtained free-standing gold pattern (Figure 20.10 (7)), respectively. Resolution of the free-standing zone plate is estimated as 0.6μm.

Another type of a zone plate, fabricated by ion beam lithography is under construction. In Figure 20.12, a schematic view of the ion beam apparatus is shown, where gallium is used as an ion-source. Beams are generated spherically. Figure 20.13 demonstrates the sharpness of the ion beam lithography, with which the resolution of the secondary image obtained by ion beam is around 400Å. A zone plate pattern fabricated by the ion beam is apparatus is shown in Figure 20.14. Fabrication characteristics are shown in Table 20.3. The Au pattern was fabricated on a glass support. The process to a free-standing zone plate is being under way.

Table 20.3. Specifications of free-standing Fresnel zone plates fabricated by electron and ion beam lithographies.

	Electron beam			Ion beam
Wave length	80Å		40Å	40Å
Focal length	150mm		75mm	1.56mm
Diameter	495μm	1997μm	498μm	40μm
# of transparent zones	26	416	104	35
Outermost zone width	2.43μm	0.60μm	0.60μm	0.28μm

The zone plate fabricated by electron beam lithography was tested at UVSOR (synchrotron radiation facility at the Institute for Molecular Science at Okazaki, Japan) with a beam line BL6A2 (Watanabe, 1986; Seki *et al.*, 1986). A Schematic view of a testing camera is shown in Figure 20.15. The beam is focussed to 1x1mm^2 at a pinhole by the monochromator at upstream, and is expanded with 30 mrad.

(1) Focal test

The focussing characteristics were tested by shifting slit and changing wavelength. Althrough data analysis has not finished before the conference of X-ray Microscopy '86 in Taiwan, Republic of China. Preliminary analysis indicates the zone plate gave good focussing.

(2) Resolution test

With a smaller pinhole and a slit, a resolution test was planned. However, with a pinhole of 0.05mm diameter, beam intensity was too low to perform a quantitative analysis. To get higher resolution, an improvement should be done, probably without the monochromater.

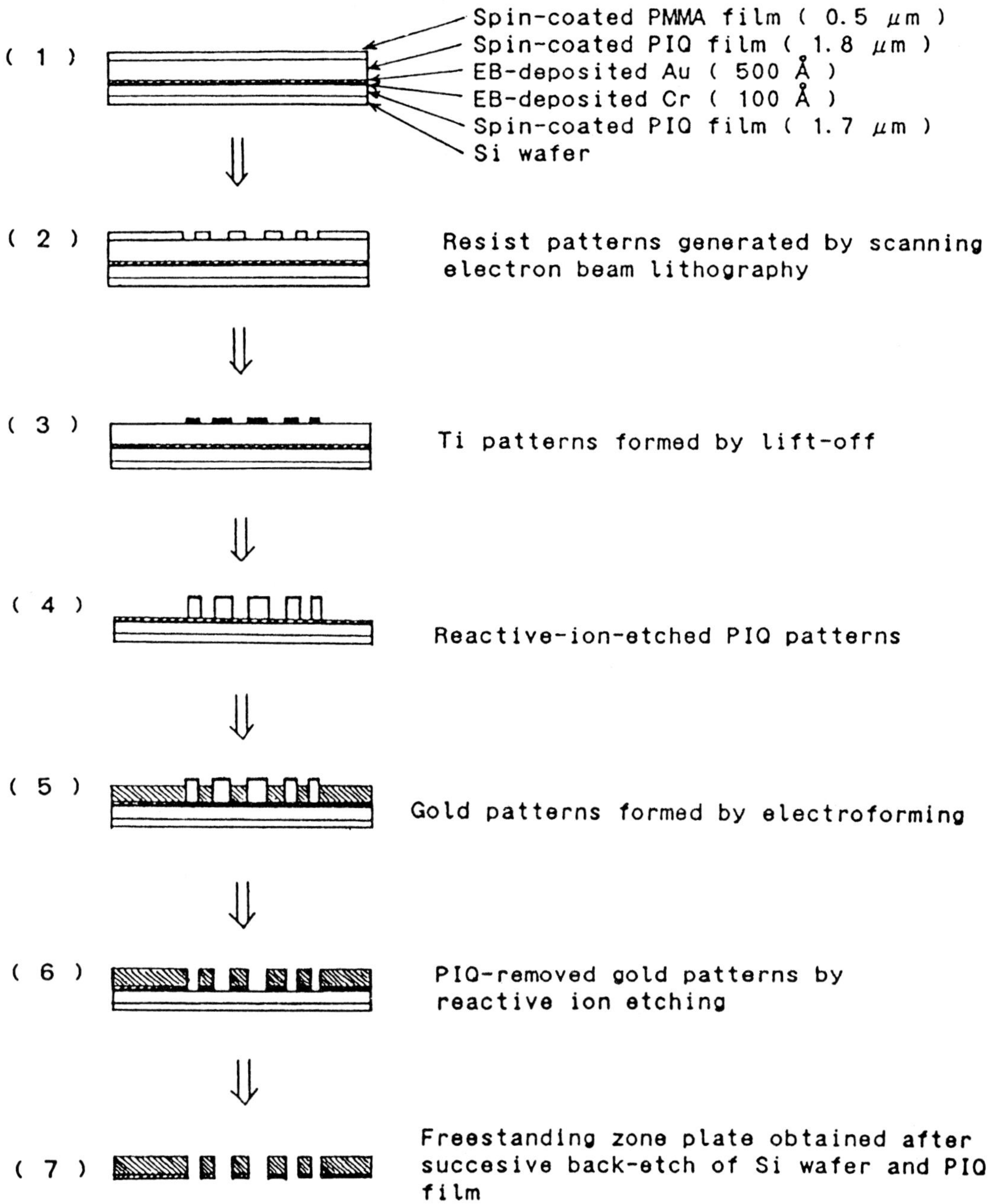

Figure 20.10. Fabrication procedure for free-standing zone plate by electron beam lithography. PIQ: a polyimide film, PMMA: polymethylmathacrylate, EB: electron beam.

a
50 µm
b
6 µm
c
20 µm
d
50 µm

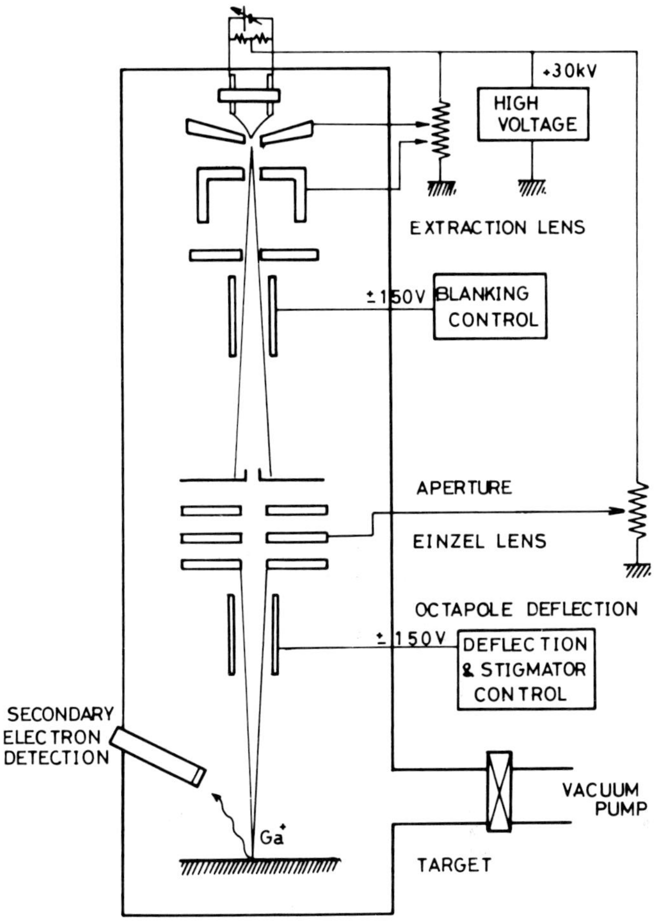

Figure 20.12. Schematic view of ion beam apparatus. Gallium ion was used as a probe ion. Acceleration voltage: 0-30 kV, ion emission current: 1-50 μA (Ga^+). Einzel lens: f = 40mm, C_s = 4600mm. C_c = 131mm. Gas presure: $p_{gun} = 10^{-7}$ Torr, p(spe. chamber) $\sim 10^{-6}$ Torr.

Figure 20.11. (a) Zone plate pattern on PMMA resist produced by scanning electron beam lithography. (b),(c) Reactive-ion-etched PIQ patterns. (d) Free-standing gold patterns after electroforming.

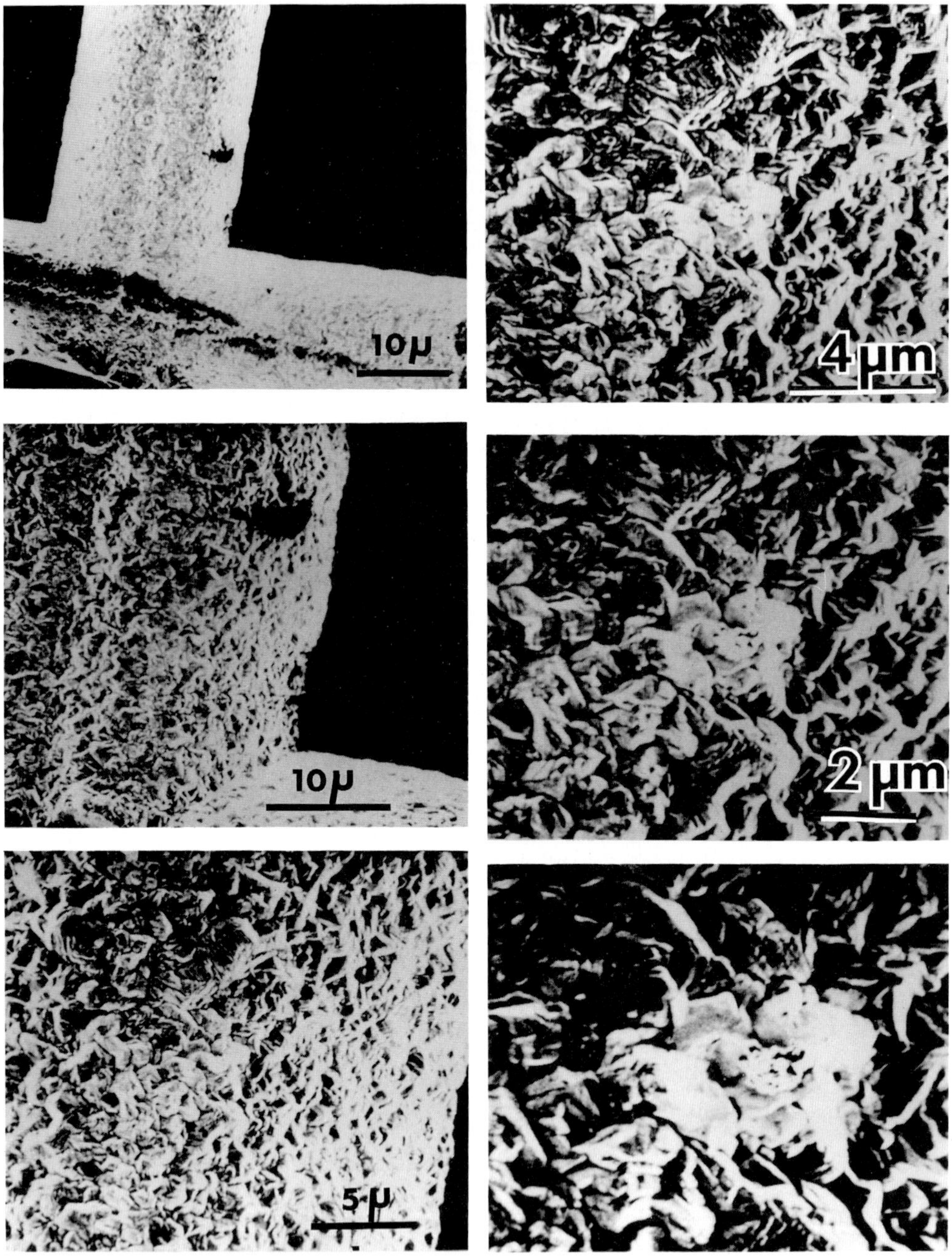

Figure 20.13. Scanning ion micrograph of Ni mesh images. Resolution of the secondary electron image by the ion beam was estimated as 400Å from the figure. Ga^+ ion current: 9μA. $V_{(accleration)}$ = 25kV.

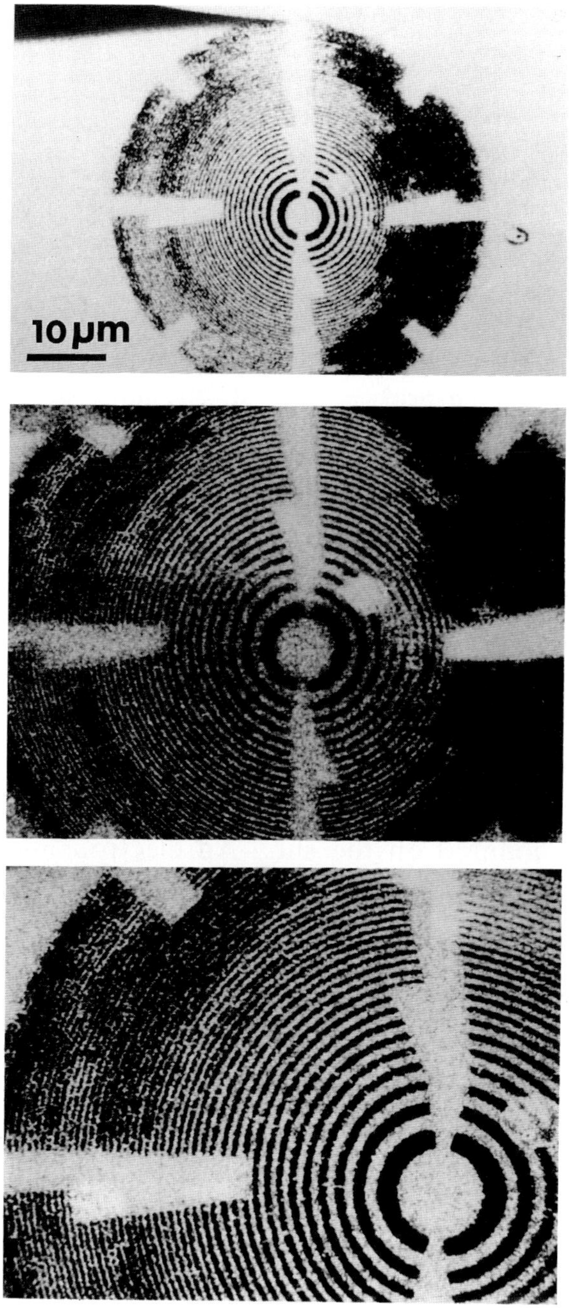

Figure 20.14. Fresnel zone patterns fabricated by ion beam lithography. $V_{(acceleration)} = 25kV$. $I(Ga^{+}) = 7\mu A$.

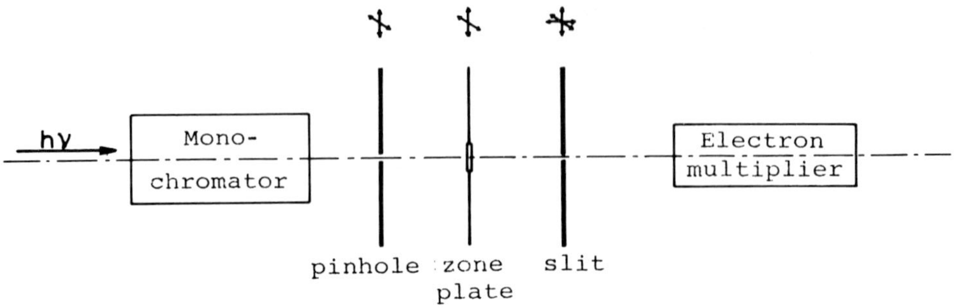

Figure 20.15. Schematic view of the test camera used at the Institute for Molecular Science, UVSOR. Beams from the upstream are converged at a pinhole position of 1 x 1mm^2 with 30mrad.

(3) Mesh pattern

Another focal test with Ni mesh (No. 1000) was taken. Quantitative analyses are under way.

(4) Tooth specimen

As a feasibility study to check the applicability of x-ray microscopy in VUV region to dental and bone medicine, a sliced tooth specimen was mounted on the slit. An electron micrograph of the specimen is shown in Figure 20.16. The transparent part in the figure is a hole. Snailing traces from the hole are cracks. It becomes thicker from the side of the hole to the edges. In our experiments at UVSOR, the tooth was found to be too thick. Thinner specimens would be required, although detailed speculations will be given after data analysis.

At the next experiment, imaging will be tested with films, which will hopefully give us more informations, especially about resolution.

Figure 20.16. Transmission electron micrograph of an ion beam thinned transverse section of human dentine. This section as never uniformly thinned, but a considerable portion near to the central hole caused by ion beam erosion was thin enough to permit transmission electron microscopy. The large number of small holes are due to the tubles seen in transverse section. Lines are cracks from the hole at the center, which were artifacts produced during dehydration.

Conclusion

(1) Observation of biological materials by x-ray microscopy in VUV regions will open special and unique fields, differing from soft x-ray region (24Å - 45Å). It shows better contrast than soft x-ray microscopy in subjects such as DNA distribution in proteins, proteins in lipids and microheterogeneity of hydroxy apatite in fossile, tooth, shell and bone.

(2) With the use of L-edge of phospher, 1M of ATP is detectable. Although it is not feasible in normal living cell (ATP concentration is ~ 10mM), it will be applicable to cases with specially condensed phosphor compounds such as ATP.

(3) Electron beam-fabricated zone plate with the resolution of 0.5μm has been contructed and tested at UVSOR.

(4) Ion beam-fabricated zone plate has been under construction with the resolution of 0.3μm.

21. Sample Preparation for X-ray Imaging and Examples of Biological X-ray Images

Ping-chin Cheng

IBM T. J. Watson Research Center, Yorktown Heights, NY 10598 USA

Introduction

In the past decade, despite the rapid advancement of x-ray imaging technology, many x-ray micrographs which appeared in published articles were poor from a biological perspective; most of the specimens were prepared probably by scientists who had little knowledge of biological specimen preparation for microscopy. For example, air-drying was generally used to "prepare" biological specimens. It is known that air drying introduces gross morphological changes and generally not acceptable for preparing biological specimens for microscopy. Further, the interpretation of x-ray images requires knowledge of image formation physics. In the case of modern x-ray microradiography, the x-ray contact images are generally magnified by means of electron microscopy, understanding the characteristics and operation of electron microscope is essential.

It has been wildly suggested that x-ray microscopy produces images of biological material otherwise unattainable, including the imaging of living or hydrated biological specimens without the use of stains. However, it would be a mistake to ignore the vast amount of knowledge histological staining techniques have contributed to biology. Recent advances in immuno-cytochemistry using either polyclonal or monoclonal antibodies conjugated with colloidal gold provides powerful tools for probing cellular structures. The fluorescein conjugated and colloidal gold conjugated antibodies are "stains" which are useful

X-ray Microscopy
Ed. by P. C. Cheng and G. J. Jan

for studying specific proteins and their functions. "Stains" are also useful for probing intracellular pH and ionic concentration (*i.e.* aequorin, Allen *et al.*, 1977) in biological research. X-ray microscopy, without using the existing techniques, will not likely make significant contributions to biological research. Besides the potential of studying living or hydrated biological specimens, x-ray microscopy should be equally useful for studying dehydrated materials. Due to the high penetration ability of x-rays, x-ray microscopy offers unsurpassed ability for examining thick specimens, otherwise not possible by electron microscopy.

The intention of this article is to review the existing technology of specimen preparation for electron microscopy. It is clear that most of sample preparation techniques for EM are directly applicable to x-ray microscopy. Examples of x-ray contact images are provided for the discussion of image interpretation.

Sample Preparation for X-ray Imaging

For any type of microscopy, specimen preparation represents a key step for success. In biological microscopy, specimen preparation involves the initial culture of organisms or cells, and the subsequent processes to ensure preservation of ultrastructures during imaging.

Cell Culture

Preparation of cultured animal cells for x-ray imaging are very simple. Similar to electron microscopy, cells can be grown directly on the surface of various thin film substrata such as polystyrene, Formvar, and silicon nitride (Si_3N_4). Among many types of plastic supporting films, polystyrene shows excellent x-ray stability when subject to high dose x-ray irradiation (Cheng *et al.,* 1986). Identification of individual cells is facilitated by cultured the cells on UV-sterilized Formvar-carbon coated gold or nickel index grids (*e.g.* Ernest Fullam,

Schenectady, N.Y.). For contact imaging, cells can be grown directly on the surface of polymethyl methacrylate (PMMA) resist. The polystyrene and Formvar surfaces require treatment to facilitate cell attachment. These surfaces can be prepared by vacuum evaporation of a thin layer of carbon or by dip-coating with polylysine. Silicon nitride generally requires no surface treatment for cell attachment. However, it has been our routine practice to coat the surface with a thin layer (~100Å) of carbon to facilitate cell attachment. Polystyrene film shows higher x-ray stability than Formvar film (Cheng, 1985). For x-ray contact imaging, the PMMA resist surface requires coating with polylysine to facilitate cell attachment. Carbon coating cannot be used since it interferes with the chemical development of the PMMA resist.

Chemical Fixation

Chemical fixation of living tissue helps to minimize alteration of cellular structures after killing biological specimen. Tissues must be fixed promptly after removal from an organism or cessation of circulation to prevent postmortem decomposition. Fixation preserves cellular structures by cross-linking proteins and stabilize lipid structures within the tissue. Fixation is generally achieved by immersing the tissue in fixative or subjecting the tissue to heat (by convention heating or microwave energy). Since no single reagent is available to achieve all the requirements of a "good" fixative, two or more reagents are generally combined or used consecutively.

In routine electron microscopy, fixation is generally carried out by an aldehyde pre-fixative followed by a post-fixative such as OsO_4. Aldehydes, such as formaldehyde, glutaraldehyde and acrolin, are used to cross-link proteins to form an insoluble "gel". The OsO_4 is used to stabilize lipid constituents in a cell such as membranes. Therefore, OsO_4 is not only a heavy metal stain, but also an important fixative.

Dehydration

After fixation, cellular water is removed. The dehydration process can be carried out by immersion fixed tissue in a graduated series of hydrophilic solvent. Typical solvents used in tissue dehydration are ethanol, acetone, methanol, and butyl alcohol. Tissue can be dehydrated in a discrete series of solvent (*e.g.*, 20%, 40%, 60%, 80%, 90% and finally 100%), or in a continuous gradient (Cheng *et al*, 1979). Ethanol or butanol should be used for dehydrating cultured cells supported on thin film of Formvar or polystyrene substratum, since acetone dissolves both. Certain degree of shrinkage and extraction of cellular components is expected in the dehydration process.

After exchanging all the cellular water with pure solvent, the tissue is embedded or critical point dried. Bulky specimens are generally embedded in a suitable embedding medium and subsequently sectioned. Thin specimens, such as cultured cells, can be critical point dried and used for whole mount studies. The acceptable thickness of a specimen depends on the desire resolution, the type of imaging method (contact, scanning *etc.*), and the wavelength used.

Drying

The force generated by the receding surface of the evaporating water as it passes over the specimen can seriously destroy surface features. The surface tension forces exert to the specimen surface are in the order of $140 kg/cm^2$, can cause both volume and surface changes. Figure 21.1 shows the difference between critical point dried and air dried anther epidermal cell of *Zea mays*. The distortion can be reduced by replacing the tissue water with a low surface tension solvent, such as ethanol or propylene oxide. The specimen distortion can be further minimized by processing the specimen via a critical-point drying process.

(a) Critical-point drying: If a liquid is placed in a seal container and the container is heated, the vapor pressure will increase, hence the

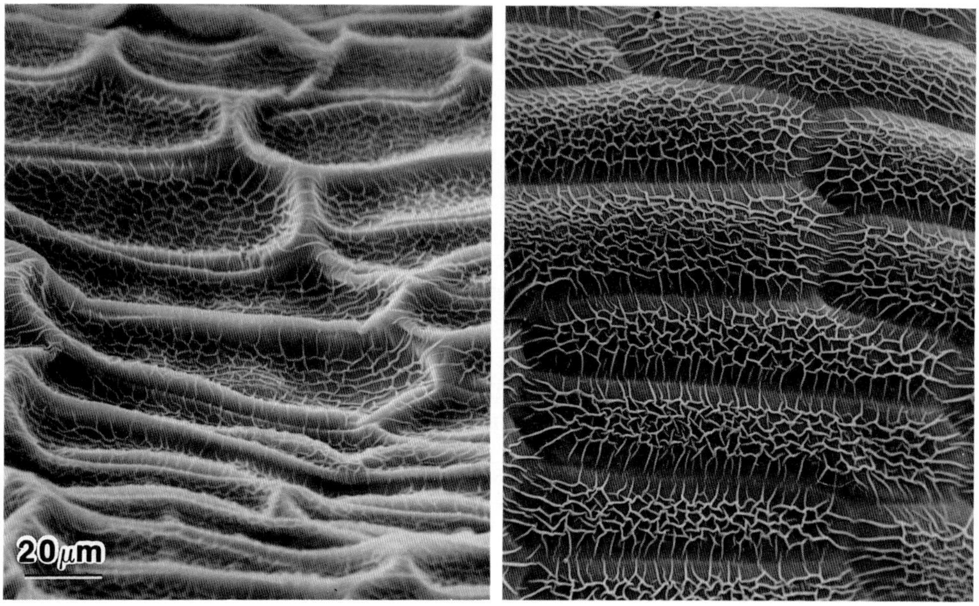

Figure 21.1. Air dried (a) and critical point dried (b) maize anther. Note air drying causes all the epidermal cells to collapse.

density of the vapor. The thermal expansion of the liquid causes the density of the liquid to decrease. There comes a point, where the density of the gas phase and liquid phase coincide. In other words, above such point the liquid phase becomes indistinguishable from the gas phase, hence the phase boundary disappears. The temperature and pressure while the phase boundary disappear is known as the critical temperature and critical pressure, and is generally referred as the critical point. Figure 21.2 is a phase diagram of carbon dioxide, the critical point is 31° and pressure at 75kg/cm^2. Above the critical point, the pressure can be reduced to zero (while maintaining the temperature) without the reappearance of liquid phase, hence the possibility of structure distortion by a receding liquid phase is avoided. Figure 21.2 shows the phase diagram of the critical point drying process. It is important to note that specimens processed by critical point drying are not artifact-free; specimen shrinkage can occur in various steps of the specimen preparation, such as in the process of fixation and dehydration.

(b) Freeze-drying and freeze substitution: Freeze drying offers less overall shrinkage of the specimen, however, specimens generally suffer from serious ice crystal damage. Due to the thermal conductivity and nucleation temperature of water, even with modern freezing techniques, such as slam freezing, propane jet freezing and high-pressure freezing (Moor *et al.*, 1980), the ice crystal-free region in a specimen is restricted only to the very surface layer of the sample. By applying high hydrostatic pressure (2100 bar) to the specimen during freezing, the melting point of water is lowered (from 0° C to -22°C); the transition temperature from heterogenous to homogeneous nucleation is lowered from -40°C to -90°C. The rate of nucleation and the rate of ice crystal growth are considerably reduced. The cumulation of these alterations lowers the critical freezing rate in animal tissue from about 10000 to about 200°K/sec.. Hence, high pressure freezing provides ice-crystal-free freezing to a depth of 500μm from the specimen surface (Moore *et al.,* 1980). Ice within tissue can be sublimed in vacuum. One should aware that the recrystalization occurs above -130°C.

Commercial apparatus such as Balzer Freeze-drying unit (FDU 001) could be used for freeze-drying operation. Freeze-drying of cultured cells supported on a Formvar-coater grid is quite simple. Cell-bearing grids are rapidly frozen in a liquid-solid N_2 cocktail (liquid nitrogen at melting temperature), and then placed in a multi-grid holder placed on the cooling stage of a freeze-drying machine. The cells are then freeze-dried under vacuum. In the case of freeze-substitution, frozen grids can be placed in low temperature methanol for 2 days, and the samples then critical-point dried with CO_2. The major technical difficulty is that the supporting film tends to break

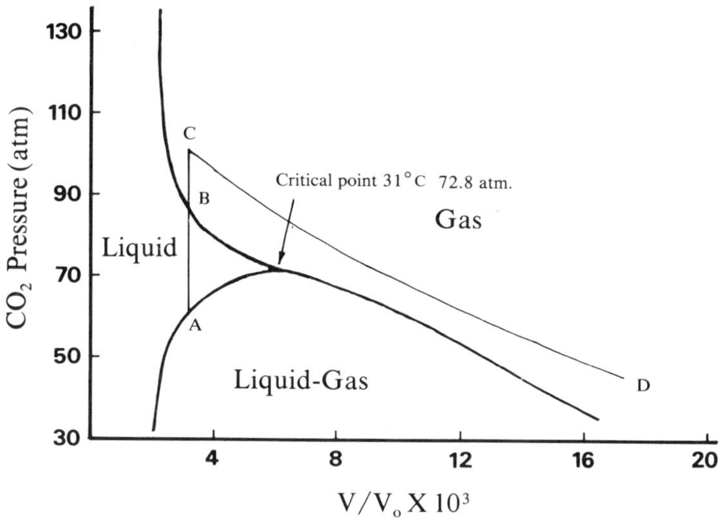

Figure 21.2. Phase diagram of carbon dioxide.

during the freezing process. Therefore, the yield of useful specimens is quite low.

Tissue Sections

Due to the similarity in x-ray absorption between biological materials and most plastic embedding matrices, plastic embedded tissue sections will not provide satisfactory image contrast unless the tissue is fixed and stained with proper stains (Figure 21.3). Recent development of a resinless sectioning technique could be quite useful for the preparation of samples for x-ray microscopy. Both polyethylene glycol (PEG)(Wolosewick, 1980) and poly methyl methacrylate (PMMA) can be used as embedding media. A typical procedure using PEG embedding medium involves: Direct or perfusing fixation with fixative. A typical bone marrow fixative for prefusion, for example, consists of 2.5% glutaraldehyde, and 1.5% paraformaldehyde in 0.1M sodium cacodylate buffer, pH 7.4. Then, the tissues are dissected out and postfixed in the same fixative. After fixation, the tissue is washed in a suitable buffer (*e.g.* 0.2M sodium cacodylate buffer pH7.4), and dehydrated in EtOH, infiltrated with 1:1 ethanol/polyethylene glycol (EtOH/PEG) for 30 min. at 60°C. Finally, the tissue is embedded in a beam capsule, and cooled rapidly in liquid nitrogen. For fixing plant material, it is preferable to use high penetration fixative (such as Cheng *et al.,* 1979) to penetrate the epidermal cuticles. Our experience indicates that it is easier to section PMMA embedded plant tissue than PEG embedded materials.

Sectioning is performed on a ultramicrotome with a dry diamond knife. The sections are picked up in a drop of 40% PEG on a wire loop, and transferred to a polylysine-Formvar coated grid or window. The grids or windows are soaked in water for 1 hr to remove the PEG embedding medium, dehydrated in EtOH and critical point dried with CO_2.

Post-sectioning labeling with immunocytochemical probes can be easily performed on both PEG and PMMA embedded tissue sections. Immunocytochemical labeling on plastic (*i.e.*, LR White) embedded section is also possible (Timms, 1986).

Examples of Biological X-ray Images

A set of selected biological x-ray images is provided in this article as an example of the x-ray imaging technique. The use of cultured *Xenopus laevis* embryonic cells, human fibroblast and mouse 3T3 cells as testing materials for x-ray lithography is advantageous, because the ultrastructural features have been extensively studied and provides basis for a detailed interpretation of x-ray images. Furthermore, the *Xenopus* cells, human fibroblast and mouse 3T3 cells can be cultured on a surface made of Formvar, polystyrene or Si_3N_4 . All of these allow direct observation of the original cells by TEM. Under appropriate culture conditions, the cells spread out onto the cultural substratum with cell thicknesses ranging from 10μm in the nuclear region to a micron or less at the margin. This variation in specimen thickness provides an opportunity to study the effect of sample thickness on image quality.

Images of Aldehyde-osmium Fixed Tissue and Cells

Aldehyde-osmium fixation is the common method for the preparation of electron microscopy specimens. Such specimens generally produce high-contrast x-ray images. The high image contrast is primarily due to the high x-ray absorption of the osmium-containing specimen. Figure 21.3 is an image of the anther tissue of *Caltha palustris*, L., fixed in Cheng's fixative (Cheng *et al.,* 1979) and post-fixed in OsO_4, embedded in Spurr's medium and sectioned (2000Å). Figure 21.4 is x-ray micrographs of *Xenopus laevis* myotomal cell. In Figure 21.4a, organelles can be clearly identified, such as the nucleus (N), and the nucleolus (No). Myofibrils are characterized by regularly spaced sarcomeres in which both the Z line (Z) and A-I bands are clearly visible (Figure 21.4b). The resulting x-ray micrograph is an image formed by superimposing all the cellular compartments. In light microscopy, observers tend to use the focusing mechanism to focus the microscope across the thickness of the specimen, and the three dimensional structural information is reconstructed by the human brain. However, in x-ray contact microscopy, it is clear that a three dimensional imaging technique is required, because the nature of the imaging technique allows neither real time observation nor focusing on different planes of the specimen. Instead, every structure is projected onto the resist to form a complicated projection image.

In Figure 21.5, a depression on the resist (D) is found around the nucleus (N). This depression is sometimes mistakenly interpreted as being an x-ray translucent zone by various authors, however, after careful study of the physical properties of contact imaging, it is believe

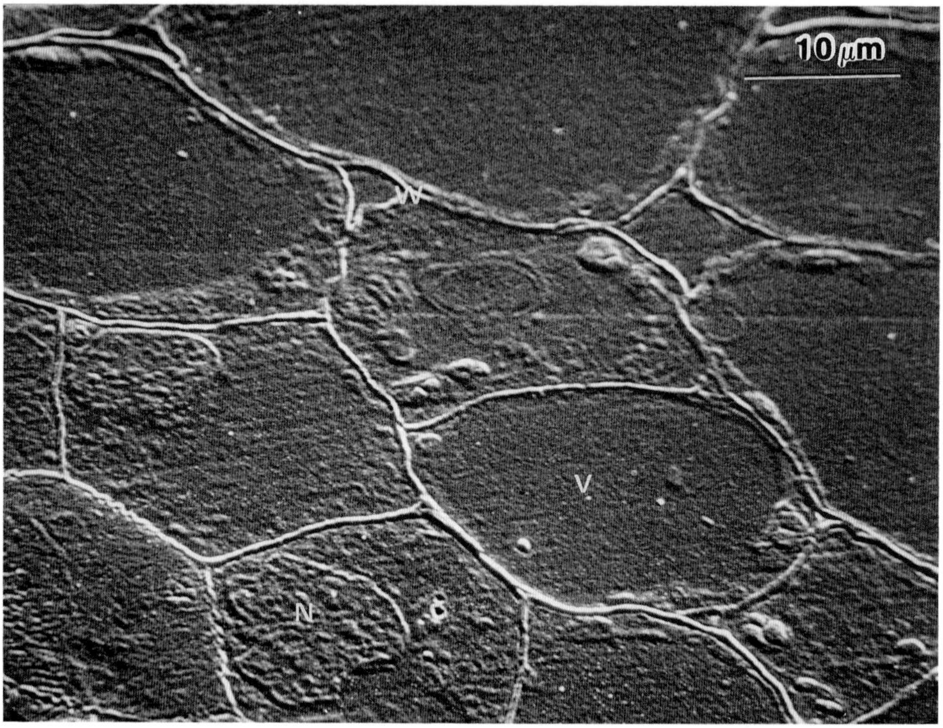

Figure 21.3. X-ray contact image of the anther tissue of *Caltha palustris* L.. The tissue was fixed in Cheng's fixative (Cheng *et al.*, 1979), post-fixed in OsO_4, and embedded in Spurr's medium.

that such an image is due to the Fresnel-type diffraction of the edge of the nucleus.

Images of Aldehyde Fixed Cells

For the aldehyde-fixed cells, the image contrast is produced by the difference of photo absorption cross section of the low Z elements of the biological specimens. If the specimen is embedded in an epoxy matrix, very low image contrast can be obtained, since the difference between protein absorption and matrix absorption is very small. However, a glutaraldehyde- fixed, critical point dried cell should pro-

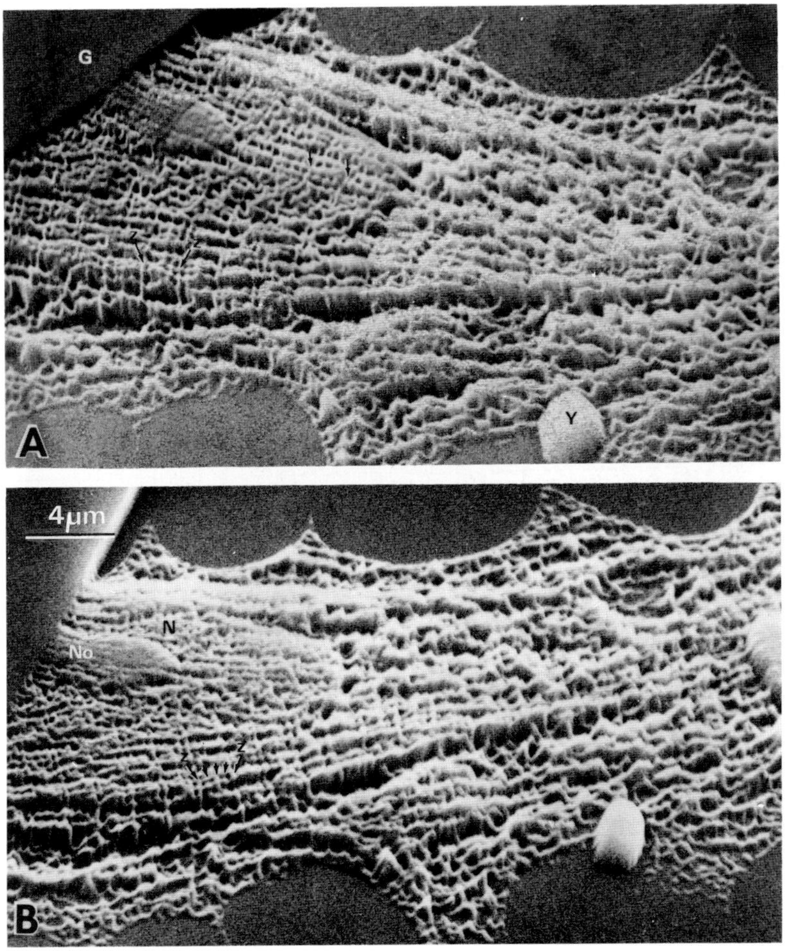

Figure 21.4. SEM-magnified x-ray contact images of *Xenopus laevis* muscle cell. A: SEM-magnified x-ray contact image of an osmium- fixed *Xenopus laevis* muscle cell. Note the regularly spaced Z lines of the sarcomeres. Within the cell nucleus, superimposed myofibrils are evident (arrows). However, due to the viewing position, some of the sarcomeres (arrows) are not easily recognized in photo B. Note the high x-ray absorption of the nucleolus. C-K image. G: EM grid, Y: yolk granule. B: The same resist as shown in A, but viewed at a different angle. Between Z lines, three x-ray dense bands can be observed (short arrows). Note the images of the nucleolus (No) and nucleus are superimposed with myofibril's image.

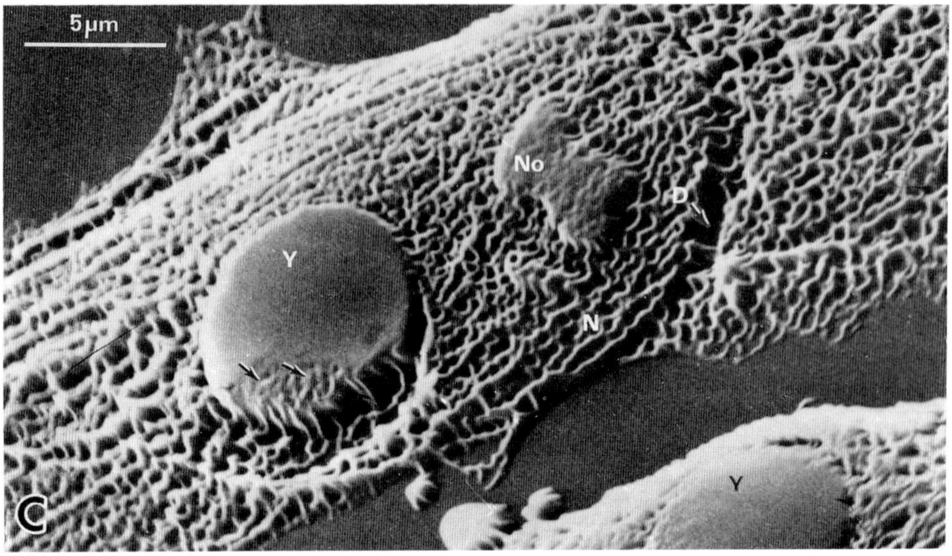

Figure 21.5. SEM magnified x-ray contact image of osmium-fixed *Xenopus laevis* myotomal muscle cell. Note the high x-ray absorption of the yolk granule (Y). The resist has been developed for a relatively long period.

vide adequate image contrast, because the cell is not embedded in any matrix. Figure 21.6 shows such an image taken with th Göttingen imaging microscope. The image is the nuclear region of a critical point dried human fibroblast.

Figure 21.7 shows a set of muscle cell images taken by C-K and V-L radiations. The nucleus and nucleolus are clearly visible in the C-K image (Figure 21.7a), However, in contrast to the V-L image (Figure 21.7b), the x-ray density of the nucleus is much higher, therefore, very few details can be observed. Figure 21.7c shows the corresponding TEM image of the same cell. Note the high electron scattering of cell nucleus (due to the thickness of the nucleus). The

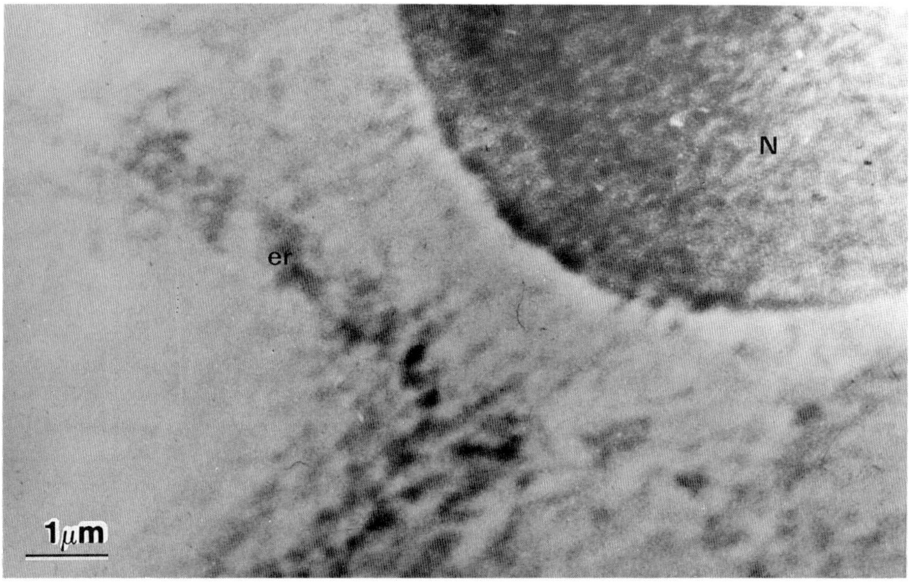

Figure 21.6. X-ray image of a human fibroblast obtained with the Göttingen imaging x-ray microscope at 45Å. The cell was fixed in glutaraldehyde and critical point dried. (W. Meyer-Ilse, G. Nyakatura, P. Guttmann, B. Niemann, D. Rudolph, G. Schmahl and P. C. Cheng)

three clear zones (L) in the cytoplasm represent the locations of lipid granules which were removed during specimen preparation.

Between Figures 21.7a, 21.7b and 21.7c, there are small disagreements on the positions of certain cytoplasmic structures which are marked by white square dots in Figure 21.7b. The relocation of cellular structures (primarily yolk granules in this case) is the result of the reloading process. It is essential to obtain intimate contact between specimen and resist, but the tradeoff is that the cell could be disturbed or even damaged by physical contact. This is especially serious when

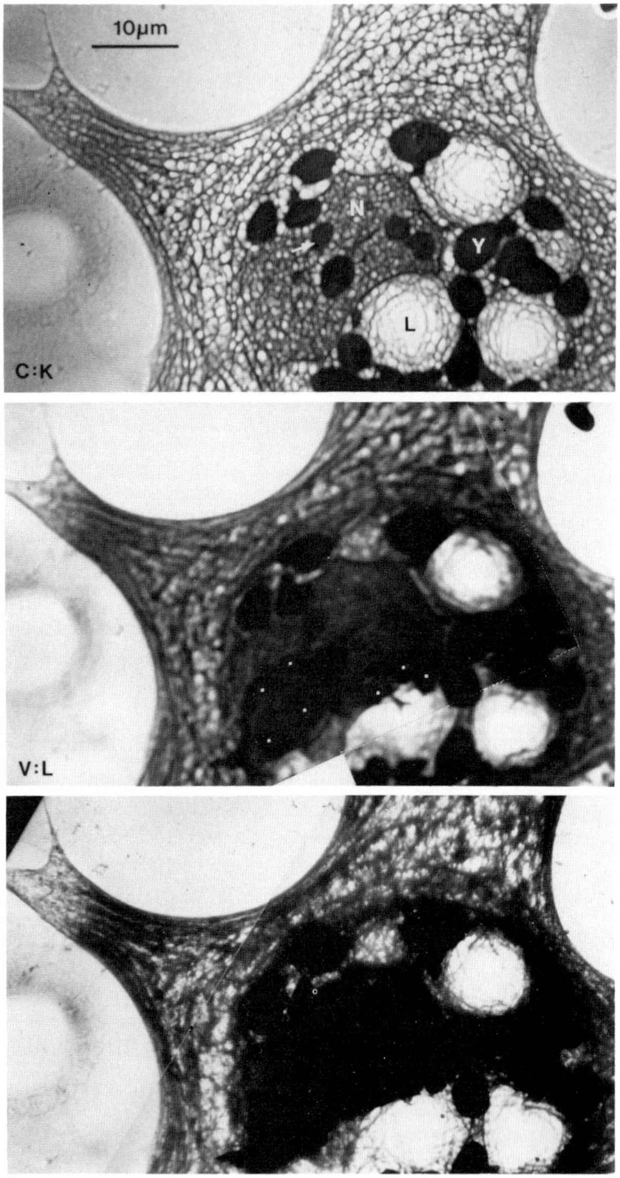

Figure 21.7. A *Xenopus laevis* muscle cell, cultured on the surface of Formvar film (~ 1500Å), fixed in 1% glutaraldehyde, dehydrated in EtOH, and critical-point dried in CO_2. Upper photo: Carbon K image on 1.2μm photo resist (PMMA) on Si_3N_4. Middle photo: Same cell, but imaged by V: L radiation. Lower photo: Electron image of the original cell. Y: yolk granule, N: nucleus, L: lipid granule (removed during dehydration process).

attempting stereo imaging and above- and below-edge imaging and image subtraction, because the above- and below-edge images are out of registration for further subtraction.

Effect of Heavy Metal Fixation and Staining

Osmium Fixation

Osmium fixation is the most common method in the specimen preparation for electron microscopy; it provides preservation of lipid structures and also increases the degree of electron scattering of the specimen. Generally, tissues are pre- fixed in aldehyde fixative, and post-fixed in an osmium tetroxide solution. Since the osmium tetroxide crystal sublimes readily at room temperature, tissues can also be fixed by osmium vapor. Since the vapor fixation technique does not involved rehydrating and re-dehydrating of the specimen under study, potential structural changes can be kept at minimum. Therefore, the vapor fixation technique provides a rare opportunity for us to study the effect of osmium fixation on the specimen. Figure 21.8 shows the nucleus of a mouse 3T3 fibroblast before and after osmium vapor fixation.

Polychromatic Images

Figure 21.9 shows a *Xenopus laevis* myotomal muscle cell imaged by Al filtered synchrotron radiation (zero order from the CSRF Grasshopper monochromator). The contact image was formed on a free-standing PMMA film back-supported by a thin layer of carbon. Therefore, the slight fuzziness of the image is primarily due to the drifting of the free-standing resist under TEM observation. The insert is the phase contrast image of the original cell. Note the high x-ray absorption of cell nucleus (N) and yolk granule (Y). The high x- ray absorption is expected because of the high absorbance of most of the

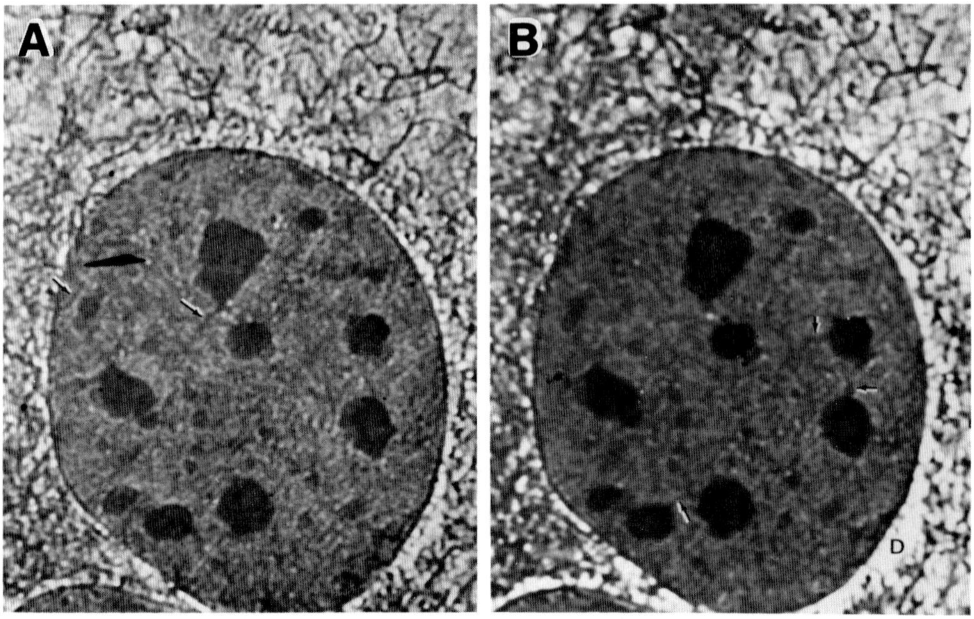

Figure 21.8. Image of the nucleus of 3T3 fibroblast before and after osmium fixation. A: Before osmium fixation. The cell was fixed in 1% glutaraldehyde, dehydrated in EtOH, and critical point dried. The x-ray contact image was obtained by using carbon K radiation. Note the x-ray dense nucleolus (No), arrows indicate structures not visible in the osmium fixed image (B). B: After osmium fixation. The above mentioned specimen was osmium fixed in OsO_4 vapor for 3 min., and the contact image was obtained at exactly the same conditions as those used in A. Note the overall image contrast is increased. Arrow shows structure whose density is greatly increased compared to what it was in A. Three- dimensional image-pair analysis suggested that this structure is likely to be the nucleoema of the nucleolus. Around the edge of the nucleus (N), a bright halo (D) is observed. This is the result of Fresnel type diffraction.

biological material in the long wavelength region of the soft x-ray spectrum. Arrows indicate bundles of stress fibers.

Monochromatic Images

Experimental runs using monochromatic synchrotron radiation have been conducted at the Tantalus storage ring, Wisconsin 1GeV storage ring and at the Brookhaven VUV ring. Papers reporting some of the Tantalus results have been published (Tan *et al.*, 1984; Cheng *et al.*, 1984; 1985; 1986).

Elementary analysis by x-ray absorption has been proposed by various authors (Cosslett and Nixon, 1960; Engström, 1962; McGowan and Malachowski, 1980). One classic example of such work demonstrated the calcium distribution of a calcified aorta by the differences between above- and below calcium K edge (Engström, 1966). Research has been done on the determination of calcium and phosphorus content in Haversian systems (Lindström, 1955). Soft x-rays have also been used for the determination of the amount of dry substance per unit area or unit volume of certain tissues (Nurnberger *et al.*, 1952; Dahl *et al.*, 1957; Omnell *et al.*, 1960). All the above mentioned work was performed using a silver-based emulsion as the photo detector. Therefore, the resolution is limited by the size of the silver grains.

Based on theoretical calculations of absorption spectra of biological materials, we have begun a series of investigations on imaging identical cells at the two energies just above and below a certain elemental absorption edge, and a preliminary attempt on above and below P-L edge imaging was published (Cheng *et al*, 1986). However, specimen damage due to loading and reloading of the photo resist presents a serious problem for subsequent registration of the two im-

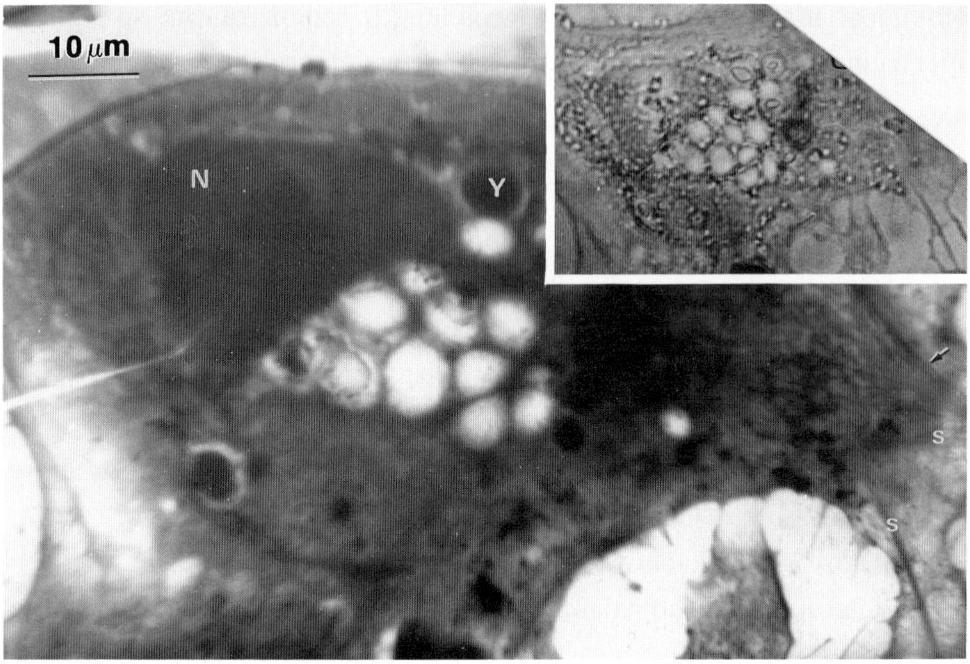

Figure 21.9. Images taken with polychromatic light. TEM-magnified x-ray contact image of *Xenopus laevis* cultured cell.
The contact image was formed on a free standing PMMA film (1.5μm) with white light synchrotron radiation (with a Al filter in the beam). Insert: phase contrast image of the original cell.
Y: yolk granule, N: nucleus, S and arrows: stress fiber.

ages (above and below absorption edge). The intensity variation of the synchrotron radiation beamline post further complication on the microchemical mapping experiment. Unless a complicated apparatus is used to ensure a even exposure to the specimen, it is technically difficult to ensure the two images (above and below an absorption edge) are exposed to the same position of the beam, hence, exposed to the same dose of x-ray radiation. A simple approach to achieve a uniform x-ray source is the use of a pinhole. However, this approach sacrifice the needed x-ray intensity.

Generally, monochromatic x-ray radiation is obtained by using grating monochromators (*e.g.*, Grasshopper or TGM), the inherit high order radiation contamination post another serious problem for quantitative work. For instance, a "monochromatic" x-ray beam (*e.g.* 90Å) is contaminated by 20% 2nd order radiation of shorter wavelength (45Å). If a biological image was taken with such x-ray beam, the resulting x-ray image will have a lower x-ray density than reality. This is because carbon based material has relatively low absorption at the 45Å radiation, but very high absorption at 90Å wavelength. Therefore, most of the radiation with the wavelength of 90Å will be absorbed by the specimen, while allowing significant amounts of 45Å radiation to pass through. The resulting image is a mixture of the 45Å and 90Å radiation with the former contributing much higher than 20%. Typical example is the $P - L_{II,III}$ absorption edges which are very close to the 2nd order of carbon K edge (Cheng, 1985). Methods to minimize the high order contamination problem are the use of filters and the use of a lower energy storage ring, so that x-ray emission at high energy (*i.e.* 45Å) is very low.

In addition to the technical difficulties, the mass fraction of certain elements in tissue, such as calcium in physiological concentration, is too low to be detected by x-ray microscopy (see calculation by Nagai *et al.*, in this volume). Therefore, by summarizing all the technical difficulties, even if it is possible to subtract the two images, it may not be practical for routine biological application. However, the wavelength-dependency of image contrast is quite useful for x-ray imaging, a suitable wavelength can be chosen to suit a particular specimen. X-rays with wavelength just longer than the carbon carbon K edge is quite useful to image thick specimens. Shorter wavelength (<15Å) can also be used. Where, for very thin specimens (critical point dried specimen or specimen with embedding matrix), x-rays with wavelengths at the maximum absorption of carbon can be used to maximize the image contrast.

Surface Replicas

Due to the sensitivity of PMMA resist to electron beam bombardment (Cheng, 1985; Cheng *et al.*, 1986; Erasmus, 1987; Shinozaki and Robertson, this volume), surface replica of the contact image on the surface of the photo resist provides the highest resolution. In interpretation of this type of image, one has to bear in mind that, due to the low angle metal shadowing, information is lost on the side of a high profile facing away from the metal source. Furthermore, because of the difficulties on precisely control the angle of shadowing, internal standard is useful for estimating feature height. Polystyrene beads with known size can be sprayed on the developed resist prior to metal shadowing, so that the shadow angle can be calculated from the shadow length of the known beads. Figure 21.10 shows a typical replica image obtained from a x-ray contact image on PMMA resist. The image is the cortex region of a human fibroblast.

Conclusion

It is not possible to provide a complete coverage on the specimen preparation techniques used today, in stead, this article is only intend to point out the vast amount technology which available to us. For further information, the readers should turn to books on histological techniques and electron microscopy. The examples of x-ray images described in this article should provide general outlines what x-ray microscopy can offer to biology at the present time. It is up to the biologists to find applications for x-ray microscopy. Examples on biological applications can be found through out the references at the

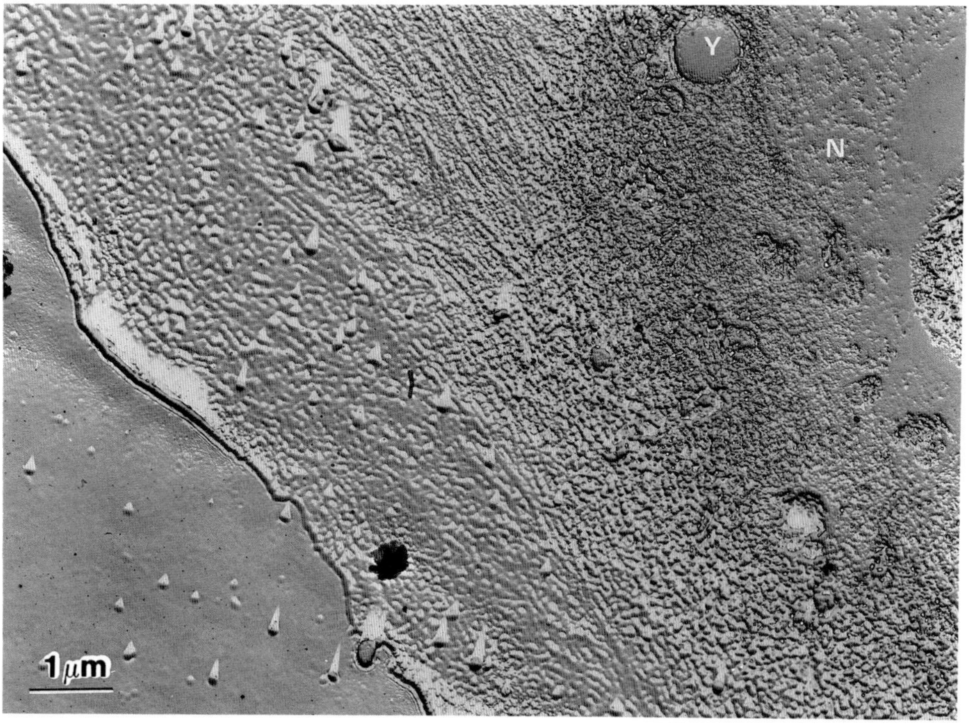

Figure 21.10. Metal replica of a contact image on PMMA resist. (osmium-fixed *laevis* myotomal cell).

end ot this proceedings. Although, x-ray contact images are described in this article, the general specimen preparation techniques should be applicable to all types of x-ray microscopy. X-ray imaging microscopy (Meyer-Ilse *et al.,* this volume), which has achieved diffraction-limited performance, should becomes a new tool for cell biology. Typical images produced by the imaging microscope are shown in Figure 3.3, in the article by Meyer-Ilse *et al.*, and in the Mini Atlas at the end of this proceedings. In conclusion, the development of x-ray microscopy is still in its infancy, not only the instrumentation needs further improvement by physicists, more important, the biological communities should develop suitable specimen preparation protocol to utilize the emerging technology. Cooperative efforts would be essential between physicists and biologists to further develop x-ray microscopy into a useful tool.

Acknowledgements

Parts of this work have been supported by NSERC and MRC of Canada. I wish to express my thanks to the Department of Anatomy, University of Illinois at the Medical Center for providing biological specimen preparation facilities, the Canadian Synchrotron Radiation Facility for providing beamline facilities, and Dr. D. Wright of the Brooklyn Botanic Garden Research Center for critical reading of this manuscript.

22. Studies of Calcium Distribution in Bone by Scanning X-ray Microscopy

F. Cinotti[1,4], M.C. Voisin[2], C. Jacobsen[3], J.M. Kenney[3,5], J. Kirz[3], I. McNulty[3], H. Rarback[1], R. Rosser[3] and D. Shu[1]

[1]National Synchrotron Light Source, Brookhaven National Laboratory, Upton, New York, 11973 USA

[2]Departement de pathologie tissulaire et cellulaire, CHU Henri Mondor, 94010 Creteil, France

[3]J.S. Guggenheim fellow, Physics Department, State University of New York, Stony Brook, New York, 11794 USA

[4]Present address: 16 rue de l'Esperance, 94000 Creteil, France

[5]Present address: Physics Department, King's College, University of London, London, WC2 R2LS, UK

Introduction

Physiological and morphological changes in bone tissue are poorly understood phenomena associated with the aging process. A decrease in bone volume and a reduction in cellular activity are most commonly noted. For research, for diagnostic purposes, and eventually to find and evaluate specific forms of treatment there is a need to be able to study the distribution of calcium in bone tissue.

A variety of techniques have been used to study calcium in bone, and to gain insight into the histopathology of metabolic bone desease. At the macroscopic level these include neutron activation analysis (Harrison *et al.*, 1975) and photon absorptiometry (Cameron *et al.*,

X-ray Microscopy
Ed. by P.C. Cheng and G.J. Jan

1968), while histomorphometry (Meunier *et al.*, 1973, Meunier *et al.*, 1979), x-ray diffraction (Smith and Smith, 1976) and IR spectrophotometry (Baud *et al.*, 1976) have been used for studies at the microscopic level. While some of these methods (e.g. neutron activation) provide an accurate, but only a global measure of the calcium content of the skeleton, histomorphometry is used to measure the percentage of mineralized tissue in trabecular bone at the microscopic level, with a resolution of a few micrometers.

Soft x-ray microscopy provides a quantitative measure of calcium distribution with a resolution of 200nm without visible specimen damage (Kirz, 1980). Concentrations as low as $2\mu g/cm^2$ have been detected in sections of non - decalcified human bone (Kenney *et al.*, 1985).

In this study we have analyzed four sets of specimens with known pathological diagnosis in order to study the correlations between the condition of the bone and our measurements of the calcium distribution.

Experimental Method

The scanning x-ray microscope uses monochromatized synchrotron radiation from the 750MeV electron storage ring at the National Synchrotron Light Source. The beam of soft x-rays is focused to a small microprobe by a Fresnel zone plate (Rarback, 1983; Rarback *et al.*, 1984). The specimen is mechanically scanned with this microprobe, and a proportional counter detects the x- rays that pass through it. The output of the counter is used to form a digitized image in our minicomputer. The resolution of the microscope is determined by the size of the microprobe and the step size used in the scan, and has been shown to be less than 200nm (Kenney *et al.*, 1985).

The specimen, a thin section of bone tissue is at atmospheric pressure. Contrast in the image is provided by variations in the fraction of the x-rays absorbed by the specimen. The absorptivity of calcium undergoes abrupt changes as a function of wavelength in the vicinity of 3.5nm: Absorption is high near 3.55nm, much lower at 3.58nm. The absorptivity of other elements in the sample does not vary appreciably over such a small change in wavelength. We form an image at each of these wavelengths, and determine the calcium content pixel by pixel by comparing these two digital images (Kenney *et al.*, 1985; Kenney 1985).

We studied iliac bone specimens of four patients, suffering from diffuse radiological demineralization. The samples were fixed in buffered formaldehyde (pH = 7), rinsed in 0.1M cacodylate buffer, post-fixed in gluteraldehyde without post- fixation in OsO_4, dehydrated and embedded in Epon 812 in the standard way for transmission electron microscope (TEM) analysis. Sections 0.2μm thick were cut with a diamond knife on an ultramicrotome, supported on copper (electron microscope) grids. No stains were used on the sections.

Tissue from each specimen was also embedded in methylmetacrylate medium. Sections 5μm thick were cut with a Jung K microtome, stained with toluidine blue at pH = 6.8 and subjected to manual histomorphometry with a Zeiss ocular integrator.

Specimen #1 obtained from a 61 year old man suffering from diffuse skeletal demineralization with femoral fracture. Histophotometry confirmed the diagnosis of severe osteoporosis without hyperosteoidosis and with significantly depressed cellular activity.

Specimen #2 was taken from a 91 year old woman with radiological bone demineralization, collapsed vertebrae, and radiological modification of the right iliac bone, where Paget's disease was suspected. In this case histophotometric analysis of the left iliac bone biopsy showed a significantly lowered bone density along with an increase in osteoid bone (7% of the total bone volume), with quite normal osteoclastic activity, and with twice the normal osteoblastic activity.

Specimen #3 was withdrawn from an 88 year old woman with radiological demineralization and collapsed vertebrae. Measurements on this sample yielded normal bone density, unaltered mineralization, and with extensive osteoclastic activity upon 14.5% of the total trabercular surface, suggesting hyperparathyroidism.

Specimens #1, #2, and #3 were necropsy material from the antero-superior spine of the iliac crest.

Specimen #4 was biopsy material taken from a 33 year old woman suffering from diffuse skeletal demineralization and had blue sclerae as observed in osteogenesis imperfecta tarda. Histological osteoporosis was confirmed by histomorphometry. The sample showed reduced bone density without hyperosteoidosis, and with total cellular inactivity (see also L. G. Sainte-Marie *et al.*, 1984).

For each of the four specimens we made at least six calcium maps. To reduce fluctuations in the data, nine point median filtered images were generated. The images included holes due to the sectioning process and at locations corresponding to burried osteocytes. There are also areas obscured by the supporting grid bars. To eliminate from the analysis these apparently calcium-free regions, areas in the specimen with less than 5% of the maximum calcium density were isolated and removed. The remaining areas were used to determine the average calcium concentration, and were also subjected to more detailed analysis.

Results

Data concerning the global features of the images analyzed are presented in Table 22.1, and the average calcium concentration for each is also shown in Figure 22.2. Although there is considerable variation from image to image for the same specimen, the difference in calcium concentration among some of the specimens is clearly noticeable. These differences are even more clearly observed when all the data for a specimen is averaged, as seen in Table 22.2.

Table 22.1. Summary of images recorded

Specimen #1

Image #	1[a]	2	3	4	5	6	7	8
pixels	140x140	120x120	120x120	120x120	140x100	140x140	100x125	125x100
steps	0.3μm	0.6μm	0.8μm	0.8μm	0.3μm	0.3μm	0.8μm	0.75μm
Ca	2.0	3.1	4.3	4.7	4.0	1.5	5.4	3.3

Specimen #2

Image #	1	2	3	4	5	6	7[a]	8
pixels	100x100	110x110	100x100	140x140	140x140	140x90	140x140	140x140
steps	0.4μm	0.9μm	1.0μm	0.3μm	0.7μm	0.7μm	0.7μm	0.3μm
Ca	5.5	4.3	5.9	2.2	5.7	7.1	5.0	5.3

Specimen #3

Image #	1	2	3	4	5[a]	6
pixels	140x140	140x140	140x140	140x140	140x140	160x100
steps	0.5μm	0.5μm	0.5μm	0.5μm	0.7μm	0.25μm
Ca	2.3	2.6	3.3	2.7	4.1	1.9

Specimen #4

Image #	1	2	3	4	5	6[a]	7
pixels	100x100	100x100	120x120	120x120	120x120	100x100	140x140
steps	0.8μm	1.0μm	0.4μm	0.4μm	0.4μm	0.4μm	0.7μm
Ca	7.8	9.3	7.6	8.5	9.1	4.7	6.7

these images were chosen for further analysis.
average calcium concentration in calcified areas, $\mu g/cm^2$.

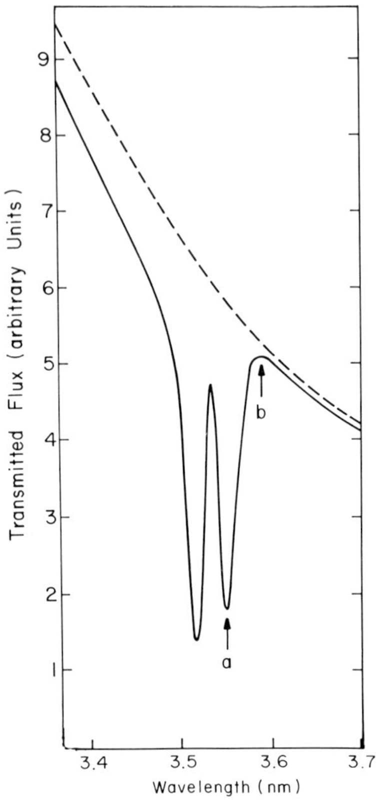

Figure 22.1. X-ray spectra near the calcium L edges as observed with our apparatus. The dashed curve was taken without the specimen, and measures the incident flux. The solid curve was obtained with a section of bone tissue in place. The arrows (a,b) point to the wavelengths chosen for imaging.

Figures 22.3 - 22.14 show a series of images and their analysis. We chose one area per specimen, and followed the same procedure for each. We shall therefore only describe the contents of Figures 22.3 - 22.5 corresponding to Specimen #1, to indicate the pattern followed in each case.

Figure 22.3a is an optical micrograph (vertical illumination) of the section mounted on a standard electron microscopy finder grid. The sketch in Figure 22.3b locates the area on this section that was sub-

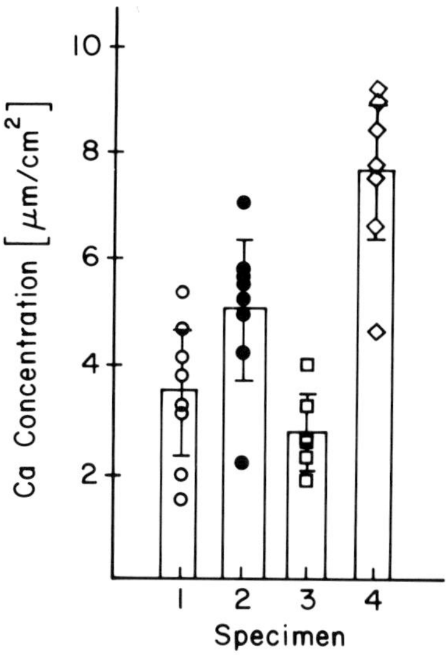

Figure 22.2. Average calcium density in the calcified areas in each of several images examined. Values corresponding to the same specimen are represented by the same symbol, and the overall avarage for a given specimen is represented by the horizontal lines. See Tables 22.1 and 22.2.

Table 22.2. Average calcium concentration ($\mu g/cm^2$)

Specimen	number of images	average concentration
1	8	3.5 ± 1.2
2	8	5.1 ± 1.3
3	6	2.8 ± 0.7
4	7	7.7 ± 1.3

jected to x-ray imaging. The x-ray micrograph of this area, taken at 3.55nm, where calcium is heavily absorbing is shown in Figure 22.4a. Different levels of specimen absorptivity are represented by the different colors, with darker shades corresponding to higher absorption. The corresponding image taken at 3.58nm is shown in Figure 22.4b. The calcium distribution based on the analysis of these images is shown in Figure 22.4c. Here lighter shades correspond to higher calcium concentration. The nine point median filtered image obtained from this calcium distribution is shown in Figure 22.5a. We marked two types of areas of interest, one group with low calcium density, and another of high density. These areas are indicated on Figures 22.5b and 22.5c.

We calculated the mean values of the calcium concentration in each of these small areas, and present the results in Table 22.3. For each type of region the dispersion is small. The average of these means is given in Table 22.4. We see that there are large variations from specimen to specimen in the difference between high density and low density areas.

Discussion

Analysis of the scanning x-ray micrographs provides detailed, quantitative information on the distribution of calcium in bone at a sub micrometer scale. Localization of the calcium is possible even in highly demineralized specimens. The results obtained show considerable variations from specimen to specimen.

What is determined in our measurements is the area density of calcium. Before we can draw conclusions concerning calcium concentration and mineralization, we must take into account possible variations in the thickness of the sections, and possible differences in specimen preparation that may lead to changes prior to the measurement. To verify that the specimens were of equal thickness, we examined them under identical tilt conditions under a scanning electron microscope. We found no differences exceeding the measurement error of 10%. Preparation procedures were similar in all four cases, with the exception that the necropsy material from patients #1 and #3 were stored in the fixative for 1 month, while that from #2 and #4 for only a few days. The extra time may have lead to some leaching of calcium into the solution resulting in some non physiological reduction in calcium concentration.

The youngest patient (#4) was found to have the highest calcium concentration as expected from the generally recognized trend of decreasing calcium concentration with age (Baud and Gossi, 1978; Wall *et al.*, 1979). This specimen also showed the largest contrast between calcium rich and calcium poor areas. This result suggests a highly non-uniform distribution of calcium in spite of a high calcium concentration in this particular case of osteogenesis imperfecta tarda. This preliminary finding needs to be confirmed and compared with calcium

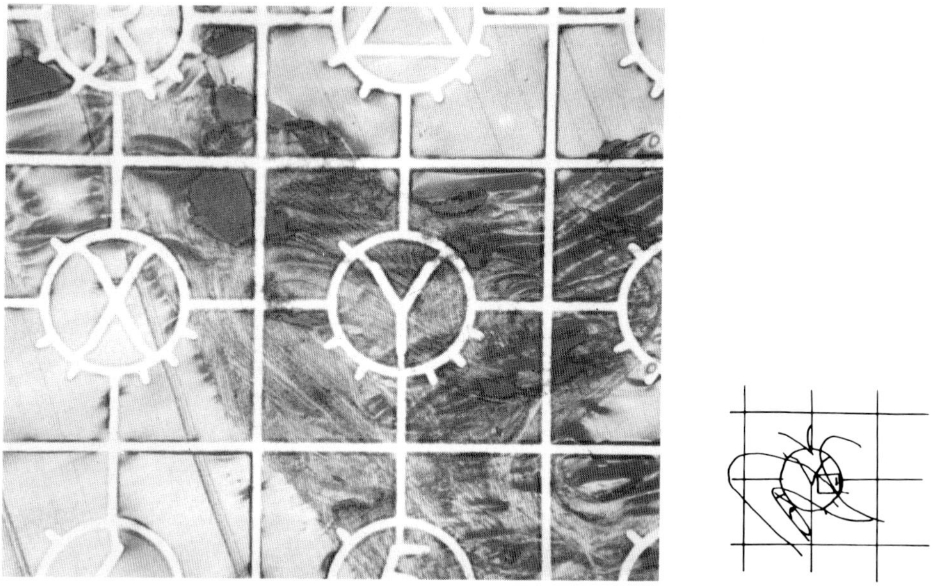

Figure 22.3. a: Optical micrograph (vertical illumination) of a section from Specimen #1 mounted on a finder grid. b: Sketch locating the area chosen for x-ray analysis.

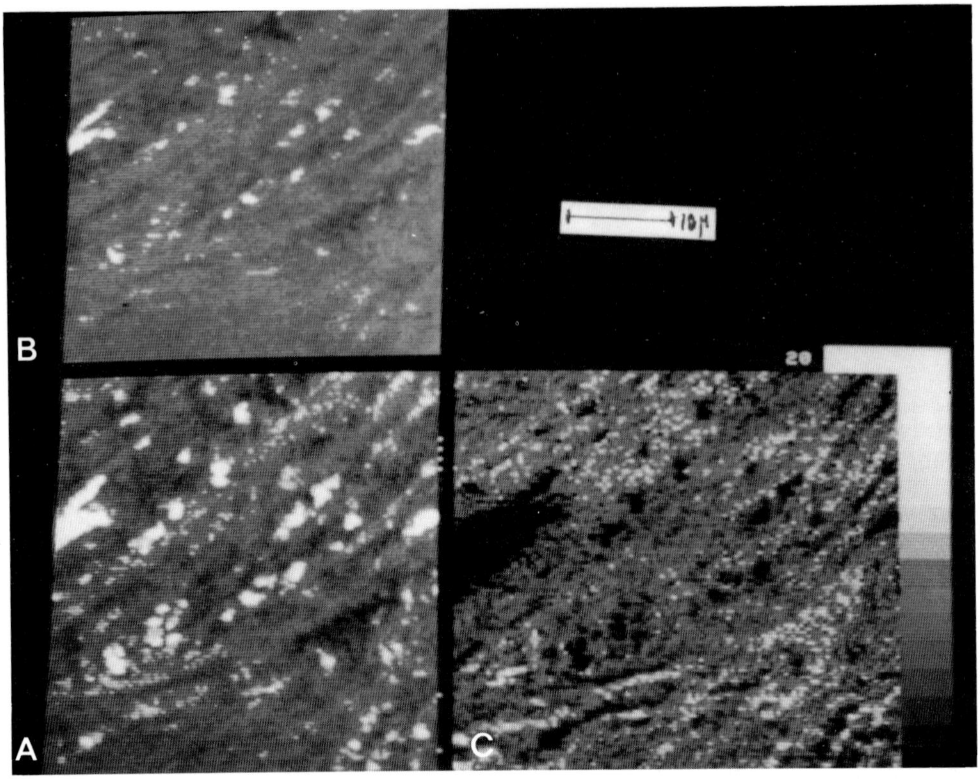

Figure 22.4. a: Scanning x-ray micrograph of the 42 x42μm^2 area indicated in Figure 22.3b (Specimen #1) taken at a wavelength of 3.55nm. Lighter shades correspond to higher transparency. b: same area imaged at 3.58nm. c: calcium distribution map determined from the data of a and b. Lighter shades correspond to higher concentrations. Lines of calcification are clearly seen. The larger black area on the left corresponds to a hole in the specimen. Smaller black areas are not osteocytes but areas of very low calcium concentration, as determined by histomorphometry.

distribution in control specimens. Nevertheless, in osteogenesis imperfecta tarda, connective tissue disorders could be associated with a modification of calcium distribution in the tissue. This could partly explain the fragility of bone observed in this disease.

Results from specimens #1 and #3 are not significantly different, although based on age and sex one might have expected to find higher

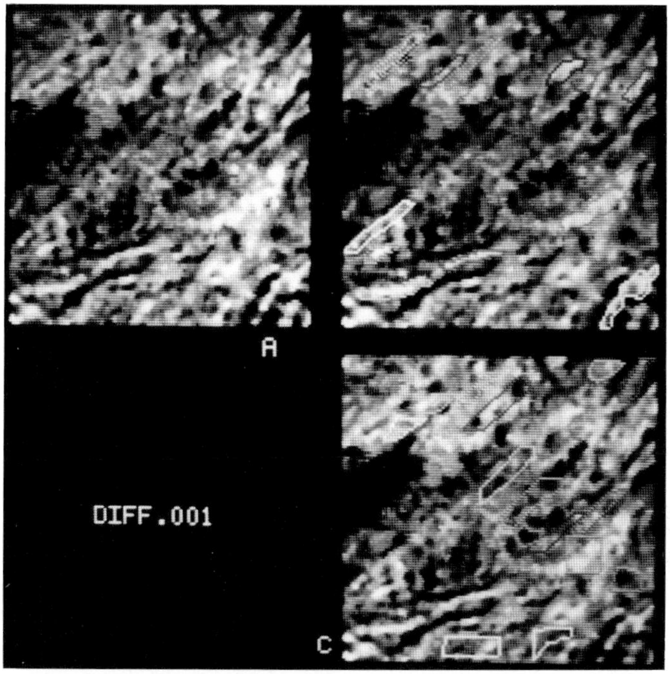

Figure 22.5. a: Nine point median filtered image derived from Figure 22.4c. (Specimen #1) b: the same image showing the high density areas chosen for analysis. c: same image showing the low density areas chosen for analysis.

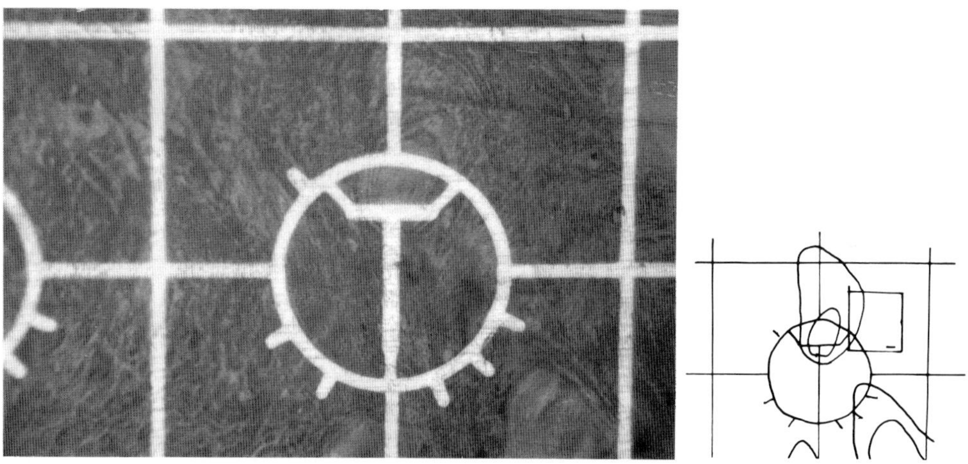

Figure 22.6. a: Optical micrograph (vertical illumination) of a section from Specimen #2 mounted on a finder grid. b: Sketch locating the area chosen for x-ray analysis.

concentrations in #1. Results of the histomorphometric examination of specimen #1, indicating a drastic decrease of bone density (while the bone density of specimen #3 was normal), might help nevertheless explain our findings.

That specimen #2, from the oldest patient in the group, shows a calcium concentration higher than those of #1 and #3 may be a preparation artifact due to the different fixation time as mentioned earlier. The larger amount of calcium in this tissue could also result from altered bone remodeling associated with the high cellular activity observed in this case.

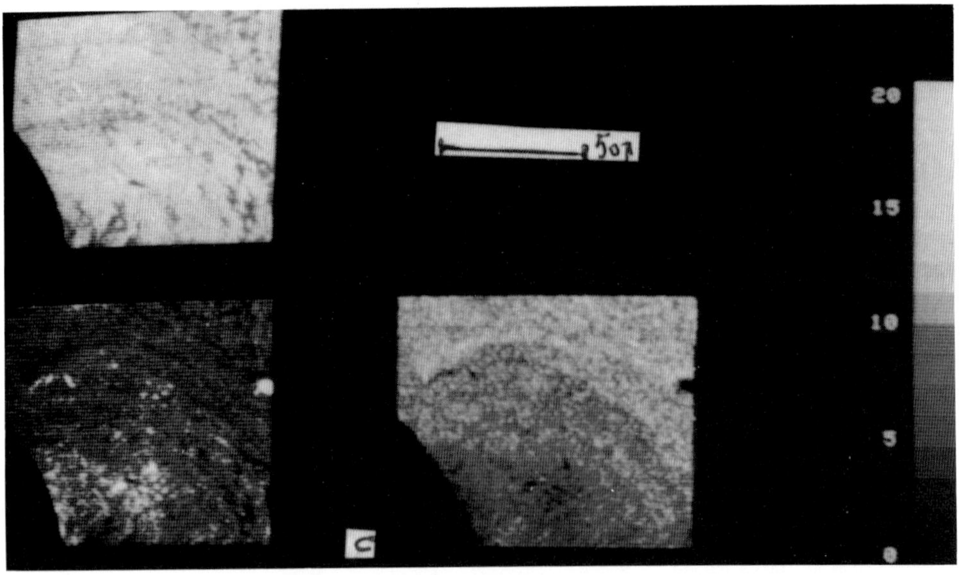

Figure 22.7. a: Scanning x-ray micrograph of the 98 x 98μm^2 area indicated in Figure 22.6b (Specimen #2) taken at a wavelength of 3.55nm. Lighter shades correspond to higher transparency. b: same area imaged at 3.58nm. c: calcium distribution map determined from the data of a and b. Lighter shades correspond to higher calcium concentrations. Two systems of calcification are present: The older one (top of the picture) is more densely calcified than the recently formed Haversian system (bottom).

Conclusions

We have attempted to take the first steps in using soft x-ray microscopy in the study of calcium distribution in mineralized tissues. This technique provides a quantitative determination of the amount of calcium present from areas as small as 0.2 x 0.2 μm^2. Under well controlled conditions the precision of the measurement is sufficient to determine the calcium concentration and distribution in bone tissue according to age, sex, and pathology. This technique could contribute to diagnosis, and give an estimation of therapeutic effects on bone mineralization in metabolic bone disease.

In view of the spatial resolution available in this form of microanalysis, the technique appears particularly well suited for the study of the correlation between cellular activity, bone matrix and mineralization on a microscopic level.

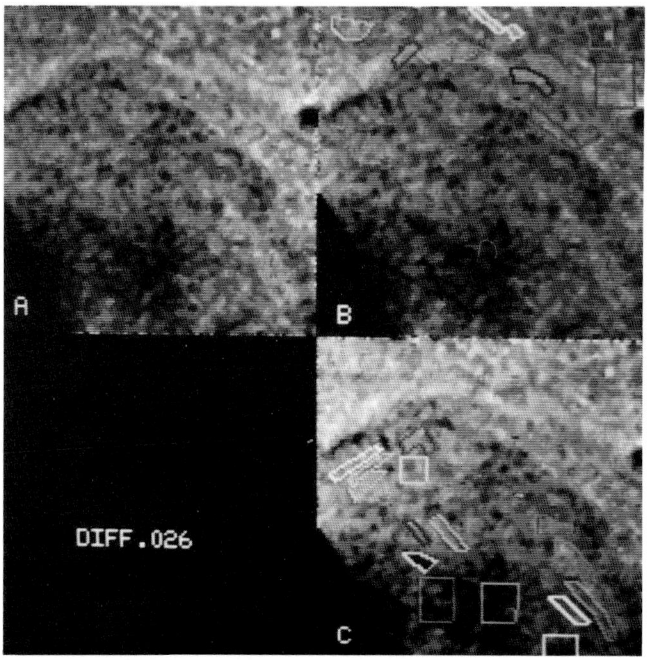

Figure 22.8. a: Nine point median filtered image derived from Figure 22.7c. (Specimen #2) b: the same image showing the high density areas chosen for analysis. c: same image showing the low density areas chosen for analysis.

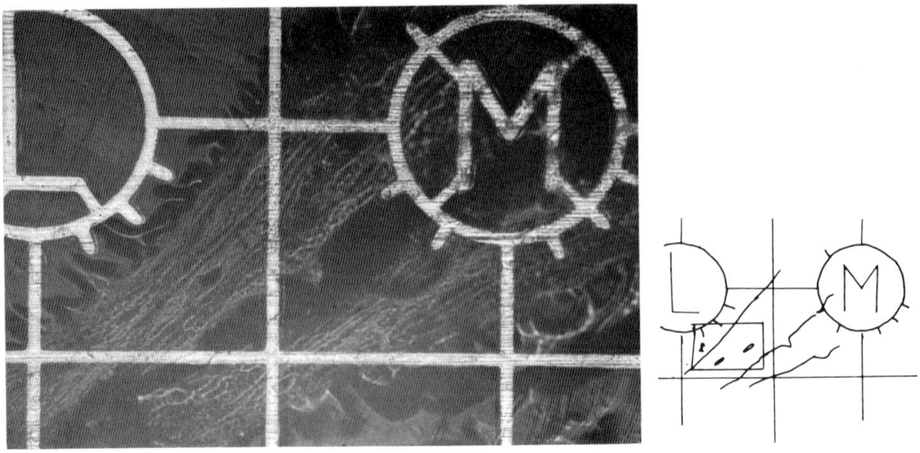

Figure 22.9. a: Optical micrograph (vertical illumination) of a section from Specimen #3 mounted on a finder grid. b: Sketch locating the area chosen for x-ray analysis.

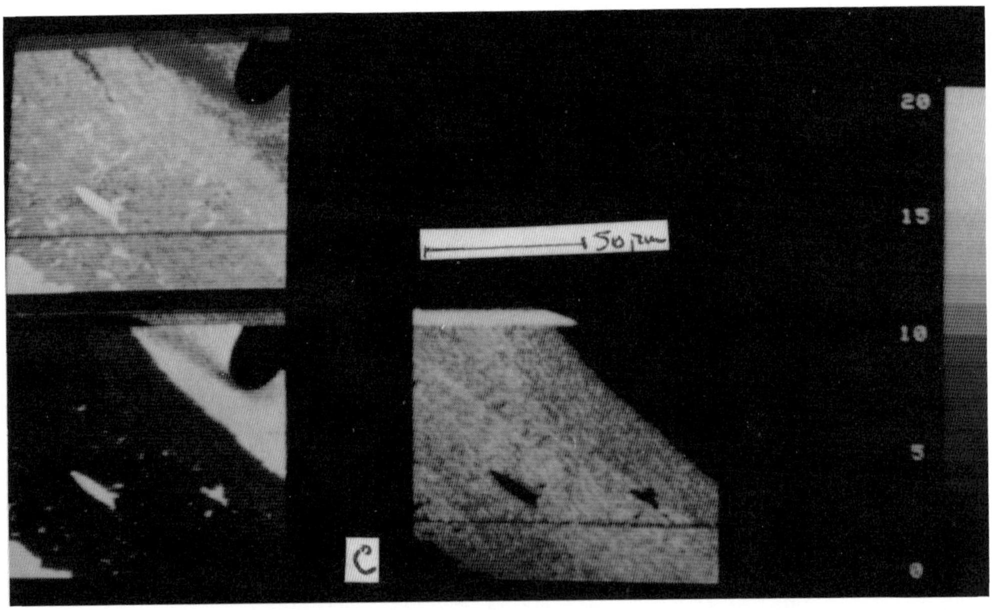

Figure 22.10. a: Scanning x-ray micrograph of the 98 x 98 μm^2 area indicated in Figure 22.9b (Specimen #3) taken at a wavelength of 3.55nm. Lighter shades correspond to higher transparency. b: same area imaged at 3.58nm. c: calcium distribution map determined from the data of a and b. Lighter shades correspond to higher concentrations. Fine lamellae are seen with decreased calcium density near the borders.

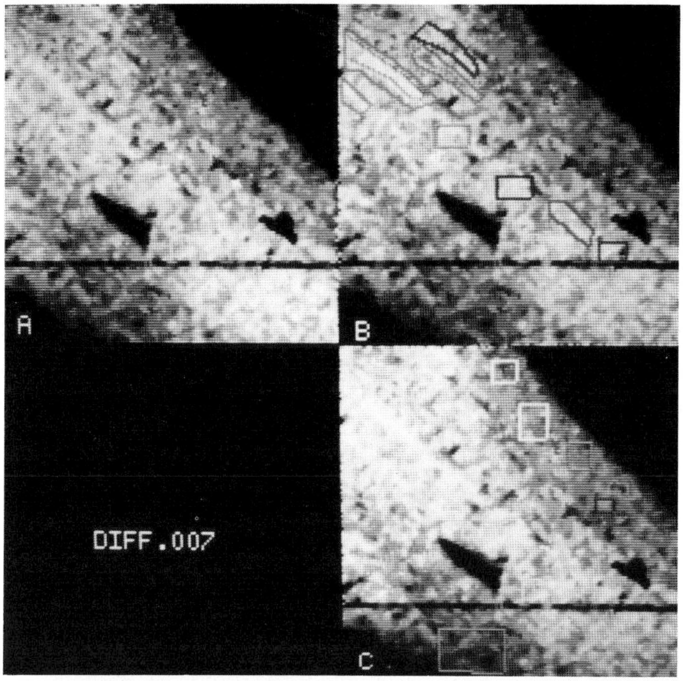

Figure 22.11. a: Nine point median filtered image derived from Figure 22.10c. (Specimen #3) b: the same image showing the high density areas chosen for analysis. c: same image showing the low density areas chosen for analysis.

Figure 22.12. a: Optical micrograph (vertical illumination) of a section from Specimen #4 mounted on a finder grid. b: Sketch locating the area, near a Haversian canal, chosen for x-ray analysis.

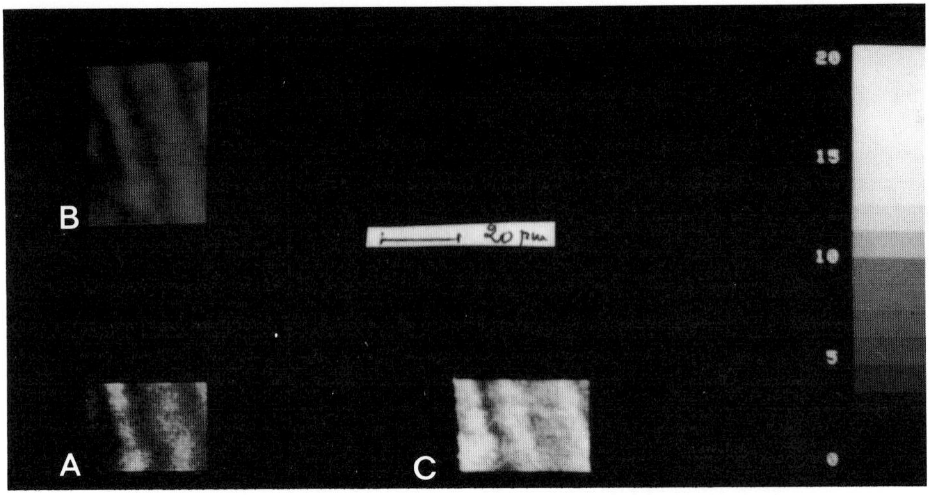

Figure 22.13. a: Scanning x-ray micrograph of the 24 x 42μm^2 area indicated in Figure 22.12b (Specimen #4) taken at a wavelength of 3.55nm. Lighter shades correspond to higher transparency. b: portion of the same area imaged at 3.58nm. c: calcium distribution map determined from the data of a and b. Lighter shades correspond to higher concentrations.

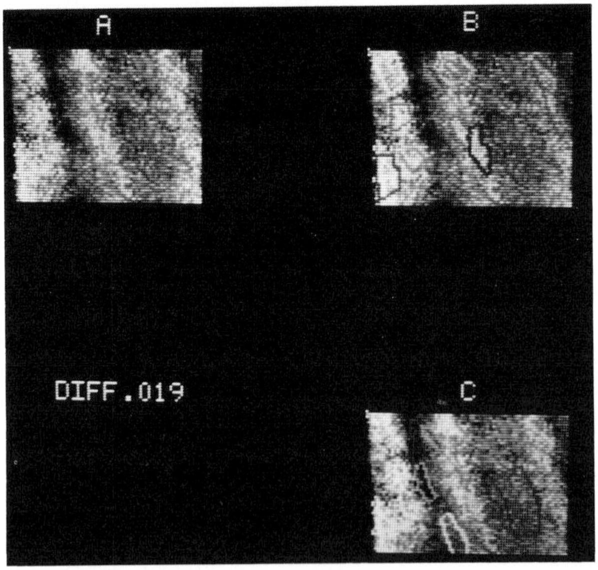

Figure 22.14. a: Unfiltered image derived from Figure 22.13c. (Specimen #4) b: the same image showing the high density areas chosen for analysis. The areas closest to the Haversian canal (11-14) have the highest calcium concentration, while the one farthest away (8) has the lowest. c: same image showing the low density areas chosen for analysis.

Table 22.3. Calcium density in selected areas ($\mu g/cm^2$)

Area#	1	2	3	4	5	6	7	8	9	10	11	12	13	14	15
Spec. hi	2.7	2.5	2.1	2.7	2.6	2.9	3.2	3.2	2.9	3.1	3.2	3.1	3.0	2.4	2.9
#1 lo	2.7	2.5	2.5	2.0	2.4	2.4	2.1	1.1	2.3	2.7	2.1	2.3	1.7	1.9	2.1
Spec. hi	6.5	6.7	6.9	6.5	6.6	6.5	6.4	6.4	5.8	6.6	6.5	6.2	6.9	6.9	7.1
#2 lo	5.4	5.1	5.5	5.3	5.1	4.9	5.2	5.5	5.2	3.1	3.3	3.0	5.1	4.8	4.2
Spec. hi	5.3	4.9	5.1	4.8	4.7	4.9	4.9	4.7	4.6						
#3 lo										3.3	4.0	3.5	3.8	4.1	3.8
Spec. hi	5.8	6.9				6.1	5.6	5.5			7.5	6.7	6.9	6.1	
#4 lo			4.1	2.4	2.7				4.3	4.3					4.3

Area numbers are as marked on Figures 22.5, 22.8, 22.11, and 22.14.

Table 22.4. Average calcium density in selected areas ($\mu g/cm^2$)

Spec #	high areas	low areas	<hi>	<lo>	<hi>-<lo>
1	15	15	2.8 ± 0.3	2.1 ± 0.4	0.7 ± 0.5
2	15	15	6.6 ± 0.3	4.7 ± 0.9	1.9 ± 0.9
3	9	6	4.9 ± 0.2	3.7 ± 0.3	1.2 ± 0.4
4	9	6	6.3 ± 0.6	3.7 ± 0.8	2.6 ± 1.0

Acknowledgments

We are grateful to the many people who contributed to this study. The zone plate used on the microscope was fabricated at IBM by D. Kern and his group. M. Di Paola of Villejuif, France contributed to the analysis of these data. Work by the Stony Brook group is supported in part by the National Science Foundation under grant #DMB 8410587. The National Synchrotron Light Source is supported by the Department of Energy under contract DEAC0276CH00016.

23. Soft X-ray Microradiography of Biological Specimens

Gwo-Jen Jan[1], Liang-Fung Chen[1], Yau-Jong Twu[2], Poh-Kung Tseng[2], Chang-Po Chen[3]

[1]Department of Electrical Engineering, National Taiwan University, Taipei, Taiwan, 10764, Republic of China

[2]Department of Physics, National Taiwan University, Taipei, Taiwan, 10764, Republic of China

[3]Institute of Zoology, Academia Sinica, Nan-Kang, Taipei, Taiwan, 10529, Republic of China

Introduction

The construction of a 1.3GeV synchrotron radiation storage ring in Taiwan (Taiwan Light Source (TLS)) began recently (Yen, 1986). The photon energy (critical energy of 0.99keV) of the TLS without insertion device will cover the near infrared, visible light, ultraviolet and vacuum ultraviolet as well as soft x-ray spectra. The synchrotron storage ring, when completed, will provide useful photons for various x-ray imaging work including x-ray microscopy. In anticipation of the commission of the synchrotron, we have built a stationary target x-ray source for some initial work on x-ray contact imaging. The stationary anode x-ray source was chosen for its simplicity and reliability. The soft x-ray source had been set up by the middle of August, 1986. After the source was tested, two biological specimens were exposed to the source.

Adult sea urchins contain a variety of mineralized tissues, these include the test, spines, teeth, jaw apparatus and in the soft tissues of the tube feet, gut and gonades. Most of the mineral in the tissue is present in the form of impure calcite rich in magnesium carbonate

X-ray Microscopy
Ed. by P. C. Cheng and G. J. Jan

(Raup, 1966). The sea urchin tooth grows continuously, and in its growing tip (plumula), calcification occurs intracellularly within the syncytial odontoblasts (Chen and Lawrence, 1986a). Therefore, the sea urchin teeth provide a good model for studying tissue calcification (Veis *et al.*, 1986; Chen and Lawrence, 1986b; 1987). However, because of the difficulty of cutting ultra-thin sections through the mineral-rich tissues, the ultrastructures between the mineral and matrix interface are poorly understood. The large penetration depth offered by soft x-rays provides a potential tool for the study of such specimens in the form of thicker sections. Therefore, thick sections of the sea-urchin tooth were chosen as a testing subject for x-ray contact imaging. In addition, mosquito's wings were also used to test our new x-ray source.

Experimental Apparatus and Setup

The stationary anode x-ray source consists of four major parts, electron guns, a power supply, a vacuum chamber and specimen holders. All of the components were made domestically. A block diagram of our soft x-ray exposure system is shown in Figure 23.1. The vacuum system is pumped by a diffusion pump which maintain a good vacuum of 10^{-6} torr. The power supply provides both high tension DC to the cathode and filament, and regulated AC for heating the filament. An ionization gauge was used for measuring and monitoring the vacuum condition of the vacuum chamber. Water cooling was used for proper cooling of the targets and the baffle plate which locates between the electron guns and specimen holders. The baffle plate prevents heating of the specimen during x-ray exposure. Figure 23.2 is a schematic of the x-ray exposure system. Figure 23.3 shows the detail design of the x-ray source which consists of a ring filament, a cylindrical cathode and a water-cooled target.

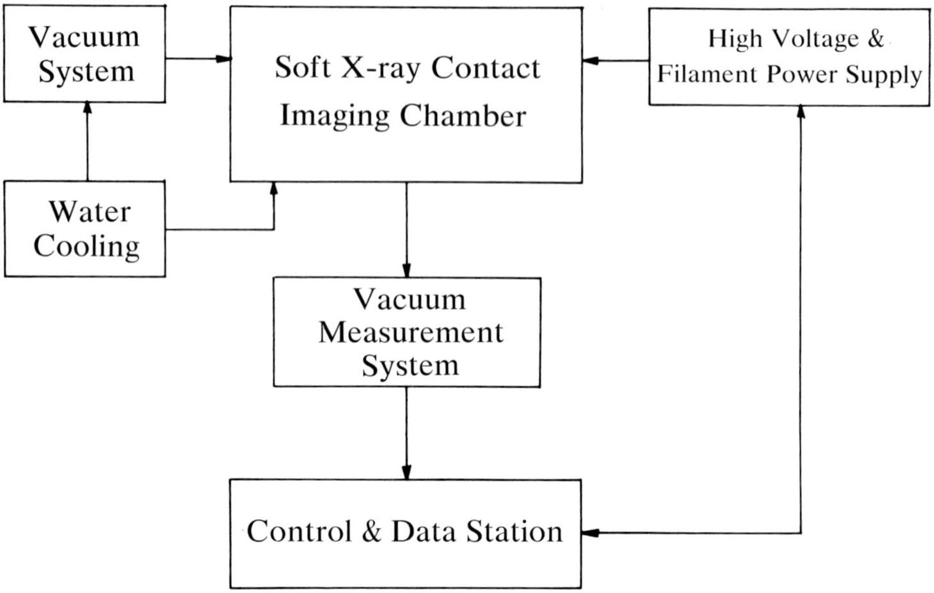

Figure 23.1. Block diagram of the stationary target soft x-ray source.

The vacuum system consists of a rotary pump and a diffusion pump. A thermocouple gauge and a Kuo ionization gauge were used to measure the vacuum conditions within the chamber. The chamber pressure is monitored by an automatic vacuum monitoring system.

The power supply of the stationary anode x-ray generator was designed and implemented in our laboratory. The detailed circuit diagram is shown in Figure 23.4. The regulated high voltage power supply provides high voltages from 0 to -10 kV. This high voltage is connected to the cylindrical cathode made of fine stainless steel screen. The ring-shaped filament is heated by a regulated AC source. The required AC voltage and current supply to the filament depend on the diameter of the filament. A 5V, 25A AC supply is currently used. The high voltage DC power supply was designed to provide a maximum current of 100mA.

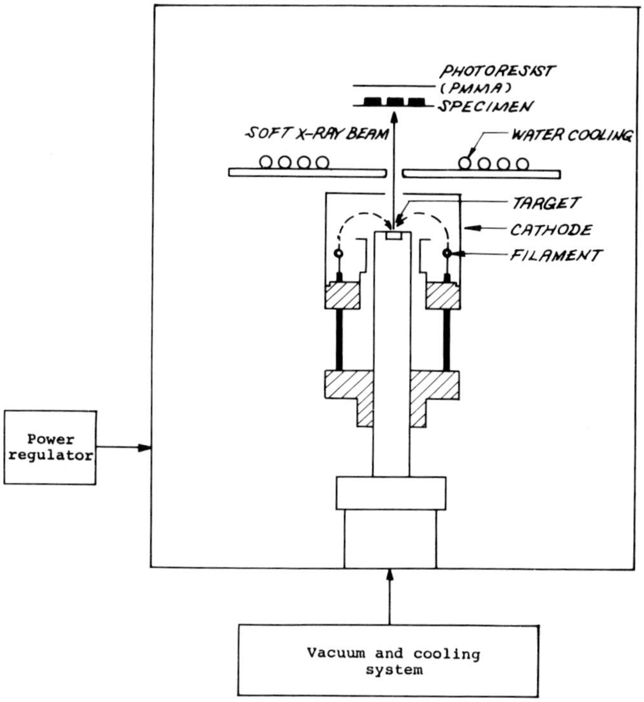

Figure 23.2. Diagrammatic representation of the soft x-ray exposure station for contact imaging.

Sample Preparation and Experimental Details

The plumula of the teeth of the sea urchin *Tripneustes gratilla* L. were fixed in cold (4°C) 4% paraformaldehyde and 5% glutaraldehyde mixture in 0.1M sodium phosphate buffer (pH 7.4) with an addition of 5% dextrose for 1 hr, then continuously fixed at room temperature (25°C) for 1 hr. After rinsing with phosphate buffer, the plumula were postfixed with 1% osmium tetraoxide in 0.1M phosphate buffer (pH 7.4) for 1 hr. After rinsing, the plumula were dehydrated through ethanol and acetone, infiltrated and embedded in low viscosity Spurr's medium (Spurr, 1969). Thick sections (0.5μm) were cut with a glass knife on a Reichert-Jung Ultracut E

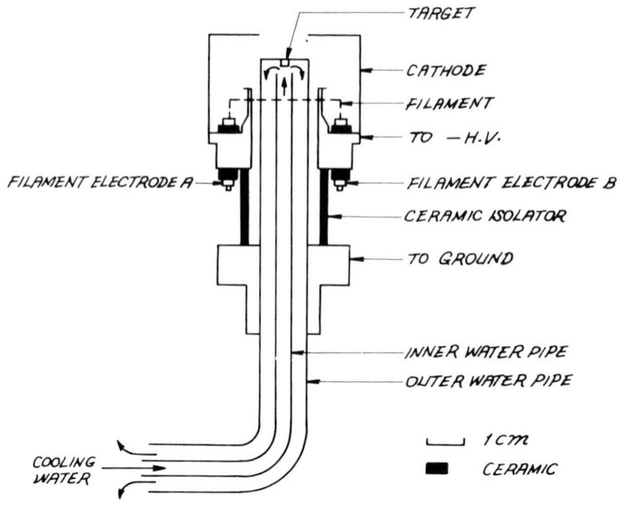

Figure 23.3. Design of the x-ray source.

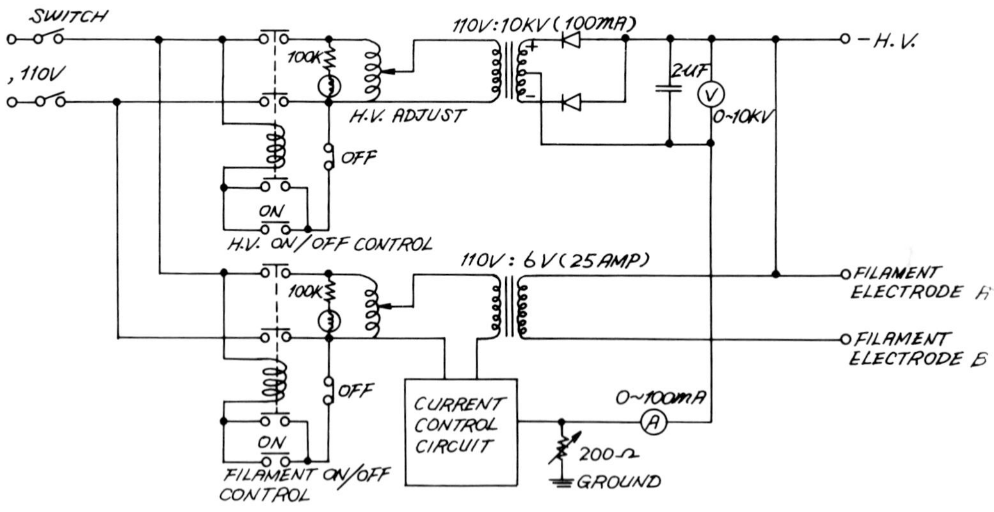

Figure 23.4. The circuit diagram of the power supply.

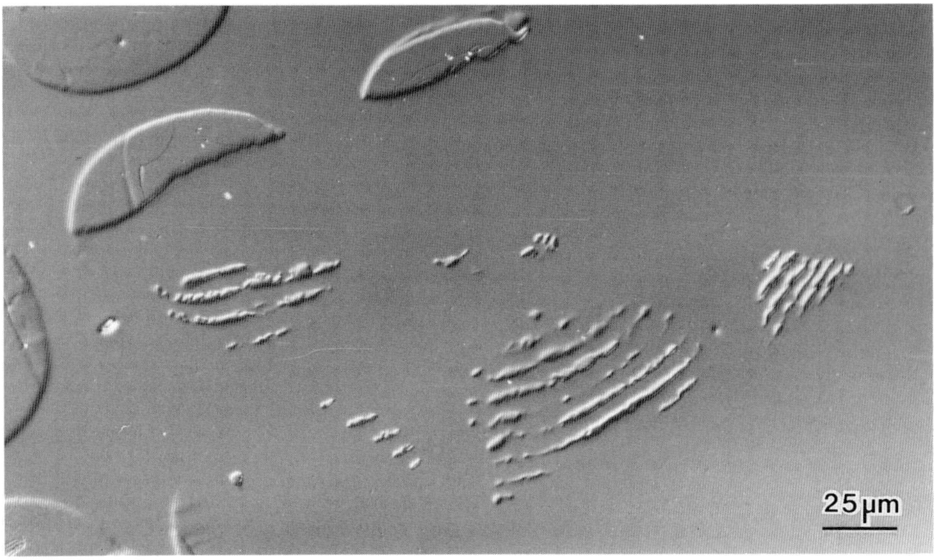

Figure 23.5. Image of the plumula skeletons of sea-urchin tooth. (Exposure condition : V = 4kV, electron current = 15mA, exposure time = 6hr, Development condition: 60sec in MIBK/IPA (1:1 v/v) at 25°C)

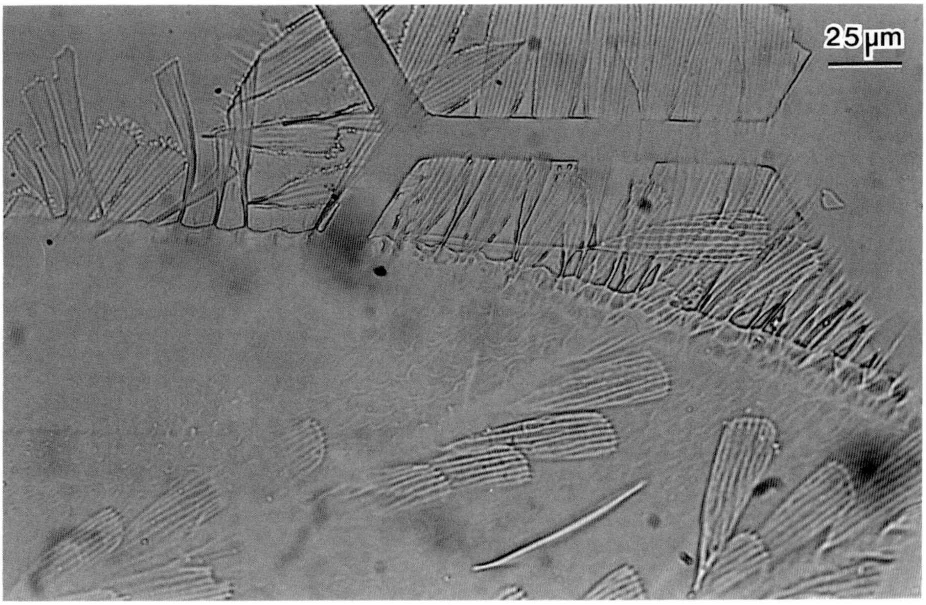

Figure 23.6. X-ray images of a mosquito's wing. (Exposure condition : V = 4kV, electron current = 15mA, exposure time = 7.5hr, Development condition: 90sec in MIBK/IPA (1:1 v/v) at 25°C.

ultramicrotome. The sections were picked up with EM grids. The mosquito's wings used in this study were air dried.

Poly methyl methacrylate (PMMA) x-ray resist was used in this study. PMMA was dissolved in chlorobenzene (6% solution). A drop of the PMMA solution was then placed on the center of a glass coverglass and spun at high speed (~5000 RPM) for 60 seconds. A thin PMMA film (~ 1μm in thickness) was formed on the surface of the glass substrate, then the resist-coated coverglass was placed in a 150°C oven for 1 hour to removed the residual solvent from the resist.

Specimens were placed in intimate contact with the x-ray resist and irradiated with a suitable dosage of x-rays. After x-ray exposure, the exposed x-ray resist was placed in the developer (methyl isobutyl ketone / iso propyl alcohol, 1:1 v/v) for 60 seconds, and the substrate was carefully removed from the developer and blown dry with dry nitrogen gas. After development, a "contact print" of the specimen was revealed. The x-ray contact images were magnified by using a phase contrast microscope.

Results and Discussion

Figure 23.5 shows the image of the mineral-rich skeleton of the sea-urchin tooth. However, the odontoblasts is not visible in this image. After being exposed to soft x-ray, odontoblasts in the section are not well stained with toluidine blue, except those cells under the grid lattice. In contrast, the calcite skeletons remain visible under polarized light. It is clear that the differences in x-ray absorption between odontoblasts and the embedding medium is small, therefore the odontoblast did not show up in our x-ray images. In order to improve the image contrast between the odontoblasts and the embedding matrix, a series of experiments on staining the tissue with various heavy metal stains and optimizing the exposure conditions are planed. The image of a mosquito's wing is showed in Figure 23.6.

Acknowledgement

The work has been supported by a grant from the National Science Council, Republic of China under contract No. NSC 75-0608-E002-05.

24. A Simple Procedure for the Fabrication of Si_3N_4 Windows

J. Pawlak[1], P. C. Cheng[1] and D. M. Shinozaki[2]

[1]IBM T. J. Watson Research Center, Yorktown Heights, NY 10598 USA

[2]Faculty of Engineering, University of Western Ontario, London, Ontario, N6A 5B9, Canada

Introduction

Various sizes and thickness of Si_3N_4 thin windows have been used extensively in x-ray imaging. This includes the use of Si_3N_4 windows as contamination barriers in synchrotron radiation beam lines, vacuum windows, supporting substrate for zone plates (Kirz, 19 ; Buckley *et al.*, this volume), target supporting material for projection shadow microscope (S. Newberry, private communication), and resist supporting material for contact imaging (Cheng *et al.*, this volume). In the past decade, Mr. R. Feder of IBM T. J. Watson Research Center has kindly provided various Si_3N_4 windows to the x-ray microscopy communities world wide. With his early retirement, the future supply of Si_3N_4 windows is uncertain. Therefore we felt obligated to write this article so that researchers can fabricate the windows in their own laboratories. Hence, this article is a technical one, and consists of no original window fabrication techniques. For a more detailed description on the topic, readers should turn to the original paper (Bassous *et al.*, 1976).

The fabrication of Si_3N_4 window takes advantage of the well-known properties of the anisotropic etching of (100) oriented single crystal silicon as well as the smoothness and rigidity of very thin self-supporting film of Si_3N_4. The anisotropic etching rate in a strong

X-ray Microscopy
Ed. by P. C. Cheng and G. J. Jan

alkaline solution decreases in the order (100), (110) and (111). If holes are etched in (100) silicon with an anisotropic etching solution, the cavities formed are pyramidal in shape, bounded by four sidewalls which represent the slowest etching (111) crystal planes forming an angle of 54.7° with the (100) surface plane (Bassous *et al.*, 1976).

The silicon wafer of (100) orientation, N- or P-type, chemically and mechanically polished on the front side and chemically polished on the back is used for the fabrication of windows covered with a film of Si_3N_4. (one can purchase this type of silicon wafer from various commercial outlets, such as Monsanto Co., and Wacker Co.).

The basic procedure in fabricating thin Si_3N_4 windows is illustrated in Figure 24.1. A silicon wafer (100) with Si_3N_4 deposited on both surfaces is used (Figure 24.1a). First, both surfaces are coated with photoresist (such as Microposit 1350J), then the mask pattern is exposed on the back surface and developed (Figure 24.1b), subsequent reactive plasma etching removes the exposed Si_3N_4 to reveal underlying silicon (Figure 24.1c). Next, the resist on both surfaces is removed by solvent (Figure 24.1d). The wafer is then etched in a pyrocatechol solution (Figure 24.1e). The etching speed depends on the crystal orientation of the silicon, and with (100) silicon, etching rate is faster normal to the surface than in any direction in the plane of the wafer. Hence a sharply defined well can be formed. Figure 24.2 shows the absorption spectrum of the Si_3N_4 window obtained at the Canadian Synchrotron Radiation Facility beamline. The small carbon K edge is the result of carbon contamination on the window.

Procedures of Window Fabrication

The procedure of Si_3N_4 window fabrication is described in the following protocol.

(1): Clean the surface of the silicon wafer with acetone and ethanol. Then, both surfaces of the wafer are coated with Microposit

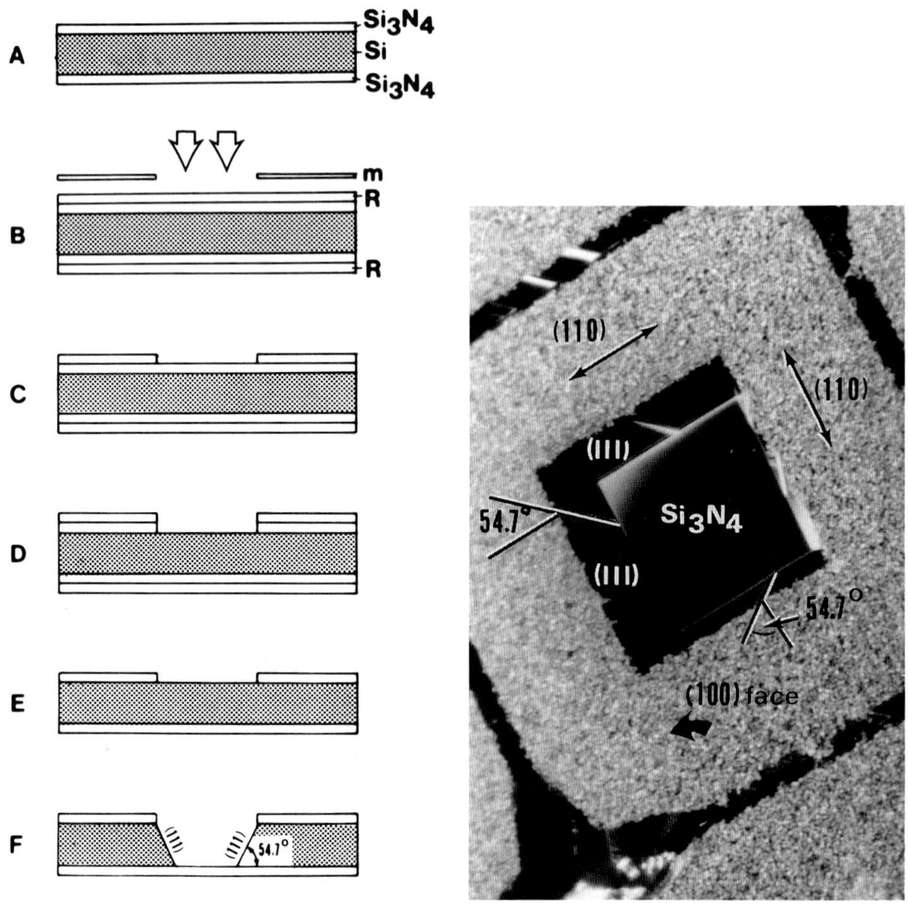

Figure 24.1. (1) Diagrammatic representation of the Si_3N_4 window fabrication processes. (2) Close-up view shows the result of anisotropic etching of (100) silicon single crystal. The sidewalls (111) forms an angle of 54.7° with the (100) surface plane.

photo resist 1350J (Shipley Co. Inc.). The wafer is coated by spinning resist on the silicon wafer at 5000 RPM for approximately 30 seconds under yellow safety light. The back surface is first coated with resist, and baked at 80°C for 30 min. Then the resist is applied to the front surface (highly polished surface) of the wafer in the same manner (Figure 24.1b).

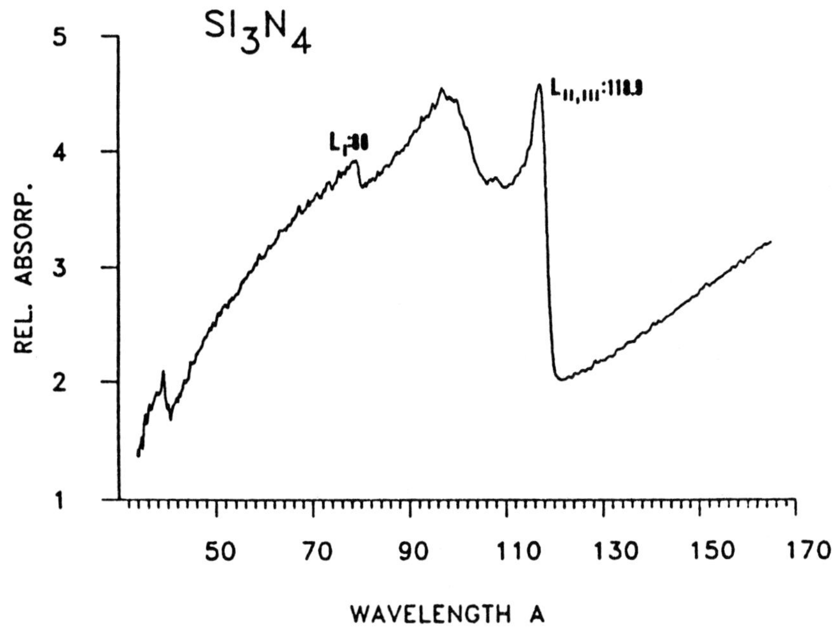

Figure 24.2. Relative absorption spectrum of Si_3N_4 window. The small carbon K edge is the result of contamination on the window. (P. C. Cheng, D. M. Shinozaki and K. H. Tan)

(2): In order to etch the desired patterns on the silicon wafer, the proper photo mask should be prepared. Figure 24.3 shows examples of various types of masks. In practice, the mask pattern can be drawn on a tracing paper with an Indian ink pen, or produced by computer using a plotter, then photographically reduced onto lithographic films (*e.g.* Kodak Kodalith Ortho 6556, type 3 and developed in Kodalith developer). The properties of anisotropic etching should be kept in mind, a window opening of 1mm x 1mm on the mask will actually produce a window opening approximately 0.7 x 0.7 mm if 0.2mm thick silicon wafer is used. Therefore, for a given window opening on the mask, the thicker the wafer, the smaller the window. Furthermore, due to the properties of the etching process, only square or rectanglar openings can be fabricated.

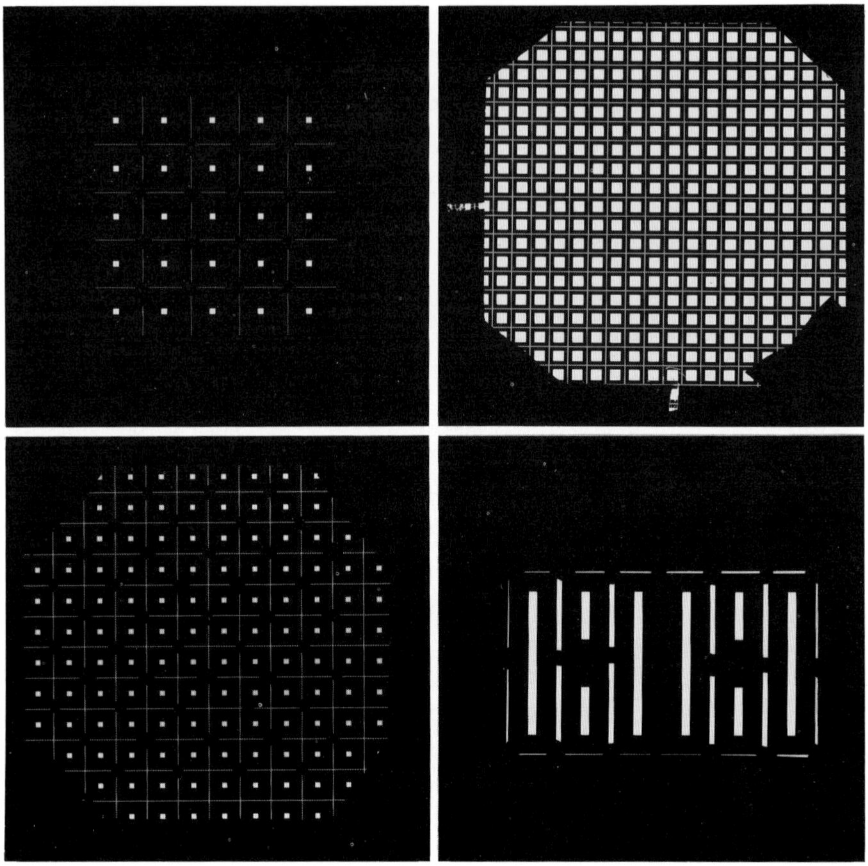

Figure 24.3. Examples of masks.

(3): Once the wafer has been coated with photo resist, masking can be performed. The first step involves orientation of the wafer so that the (110) plane is parallel with the edge of the window opening of the mask. This step is important, since the orientation of the etching plane depends on the orientation of the crystal. Misalignment results in saw-toothed window edges. Next, the desired mask is placed over the wafer surface and exposes to a 30W halogen lamp at 10cm distance for 3 minutes (Figure 24.4). After exposure, the wafer is then developed in Microposit developer concentrate until the masks pattern be-

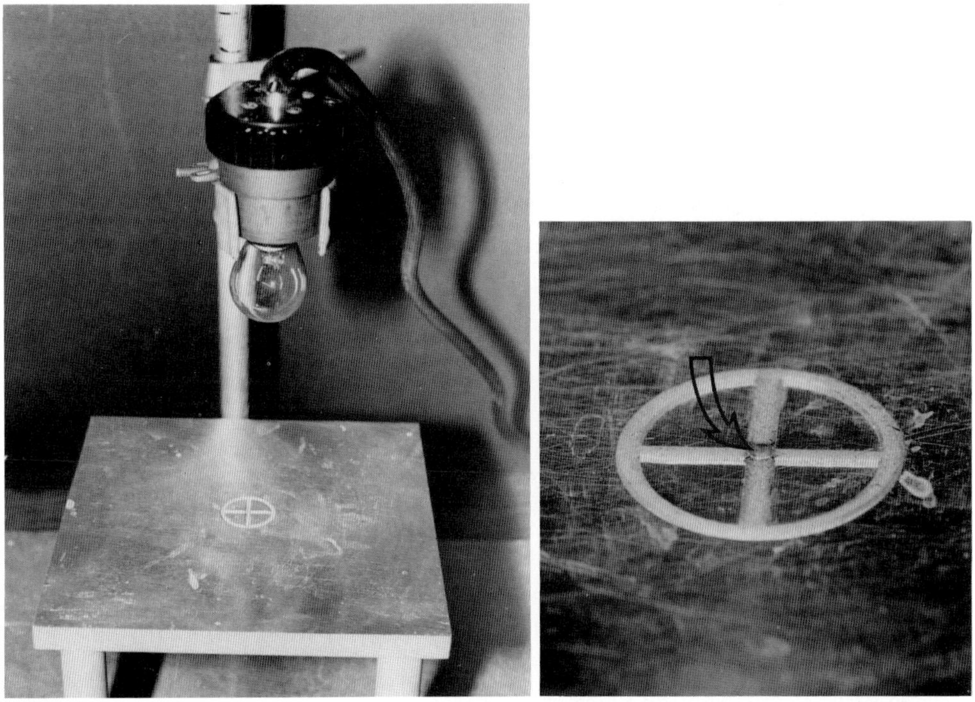

Figure 24.4. (a) Simple apparatus used to expose the resist. In order to securely attach the wafer to the exposure platform, vacuum is applied to a ring glove milled on the platform surface (arrow).

comes clearly visible. The development process takes about 20-30 seconds. After the pattern has developed to the desire condition, the wafer is immediately immersed in water to stop the development. If the mask pattern on the surface of the wafer is uneven, (*i.e.* certain parts of the pattern are not thoroughly developed), spot developing with a cotton tipped swab may be used. Once developing has been completed, the wafer is baked at 100°C for 30 minutes.

(4): The wafer is now ready for plasma etching (8.5% O_2 with CF_4 with total power of 350-400W) using a plasma etching system (LEF Co., PDS/PDE-301). The length of etching is about 1.5 ~ 2.5 minutes. When etching is complete, the surface of the wafer has a light silvery-gray appearance. After plasma etching, the wafer is again rinsed with acetone to remove trace of residual resist.

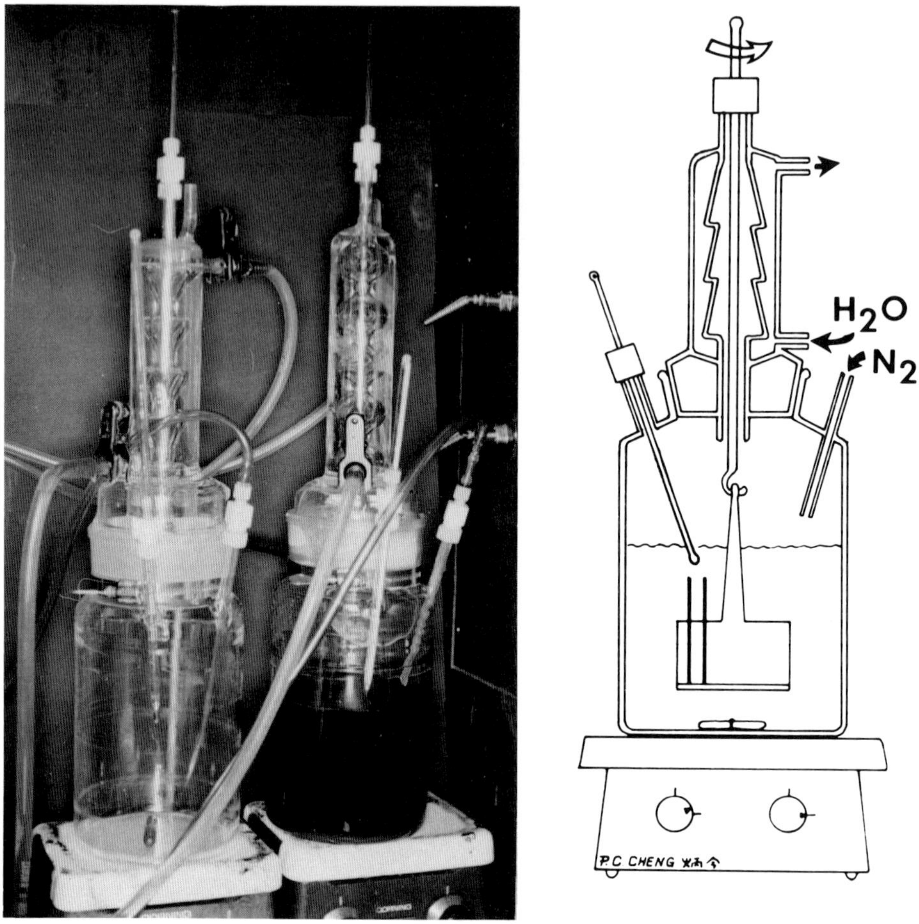

Figure 24.5. Reflux apparatus used in the etching process. The reflux apparatus consists of a water-cooled condenser with a large flat-bottom wide-mouth flask. Side ports are included to insert a thermometer and to accommodated nitrogen supply. The whole apparatus is placed on top of a hot plate equipped with a magnetic stirrer. The apparatus shown on the right contains oxidized etching solution. Fresh solution is light yellow in colour. The solution should be discarded when it turns into dark brown (as shown in the left apparatus).

(5): Finally, chemical etching using a pyrocatechol bath can now be done. A reflux apparatus (Figure 24.5) is used for the chemical etching process. The reflux apparatus consists of a large wide-mouth flask and a condenser. A magnetic stirring bar is used to stir the liquid. A nitrogen supply is connected to the flask (Figure 24.5, arrow) to maintain the bath under nitrogen atmosphere in order to extend the working life of the etching solution. A quartz "boat" (figure 24.6) is suspended from the condenser in the flask, the boat functions to immerse the wafers in the bath. The etching solution is heated to a gentle boil, about 115°C, prior to etching. Care should be taken to be sure that boiling and stirring are not too vigorous, since an overly turbulent bath could damage the wafers. A flashlight is useful to inspect the progress of etching. Approximately two hours are required to etch 2.25" wafers (0.2mm thickness) and about 3.5 ~ 4 hours for 3.25" (0.4mm thickness) wafers. Once etching is complete, the wafers are removed from the bath and rinsed in water and dried with N_2.

Figure 24.6. Quartz "boats" used for holding silicon wafers during chemical etching process.

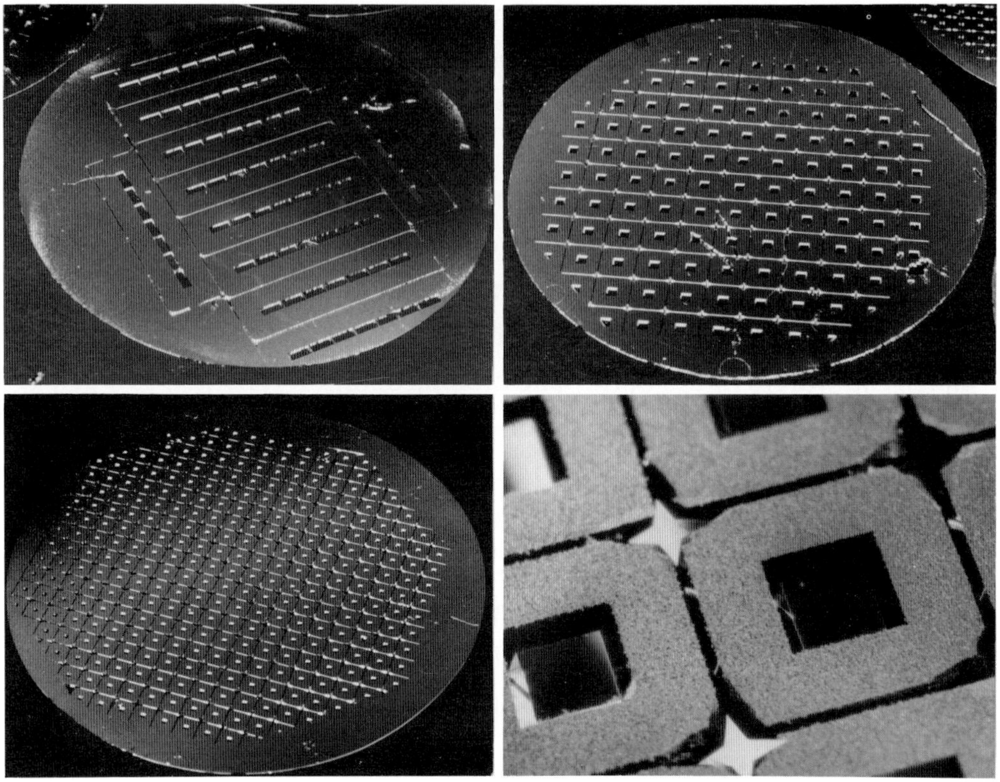

Figure 24.7. (a, b, c): Examples of various windows. (d): Close-up view of 3mm x 3mm windows.

Figure 24.7 shows various windows fabricated by the method described in this article. Individual windows can be easily broken off from the wafer. If coating of photo resist on the window is required, the coating process should be done on an intact wafer. A special chuck is required to hold the wafer onto a spinner, since the commonly used vacuum chuck tends to break the delicate windows.

The etching bath is prepared as follows:

pyrozine	6.5g
pyrocatechol	400g
distilled water	375ml
ethylene-diamine	1150ml

Protective gear should be worn when preparing the bath. The preparation of the bath should be conducted in a fume hood, and the chemicals should be mixed in the order listed.

Acknowledgements

We are gratefully acknowledge Dr. E. Spiller for helpful discussion on the window fabrication technique, and Dr. D. Sayre and Dr. Larry Herman for critical reading of this manuscript.

25. History of X-ray Microscopy

S. P. Newberry

CBI Lab., Box 11, Westcott Rd., Rd. 6, Schenectady, NY 12306 USA

When first discovered in November, 1895 by Wilhelm Konrad Röentgen x-rays created immense excitement (January 1945) because of their ability to image structures hidden within previously opaque materials. The applications to medicine were immediately obvious. Within the first year major hospitals were using x-rays for repair of fractured bones and to assist in surgery. Within this same first year Ranwez in France and Burch in England applied x-rays to botanical research to "reveal plant structure" by use of enlargements of contact radiographs, thus a form of x-ray microscopy was accomplished. By January 1896 Radiguet (1896) in France attempted to find a grainless fluorescent screen as a means of microscopy. Shortly thereafter (1898) Heycock and Neville applied contact radiography technique to alloys. These were forms of what we now call "nondestructive examination" that is they relied upon the ability of x-rays to penetrate opaque structures rather than relying upon thin sectioning or microdissection. (Incidentally Röentgen also employed a pin hole photograph to obtain an image of the x-ray source thus foreshadowing self-luminous x-ray microscopy.) It would be two decades (1895 - 1912) before an additional property of x-rays, namely their very short wavelength would bring a new reason to employ x-rays for microscopy. Here a word of caution is in order about the wavelength used for microscopy. In the first attempts at microscopes, listed above, no one knew whether x-rays were waves or not and if so what the wavelength was under various conditions of experimentation. Until the hot filament cathode tube was developed by Coolidge in 1913,

X-ray Microscopy
Ed. by P.C. Cheng and G.J. Jan

x-rays were universally continuous produced by gas discharge similar to the Crooks tubes in which they were discovered. The wavelength was of course a mixture of continum and line radiation and filtration was poorly understood. The accelerating potential and current could not be independently controlled since they were tied together by the gas pressure. As the tube was used the gas disappeared and the voltage had to be raised to maintain current. (Gas tubes were some times provided with palladium thimbles through which pressure could be restored by heating the thimble with a Bunsen burner, thus introducing hydrogen.) With the higher voltage came more penetrating radiation thus the classification of "hard" and "soft" radiation was adopted. In the early work we cannot know what wavelength was used but can reasonably assume that it ranged from a couple of Angstrom units for plant specimens to a fraction of an Angstrom for metal samples. Contrast this region of the spectrum with the 20 to 50 Angstrom wavelength most of you are employing today and one realizes that there is need for vastly different instrumentation to cover different regions of the spectrum. For example mirrors and zone plates require very long wavelength to be practical, a different type of microscope is required for one Angstrom radiation and shorter.

To understand this next adventure into x-ray microscopy we must call to mind the technological atmosphere of those times. The optical microscope and specimen preparation for its use were mature technologies. By contrast photographic emulsions were still primitive and not capable of supporting microradiography in competition to the optical microscope. There was also a bias against thinking of shadow photographs as microscopy. Photography was even rarely used with the optical microscope. The microscope was primarily a visual instrument, and was both physically and psychologically an extension of human vision. Physics was embroiled in the wave versus particle controversy and had been very reluctant to accept the indications that x-rays were electromagnetic radiation of very short wave length. William

Henry Bragg, one of the staunch opponents of the wave theory for x-rays, became converted over night by Max von Laue's prediction and the immediate demonstration by Laue's coworkers (Friedrich and Knipping) in 1912 of x-ray diffraction by a crystal.

Bragg at Leeds and his son (William Lawrence, later to become Sir Lawrence Bragg) at Cambridge immediately sought to apply the new x-ray diffraction tool to the study of the structure of matter. The elder Bragg was an experimentalist and developed a spectrometer which greatly simplified the interpretation of both the radiation and the spectra while his more theoretically inclined son accepted the Laue type of pictures and developed the fundamental equation $n\lambda = 2d \sin\theta$ for interpreting all diffraction patterns. Starting with the summer of 1913 they frequently worked together. Their work in diffraction is well known, what is not as well known is their serious contemplation of an x-ray microscope. The author learned of this work from the late Professor Jauncey who had worked under Bragg starting in 1913 as a young student (from Australia) and in the 1930's was our lead x-ray physicist at Washington University. The motivation was to escape the resolution limitation of the optical microscope which had been shown by Abbe to be due to the long wavelength of light. Their concept was a direct x-ray analogue of the optical instrument. Now Roentgen had pointed out in his first publication that the refractive index of any material for x-rays which he had tested could not be more than 1.05 and that no reflection was found in the optical sense. Knowing this background the Braggs were willing to consider the construction of a much larger instrument employing very weak lenses. This willingness to consider a project which would have been very "big physics" for that time indicates the importance to them of breaking the resolution barrier. However the best estimates they could make of the refractive power forced them to abandon the idea when they calculated that the microscope would have to be approximately one mile long (1.6 Kilometers). This result, divorced from its originators, became well known to every physics student and created a mental block in the mind of nearly all scientist against the x-ray microscope. However, the younger Bragg never completely abandoned the notion of x-ray microscopy. As ever more difficult structures were analyzed from diffraction data he sought an optical way to create the structure from the diffraction recording. He was the one to point out that diffraction reflections carry the information of the average location of electrons in the atom, which Wollan and Compton (1934) demonstrated by opto-mechanical techniques. Bragg succeeded in extracting crystal structures by optical diffraction of the x-ray diffraction images through mechanical filters. Unfortunately he was forced to guess the phase relations in the diffractogram and construct the mechanical filters from the best guess of the phase relations. He began publishing successful analyses by this method as early as 1939. How tantalizingly close he was to the concept of x-ray microscopy by holography. Knowing the history, it does not surprise us that he named this combined x-ray/optical method x-ray microscopy; however, he was not clearly distinguishing between determination of a repetitive structure and point resolution. This distinction isn't really very important when one has not yet seen the idealized structure.

The time interval around 1939 was significant in x-ray microscope history. The quarter century from the abortive attempt in 1913 had brought many changes in the technological atmosphere. X-ray physics had matured, especially with respect to interactions with matter. (Recall that Bragg was just discovering the line spectra and Mosley was discovering the orderly relation of the line spectra with respect to atomic weight in 1912-13.) An excellent review of the status of x-ray microscopy at the beginning of this period is given in a paper read before the Royal Microscopical Society (British) December 16, 1914 by Barnard (1915) (Barnard introduced the use of the Schumann plates into microradiography. He tried to repeat some of Braggs attempts to produce lenses or mirrors. He also predicted that if x-rays were found incapable of breaking the resolution barrier, that other short wave length radiations would surely be found which could be used. Apparently in common with the wisdom of his time, he did not consider the electron to have a wave nature.). Microradiography continued to grow in capabilities along with improvements in photography and discovery of the more reliable Coolidge tube. For example, Harold Sherwood at Eastman Kodak as a member of the photographic industry, was starting on a long career to improve methods and techniques for microradiography. Two additional capabilities of x-rays had been invoked to enhance the value of the microradiogram, namely stereo imaging and chemically related contrast. Golby in England applied the established radiography stereo techniques in 1925 to microradiographs. In 1929 a remarkable piece of work was published by Jentzsch of Jena at a German Physical Society meeting in Leipzig. He proposed the use of grazing incidence mirrors to make an optically magnifying x-ray microscope with calculations to support his suggestion. His work was apparently forgotten and probably unknown to many interested workers out side Germany. The next significant step was the enunciation of the principals of historadiography in 1936 by Lamarque of France. The electron microscope burst on the scene in this same decade giving a break thru of the resolution barrier but more important permanently altering our concepts of microscopy.

As evidence of this new attitude in microscopy, Zernike developed a method to employ phase contrast in the optical microscope in 1935 after 75 years of practically no innovation in the field, but manufacturers did not take up phase contrast until after the electron microscope was generally available. During this time of imaginative thinking, ushered in by the electron microscope discovery, a few people independently returned to consideration of an x-ray microscope. With the exception of Kirkpatrick of Stanford (an outstanding x-ray physicist) all of us were closely associated with electron microscope development. Trillat in France was experimenting with photoelectron recording and crystal reflections for improving the contact image. Ardenne (Manfred von,) published the concept of the shadow x-ray microscope in 1939. To Ardenne it was a direct extension of the shadow electron microscope (Böersch 1939). Zworykin and coworkers at RCA looked at the Ardenne suggestion and decided that the intensity would be too weak to be of practical concern. Marton looked at it and said if we could have a practical field emission source then we could have exposures short enough to stop motion of living organisms. The goals were by this time less oriented toward high resolution and more toward uniqueness of x-ray contrast. The desire was to get around the shortcomings of the electron microscope due to high vacuum and fussy specimen preparation requirements. A modest resolution just beyond the optical microscope would have been considered adequate. The primary urgency was in the field of biology. We did not have satisfactory microtomes for the electron microscope or surface replica techniques for tissue sections. The environmental cell was just beginning to be considered and the scanning electron microscope was not yet available. We were some what simplistic in our thinking however. (I remember well how revealing a simple calculation of an ambient air column to have equivalent mass to the 10 micrometer tissue section is. One finds that it is only 5cm. For a ten to one contrast advantage, according to this first order calculation, one would then have to restrict the total optical package to one half cm length. Because there is no x-ray window in this wavelength region to differentiate moist air from living tissue, one could not think of this x-ray region of the spectrum as giving both atmospheric penetration and specimen contrast, at least not using simple absorption. The use of x-rays to escape vacuum technology by means of an extended optical column is not realistic. The real attraction of the x-ray spectrum is the quantitative, predictable, easily varied contrast ranges available.) We were also prejudiced against shadow images, seeking methods which would produce a quotes "real image". We did not appreciate until we saw stereo views of shadow images, that the stereo image of shadow recordings provides the same three dimensional "point for point image to object correspondence" as a true lens image.

World War Two caused the advanced x-ray microscope work to be put aside. The more established Microradiography and topographic imaging were applied to practical problems in alloy segregation during Spot Welding and enjoyed wide spread acceptance and rapid improvement. It became text book technology during this time, being referred to by Barrett, Maddigan, Clark and others as the "New microscopy".

Following the war, workers began to turn again to the search for the "true x-ray microscope". The academic pace of the work converted overnight to a feverish race with the publication in 1948 by Kirkpatrick and Baez of their grazing mirror approach. The notion of an x-ray microscope was suddenly given credibility in the general scientific community and sponsors were now willing to support efforts in the field. I do not mean to imply that Kirkpatrick and Baez were the first to consider mirror x-ray lenses, in fact they cautiously suggested that the opposite might be true. We have already cited the paper in 1929 by Jentzsch and should also point out that Ehrenberg a former associate of Jentzsch, was an active worker in x-ray microscopy at the University of London and published on the formation of line images by curved mirror in 1947. It is doubtful that Kirkpatrick and coworkers knew about the earliest work until after they first published their own results. See (Kirkpatrick and Baez 1948) and (Kirkpatrick and Pattee 1957). The most important historical point is that the Kirkpatrick and Baez publications started a new renaissance in x-ray microscopy at midcentury and gave creditability to x-ray microscopy in the intellectual community. The new burst of activity was primarily by people who were instrumentalists. By contrast the serious applications people were concentrating on microradiography. We have noted the increase in metallurgical application of Microradiography, we should also note the increase in Biological application of the contact method. In 1946 (two years before Kirkpatrick's announcement), a strong school of Historadiography began taking shape under the leadership of Engstrom in Sweden. By this time Microradiography was no longer dependent upon borrowed equipment from other technologies but was practiced either by using sealed tubes designed for metals work or demountable (continuously pumped) systems designed specifically for biological contact radiography. Here we should also note the continuing contributions of Ehrenberg in England and Trillat in France.

The search for an x-ray microscope subsided appreciably with the publication by Cosslett and Nixon in 1951 of success using the Ardenne shadow projection approach. They demonstrated results, in the wavelength region of 1/2 to 8 Angstroms, comparable in resolution with the contact method, with improved depth of field and decidedly greater convenience. It is important to note that Cosslett like many of the rest of us had been working on the x-ray microscope prior to the Kirkpatrick and Baez announcement, in fact, dating back to that 1939 period when the electron microscope had just been put into useful form. He was then a student at University of London later moving his x-ray work with him to Oxford then to Cambridge. He was endeavoring to use electron microscope techniques to produce a smaller brighter source of x-rays for microradiography. Like many of of us he did not hear of the Ardenne paper for some time. On the basis of Cosslett's work he was accepted at the Cavendish and it was Sir Lawrence Bragg with his interest in x-ray microscopy who was the primary force in bringing him to Cambridge and who continued to lend encouragement to the work. The success of the shadow projection method by Cosslett and Nixon did not get generally known for nearly five years, thus considerable work continued with mirror approaches and the Kirkpatrick and Baez approach continued to give the x-ray microscope plausibility generally.

At the time of the Cosslett Nixon announcement, a very important meeting of electron optical workers was convened at the (American) National Bureau of Standards as part of a series of meetings to commemorate the Bureau's Fiftieth Birthday. All of the people who would later produced Commercial versions of the x-ray microscope were present at that meeting and both Cosslett and Marton presented papers on the shadow projection method, although Marton presented calculations only. The clarity of the stereo views shown by Professor Cosslett at that meeting dispelled any lingering doubts about the adequacy of the shadow image. The rapid development of commercial x-ray microscopes thus arose from the technology which had been developed for the electron microscope and from the cross fertilization of this meeting in 1951. Models were produced and sold by General Electric in the USA, Phillips in Holland and Microray Laboratories in England. Phillips also offered for sale a small, (sealed source), contact machine but window technology at that time did not permit any advantage in soft x-ray production compared with the much thinner windows possible in continuously evacuated equipment. Professor Gunji Shinoda and coworkers at Osaka University constructed a working model of the Ardenne type and published pictures in 1953, but to our knowledge no commercial instrument was derived from that work. In the next five years several other commercial instrument were announced.

Work continued on alternate approaches. Baez arranged for Buckbee Mears to make a micro-zone plate and tested it with x-rays. Pattee explored the direct magnification of a high resolution fluorescent screen in contact with the specimen, with considerable success. He makes no reference to earlier work which leads one to suspect that the idea was independently arrived at by him. Many groups, including Kirkpatrick and his students worked to improve the quality of the grazing incidence mirrors but with limited success. Especially noteworthy was the combination of both axes of magnification into one optical mirror (with astigmatism correction) by Marc Montel in Paris. In all the mirror types the numerical aperture has been very small and the surface smoothness requirements have been too stringent for a useful microscope to be developed. An interesting variant was an attempted in 1955 by Ladd to carry the contact method beyond the limitation of the optical microscope. He recorded on plastic which was developed by weak solvent, then viewed by replica technique in the transmission electron microscope. His exposures were very long (20 to 30 hours) and his images of carbon black were confused by the crystalline structure of the plastics used so nothing came of the work at that time. Asummaa of Finland experimented with other plastics and special emulsions shortly there after. The very earliest work in this direction was Possibly Koenig (1951).

Within ten years the great popularity of the shadow method waned and then it gradually fell into almost total disuse. It is not unusual for a new technique to be over worked initially and then abandoned only to be revitalized after the passage of many years. The reason for abandonment of this version of microscopy was due primarily to the concurrent development of the electron probe instrument of Castaign and its wedding to the scanning electron microscope (of Smith and Oatley) by Long at Cambridge. This combination could give an x-ray image of one element at a time with essentially the same resolution limit as the shadow method and relate it to a higher resolution surface structural image. It was wrongly supposed that it was an alternate approach for accomplishing the same result and did not require the purchase of a separate instrument. Actually it was a surface instrument and did not answer the need for three dimensional viewing. Shadow projection adapters for the scanning electron microscope were developed by JEOL and others, which further diminished the apparent need for a separate instrument.

In the hey day of the shadow projection instrument a few centers of productive work developed and many striking photographs were published. (Strangely enough one seldom meets a younger worker who has seen these pictures, especially not in their three dimensional versions.) Professor Saunders at Halifax University carried out extensive, significant work in the structure of synapses and microcapillaries. Wycoff and coworkers made a complete three dimensional atlas of the laboratory mouse for their work at the National Institute of Health. At General Electric in Schenectady we applied our own shadow projection microscope to a wide variety of industrial problems for many years, with many useful results obtained. Interestingly enough the most useful results did not necessarily coincide with the most striking pictures. Ernest F. Fullam Inc. included x-ray microscope research in their services for many years. Dr. Ong Sing Poen, who developed the Phillips version of the microscope under Professor LePoole, was a consistent user of the instrument and published a book containing many excellent micrographs. It is not possible to list all the active centers which published but most of them published only once or twice. Professor Saunders and coworkers stand almost alone in consistent publication over a long period of time.

The shadow method did not reach its full potential. It reached an apparent resolution barrier at about one half micrometer. Nixon (1955) was able to demonstrate that the barrier was due to scattering in the target support and showed pictures of 0.1 micrometer resolution but a practical structure was not developed to make this improvement available. Although the technology was in hand, no image intensifier was developed for the microscope. The result was a continuing requirement to dark adapt ones eyes for half an hour in order to visually focus or align specimens. Most of the successful applications were industrial, adequate specimen preparation was not developed for the then current biological needs. Biology was further slighted by lack of sufficient work with very soft x-rays. A good vacuum camera was needed (some of us made our own) and appropriate filters were needed. It should have been determined that scattering is more important than absorption with the softer region of the projection wavelength range. Also a Micro Buckey Diaphragm (also called grid) should have been developed in a concentric sphere geometry to accentuate the scattering contrast. Such a Buckey Diaphragm would have the same effect as an objective aperture has for the transmission electron microscope. Finally an automatic exposure control should have been developed which responded to the x-rays arriving at the photographic plate. These were improvements which could have been made while the technique was at the peak of attention. If revived today it could also benefit from, a host of tomography and other computer related techniques. The shadow projection technique stands alone in producing a directly magnified image in a matter of a few minutes on in house equipment. Some instant film exposures are less than 5 seconds with a processing time of 20 seconds. In the author's lab, motion pictures of live specimens were produced at 3 frames per second. Figure 25.1 shows the first set-up.

The intervening years from 1970 to 1976 was a period of neglect of the field of x-ray microscopy. X-ray optics was kept alive principally by x-ray astronomy and by attempts to see the x-ray emission from plasma implosions. The latter required development of forms of emission x-ray microscopy. The work with plasma also led to the recognition of the possibility of a brighter x-ray source which we have seen successfully applied to the construction of a new type of x-ray microscope by Feder *et al.* (1985). Even brighter flash x-ray sources using laser generated plasma have been demonstrated by Rosser *et al.* (1985a). For the benefit of both astronomy and plasma physics, attempts were made to improve mirror lenses and zone plate lenses with some success at that time in both efforts. Interestingly enough it was the improvements in electron beam lithography which made possible the next round of improvements in zone plates. We note especially the work of Möellenstedt's group at Tüebingen.

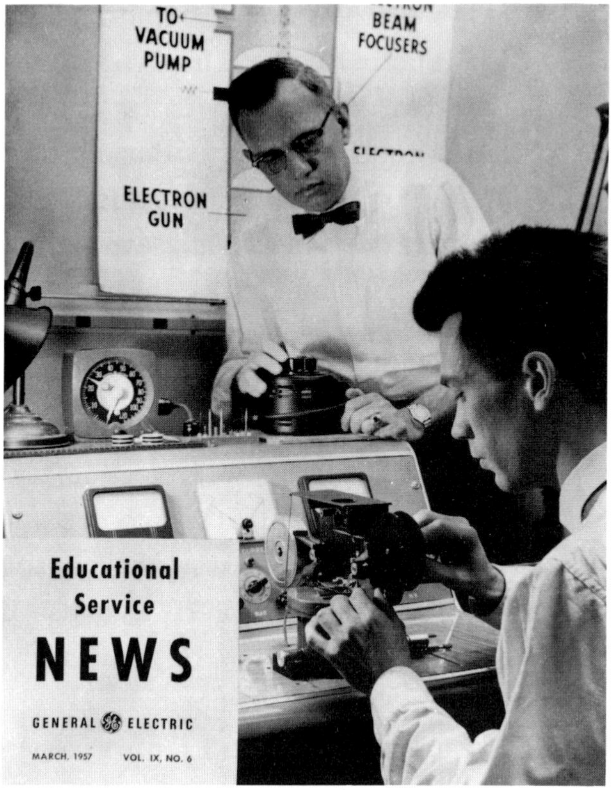

Figure 25.1. General Electric education bulletin cover photo shows the first x-ray microscope movie test set up. James F. Norton seated, Cherles Lemmond standing.

The year 1972 brought the announcement by Spears and Smith of their development of x-ray lithography for copying electron beam generated masks at submicron resolution. Just 4 years later in 1976 Feder, Sayre, Spiller and Topalian of IBM Research Center in conjunction with Kirz of New York State University, Stony Brook announced the use of x-ray lithography techniques to produce soft x-ray contact images well with in the electron microscope range of resolution (Feder *et al.*, 1976). They were able to employ wavelengths as long as 44 Angstrom units (4.4 nanometers) which not only gave adequate contrast for this resolution but also contained a window for viewing protein in the presence of water. This and subsequent publications served to ignite a new renaissance in x-ray microscopy, just as the Kirkpatrick and Baez publications had done a quarter of a century before. Furthermore Scherzer (1978) pointed out that x-ray imaging should produce less radiation damage than electron imaging. Schmahl and co-workers at the University of Göttingen (1976,1985) have built a microscope based on a step function improvement in zone plate manufacture. Their zone plates, produced holographically, have several thousand zones as compared with beam written plates with have at best, a few hundred zones. In this microscope they have demonstrated the lesser radiation damage predicted by Scherzer and calculated by Sayre *et al.* (1977) to range as high as four orders of magnitude for very thick specimens. The holographic zone plates have reached diffraction limited performance. In the past 10 years we have seen rebirth of many of the former approaches, but with renewed possibility of success, because of the continuing advances in supporting techniques. Surely as this conference convenes, we can sense the exciting possibility that within a few short years one or more of these approaches will bring forth a useful form of microscope for wide application in ultra structure research.

It has been both a great pleasure and at the same time an ominous responsibility to prepare this history for you. It should be obvious that a review of nearly a century of work must choose what is left out as well as what should be recalled. I have tried to reconstruct the attitudes and state of science at each period but inevitably this introduces some personal judgement and bias. As in all developments, much x-ray microscope work is inadequately reported or not at all. Jauncey (1945) complains that Röentgen's use of obscure journals and reluctance to publish until he had thoroughly worked through a discovery was counter productive. Röentgen was not the only x-ray worker with this human failing. By relating only unrecorded history known to me you miss out on that part known only to others from those times. We must be mindful that in oral history I or they could be in error, at least in details, nor forget that written records may be so devious as to mislead the reader. I should welcome comments on omissions or suspected errors because I feel that we should continue to complete the history of our science to the best of our ability.

MINI ATLAS

This Mini Atlas provides examples of biological x-ray images obtained by various x-ray imaging techniques. Provision has been made to ensure that the specimens were properly prepared (by electron microscopy standards) unless otherwise stated. The contributors are listed at the end of each figure caption.

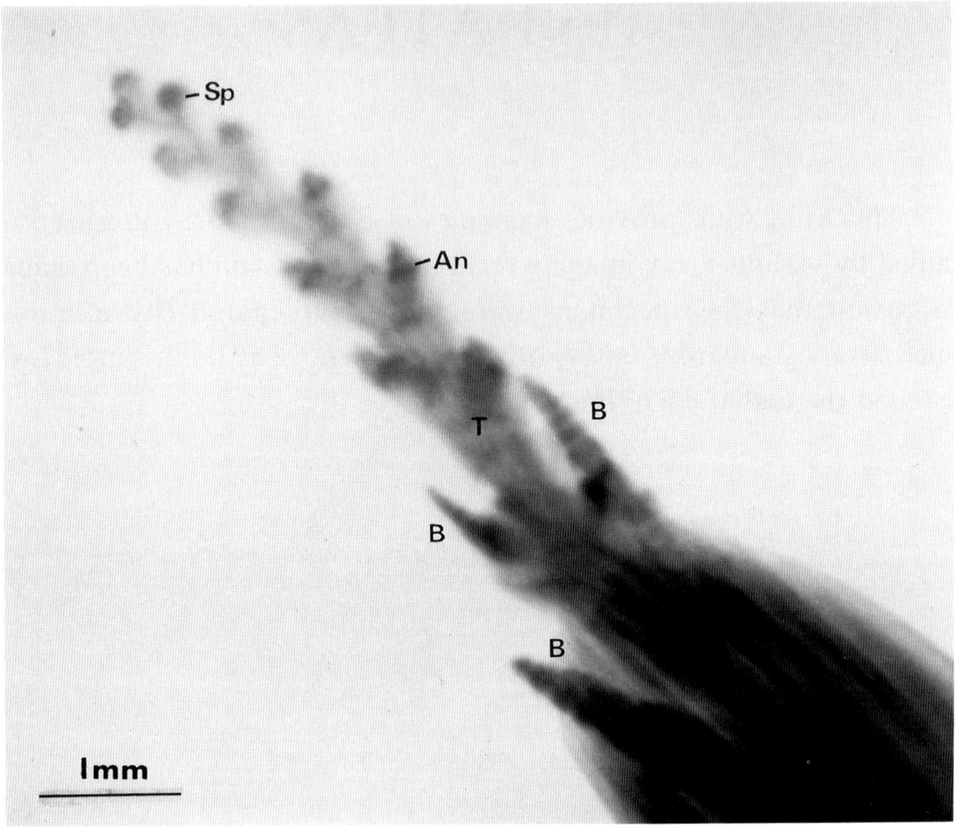

PLATE 1

X-ray projection image of a terminal inflorescence (T) and three branch inflorescences (B) of Northern teosinte (*Zea mays* subsp. *mexicana*, race Norbogam). The specimen was fixed in Cheng's fixative (Cheng *et al.*, 1979), dehydrated in acetone, critical-point dried in CO_2. Note the high x-ray density of the meristematic regions. An: stamen primordia, Sp: spikelet primordia. The image shows that the meristematic tissue has higher mass (higher x-ray density) than the surrounding tissue. (S. P. Newburry and P. C. Cheng)

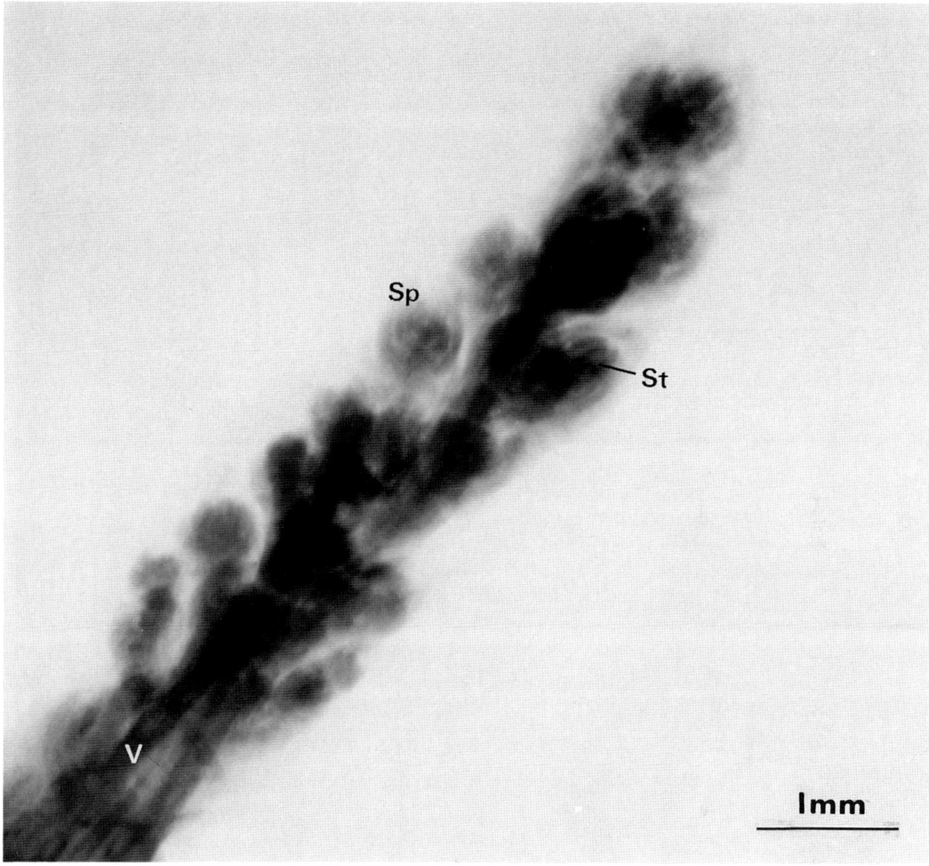

PLATE 2

X-ray projection image of the inflorescence of rice (*Oryza sativa*). The specimen was fixed in Cheng's fixative (Cheng *et al.,* 1979), post-fixed in 1% OsO_4, dehydrated in acetone, and critical-point dried in CO_2. Note the developing stamens (St), and x-ray dense vascular bundle (V). Sp: spikelet. (P. C. Cheng, S. P. Newberry and P. B. Kaufman)

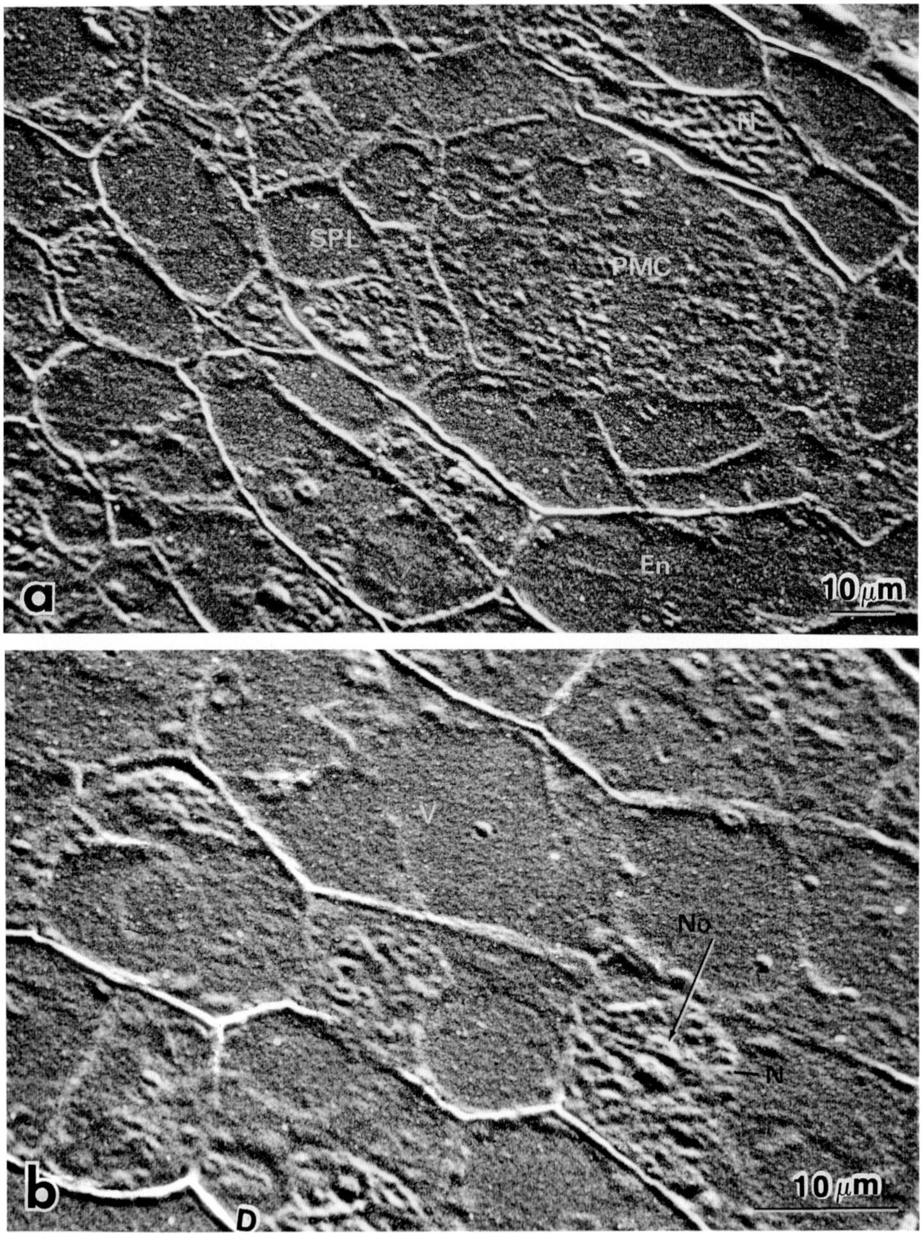

PLATE 3

Soft x-ray contact images of the floral primordium of *Zea mays* magnified by scanning electron microscopy. (a): X-ray image shows large pollen mother cell (PMC) and surrounding endothecial layer (En) and secondary perietal (SPL) of the developing anther primordium. (b): High magnification view of a developing glume. Nucleus: N, Nucleolus: No, Vacuoles: V. Both x-ray images were taken with C-K radiation on PMMA resist. (P. C. Cheng)

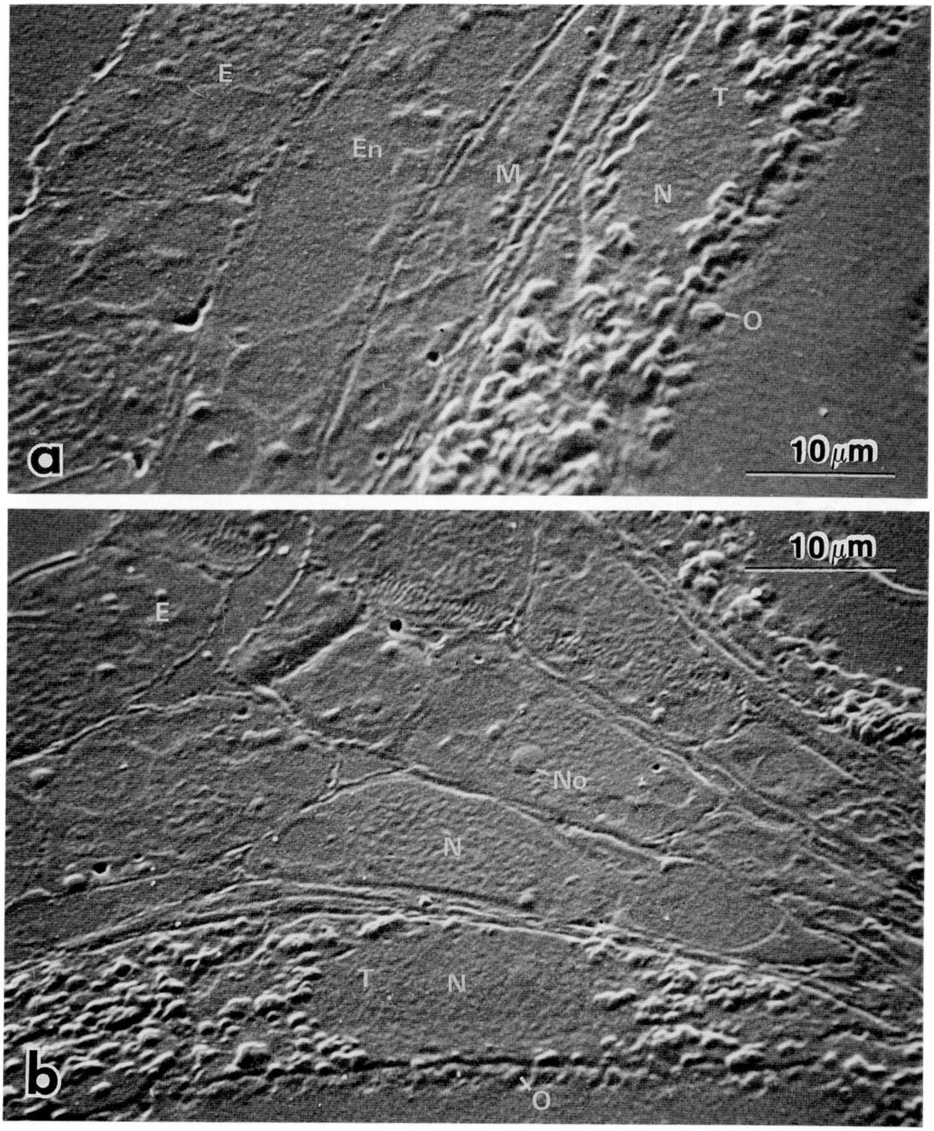

PLATE 4

SEM magnified soft x-ray contact images of the anther tissue of *Caltha palustris* L.. (a): The anther wall is composed of four layers of cells. Note the x-ray dense lipid granules in the tapetal cells. The high x-ray density of lipid granules is due to its high osmium content. (b): Anther wall near the intermicrosporangial stripe. Note the tapetal cell with numerous orbicules (O). Epidermis: E, Endothecium: En; Middle layer: M, Tapetal cell: T, Nucleus: N, Nucleolus: No. The tissue was fixed in Cheng's fixative (Cheng *et al.,* 1979), post-fixed in OsO_4, and embedded in Spurr's medium. The images were taken with C-K radiation on PMMA resist. (P. C. Cheng)

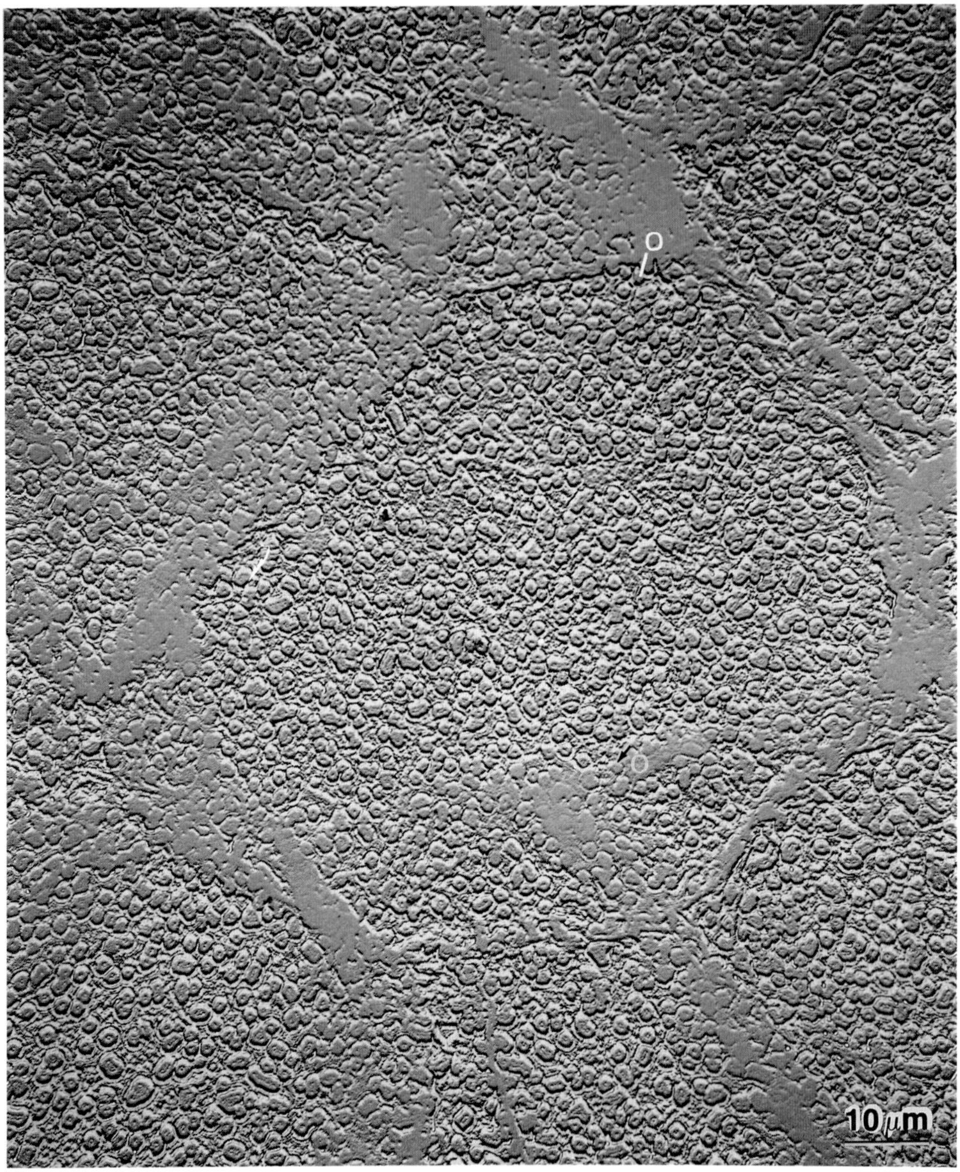

PLATE 5

Soft x-ray contact image of the tapetal film of *Zea mays*. The tapetal film was isolated from the locular surface of anther tapetum by $ZnCl_2$-HCl treatment. The isolated tapetal film was supported on an EM grid, and imaged with 35Å monochromatic x-rays on PMMA resist. After shallow development with MIBK/IPA (1:1, v/v) for three seconds, a metal replica was made from the contact image, and examined with a TEM. Note the numerous orbicules (O). (P. C. Cheng and D. M. Shinozaki)

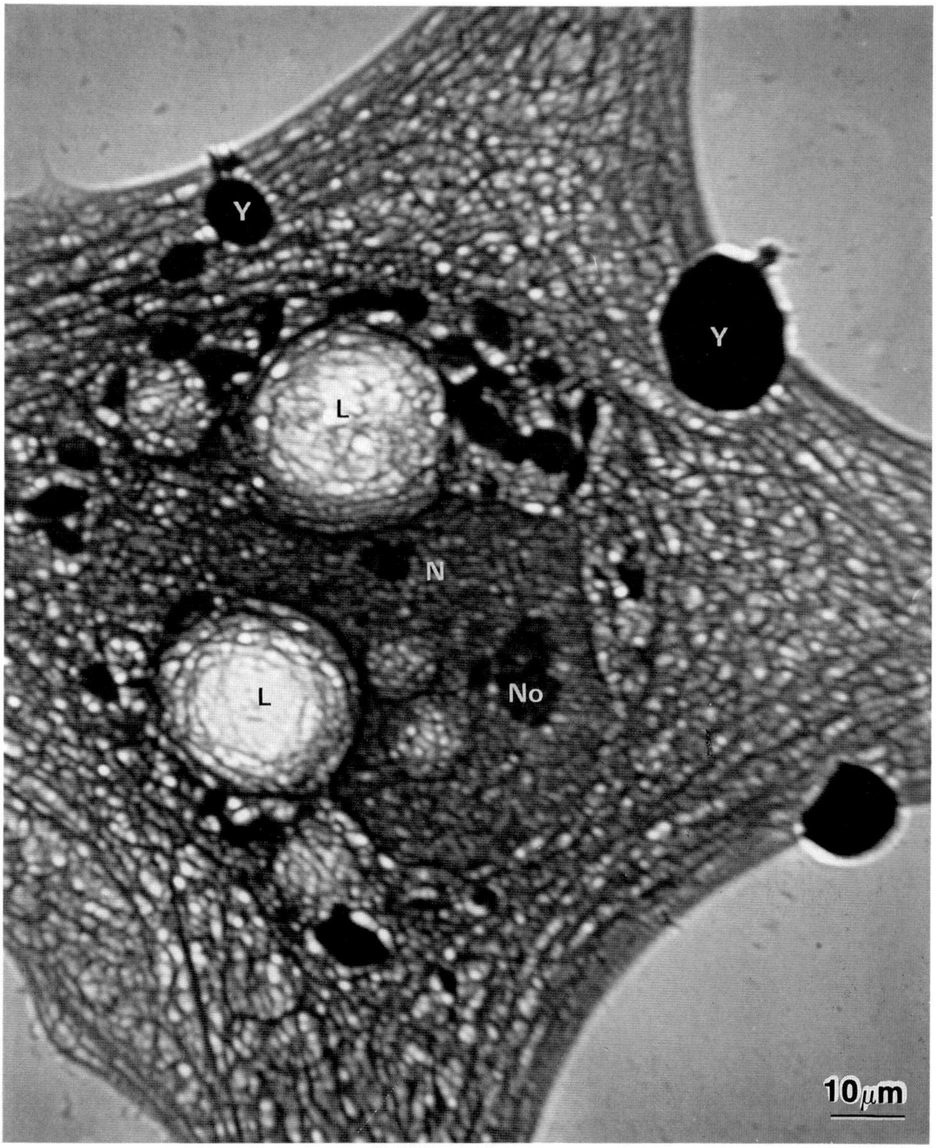

PLATE 6

TEM magnified soft x-ray contact image of *Xenopus laevis* myotomal muscle cell. The contact image was taken on a Si_3N_4- supported PMMA resist with C-K radiation. The cell was cultured on Formvar-coated EM grid, fixed in 1% glutaraldehyde, dehydrated in EtOH and critical-point dried. The dark granules (Y) are yolk granules. Note some of the yolk granules are extracellular structures (from another cell which have attached to the surface of the cell being viewed). L: lipid granules (removed during dehydration process). Nucleus: N, Nucleolus: No. (P. C. Cheng)

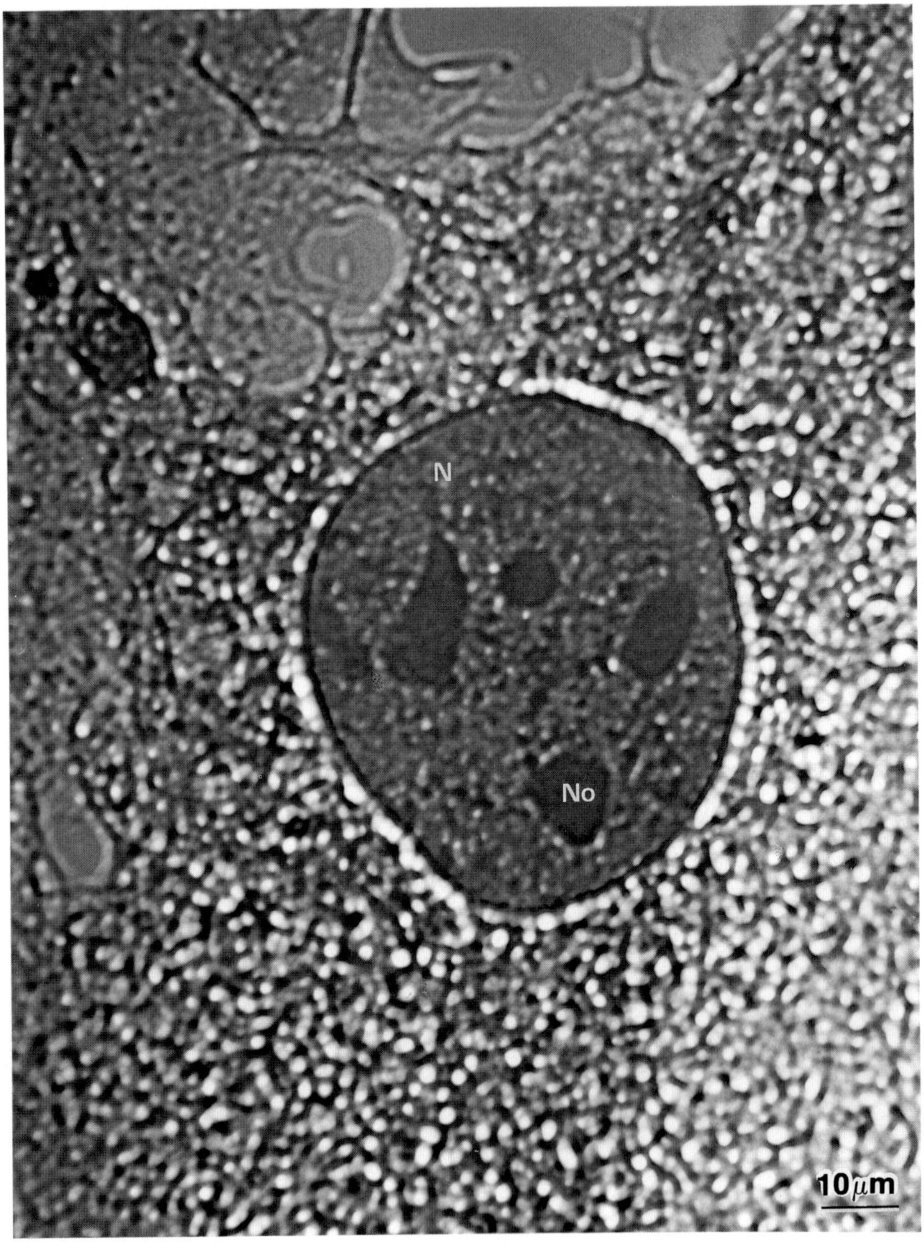

PLATE 7

TEM magnified soft x-ray contact image of a human fibroblast. The cell was cultured on a Formvar-coated EM grid, fixed in 1% glutaraldehyde, dehydrated in EtOH, critical-point dried in CO_2. The contact image was taken with 77 Å monochromatic x-rays on PMMA resist. Nucleus: N, Nucleolus: No. (P. C. Cheng, K. H. Tan and D. M. Shinozaki)

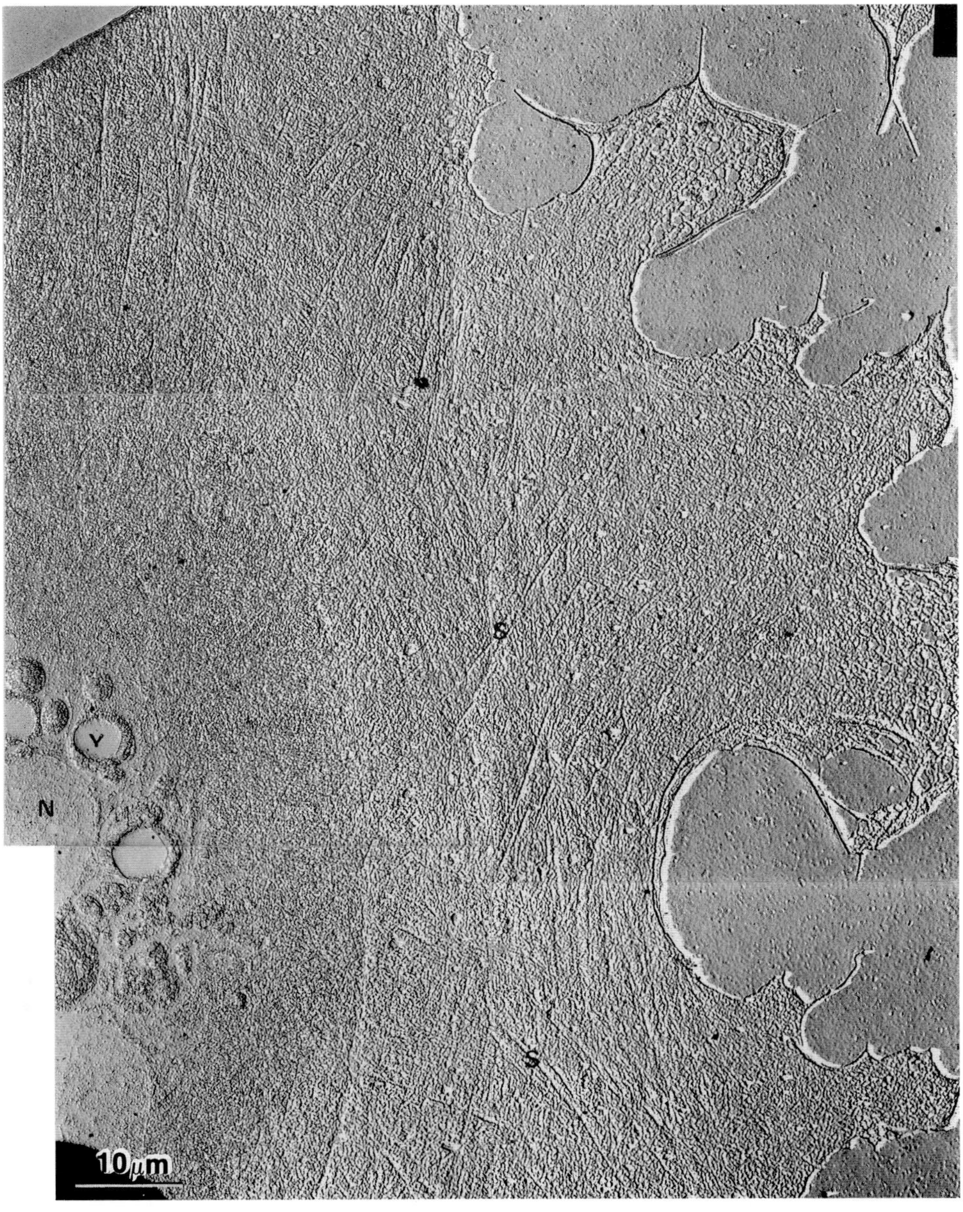

PLATE 8

Soft x-ray contact image of a embryonic *Xenopus laevis* fibroblast. The cell was cultured on a Formvar-coated EM grid, fixed in 1% glutaraldehyde, post-fixed in OsO_4, dehydrated in EtOH, and critical-point dried. The contact image was taken with 35 Å monochromatic x-rays on PMMA resist. A metal replica was made from the contact image, and viewed under a TEM. S: stress fiber, Y: yolk granules, N: nucleus. (P. C. Cheng and D. M. Shinozaki)

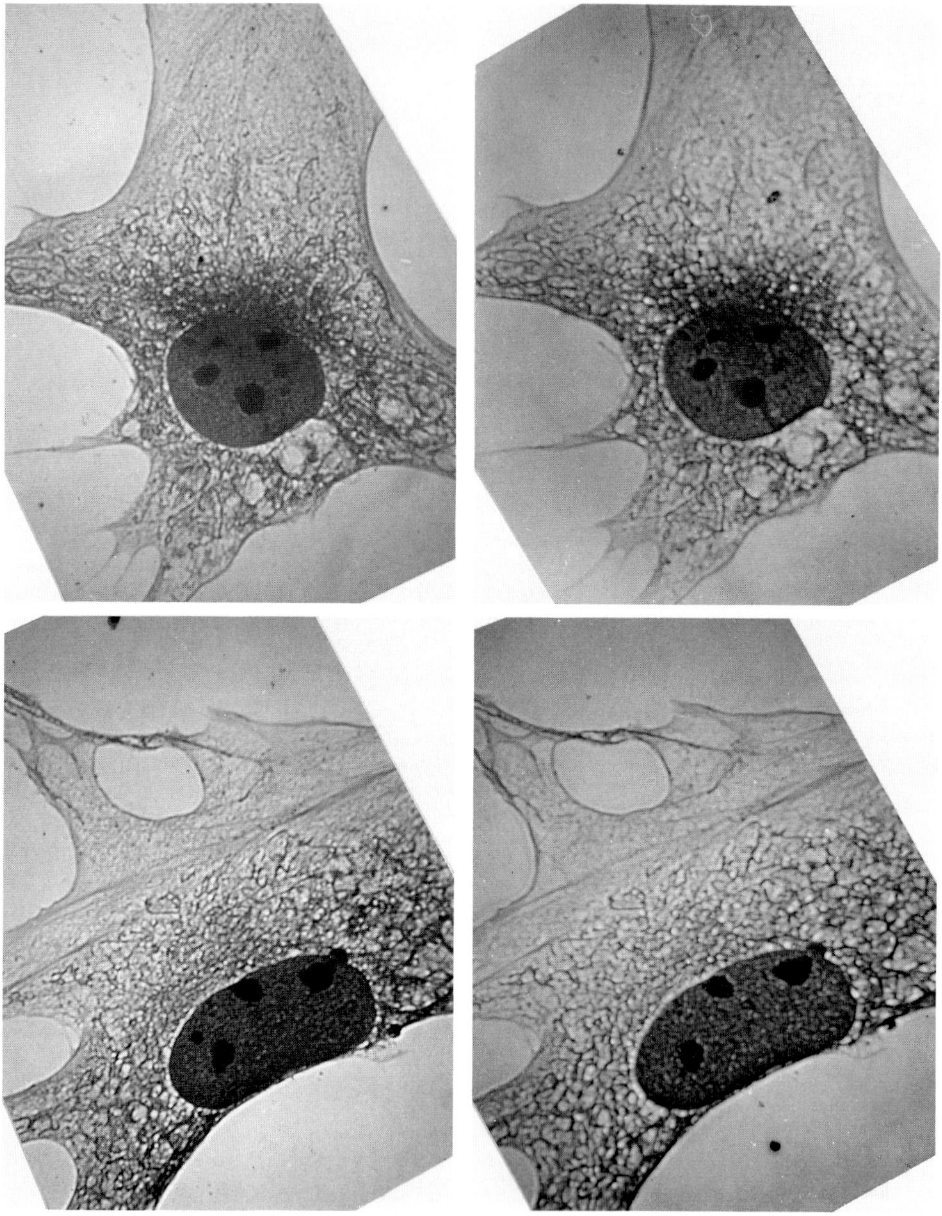

PLATE 9

Stereo-pair of mouse 3T3 fibroblasts. The cell was cultured on a Formvar-coated EM grid, fixed in 1% glutaraldehyde, dehydrated in EtOH, and critical-point dried. Due to the flatness of the cell, the image pair were taken at $\pm 20^\circ$ with C-K radiation on PMMA. (P. C. Cheng)

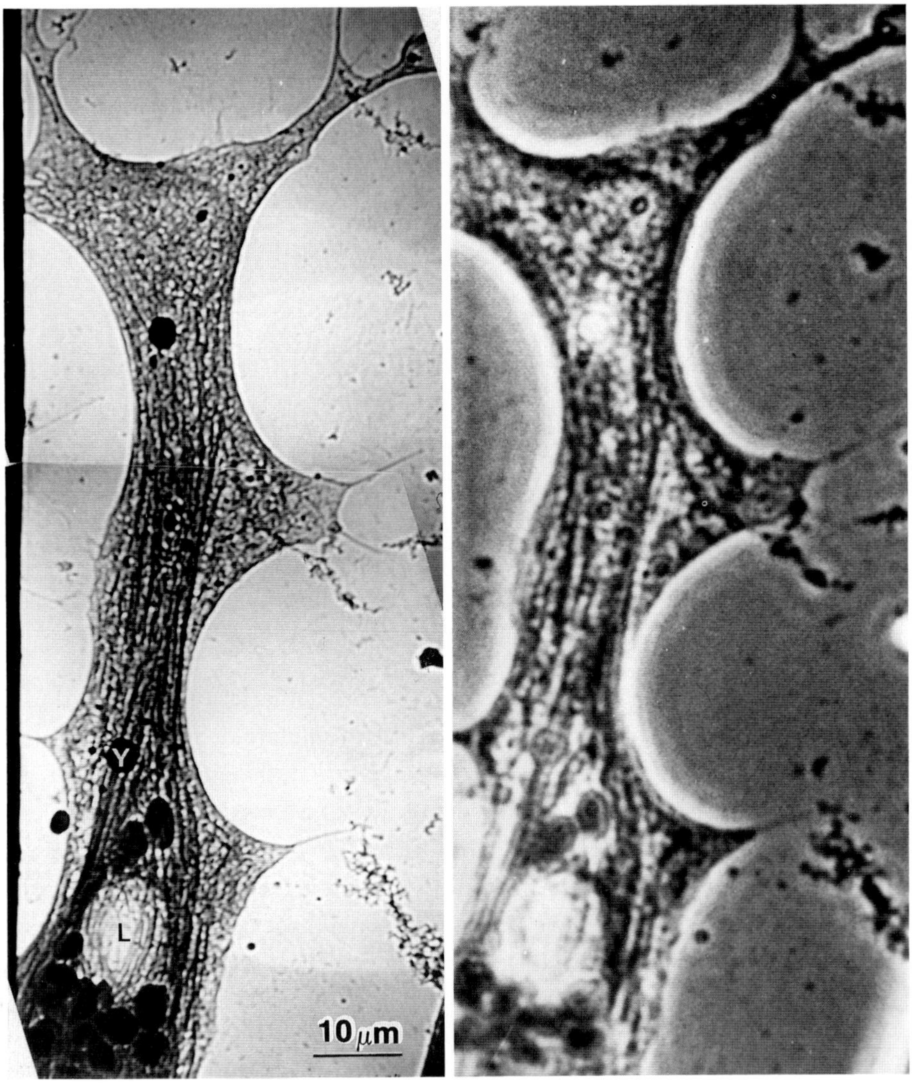

PLATE 10

Comparison of x-ray image with phase contrast image of the contact image. (a): TEM magnified x-ray contact image of *Xenopus laevis* myotomal muscle cell taken with C-K radiation on PMMA resist. The dark granules are yolk granules. Individual sarcomeres can be seen on the myofibrils. (b): Phase contrast micrograph of the contact image. The cell was cultured on a Formvar-coated EM grid, fixed in 1% glutaraldehyde, post-fixed in OsO_4, dehydrated in EtOH, and critical-point dried. (P. C. Cheng)

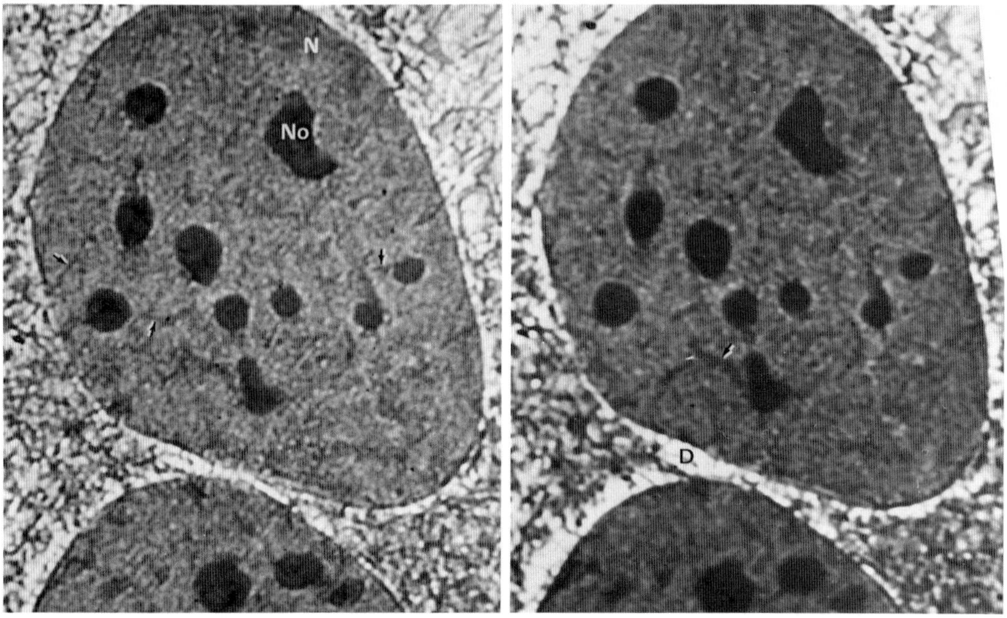

PLATE 11

Image of the nucleus of a 3T3 fibroblast before and after osmium fixation. A: before osmium fixation. The cell was fixed in 1% glutaraldehyde, dehydrated in EtOH, and critical-point dried. The x-ray image was obtained by using C-K radiation on PMMA resist. Note the x-ray dense nucleolus, arrows indicate structures not visible in the osmium fixed image (B). B: After osmium fixation. The above mentioned specimen was osmium fixed in OsO_4 vapor for 3 min., and the contact image was obtained at exactly the same conditions as those used in (A). Around the edge of nucleus, a bright halo (D) is observed. This is the result of Fresnel diffraction. (P. C. Cheng)

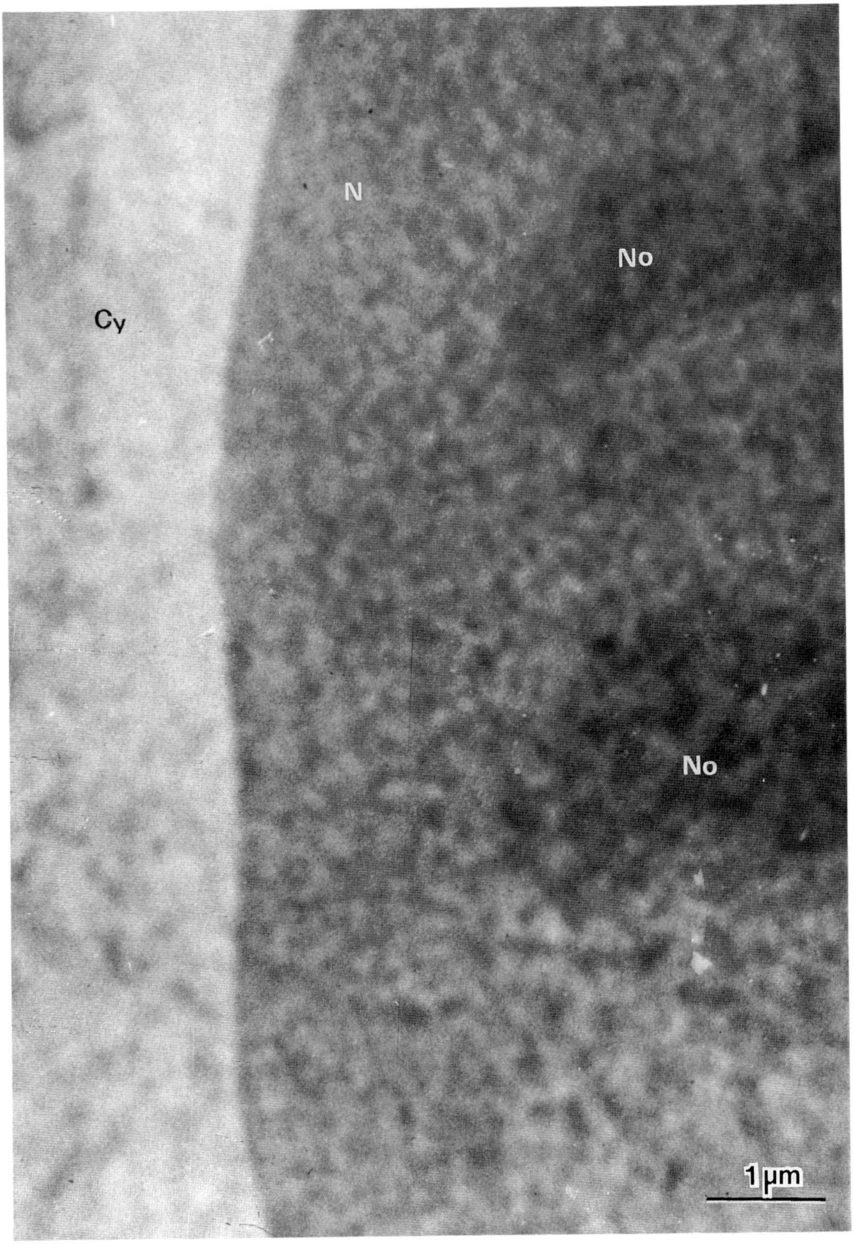

PLATE 12

Soft x-ray image of human fibroblast taken by the Göttingen x-ray microscope at 45Å The image shows the nuclear (N) with two x-ray dense nucleoli (No). Note the x-ray dense heterochromatin. Cy: cytoplasm. The cell was cultured on a Formvar film, fixed in 1% glutaraldehyde, dehydrated in EtOh and critical-point dried. (W. Meyer-Ilse, G. Nyakatura, P. Guttmann, B. Niemann, D. Rudolph, G. Schmahl and P. C. Cheng)

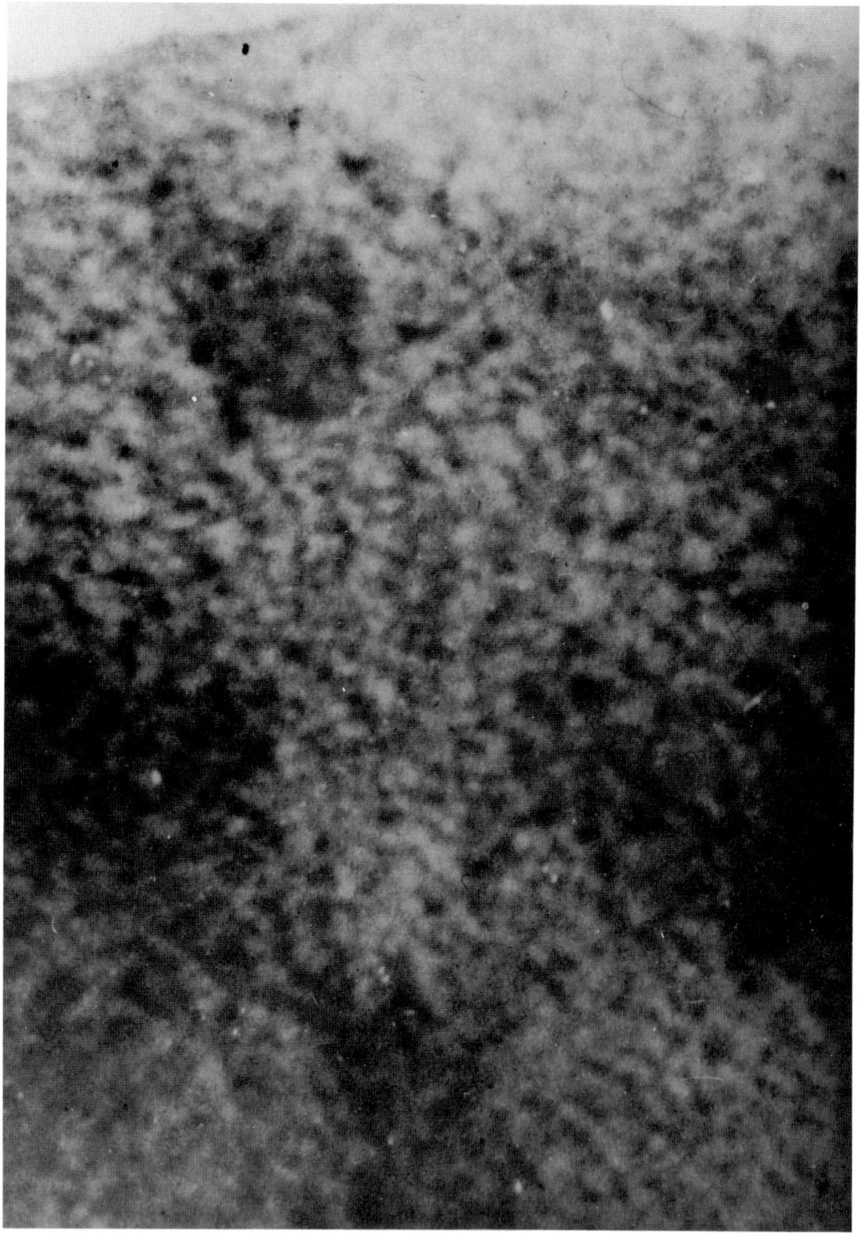

PLATE 13

Nuclear region of a human fibroblast imaged by the Göttingen x-ray microscope at 45Å The cell was cultured on a Formvar film, fixed in 1% glutaraldehyde, dehydrated in EtOh and critical-point dried in CO_2. N: nucleus, No: nucleolus, S: stress fiber. (W. Meyer-Ilse, G. Nyakatura, P. Guttmann, B. Niemann, D. Rudolph, G. Schmahl and P. C. Cheng)

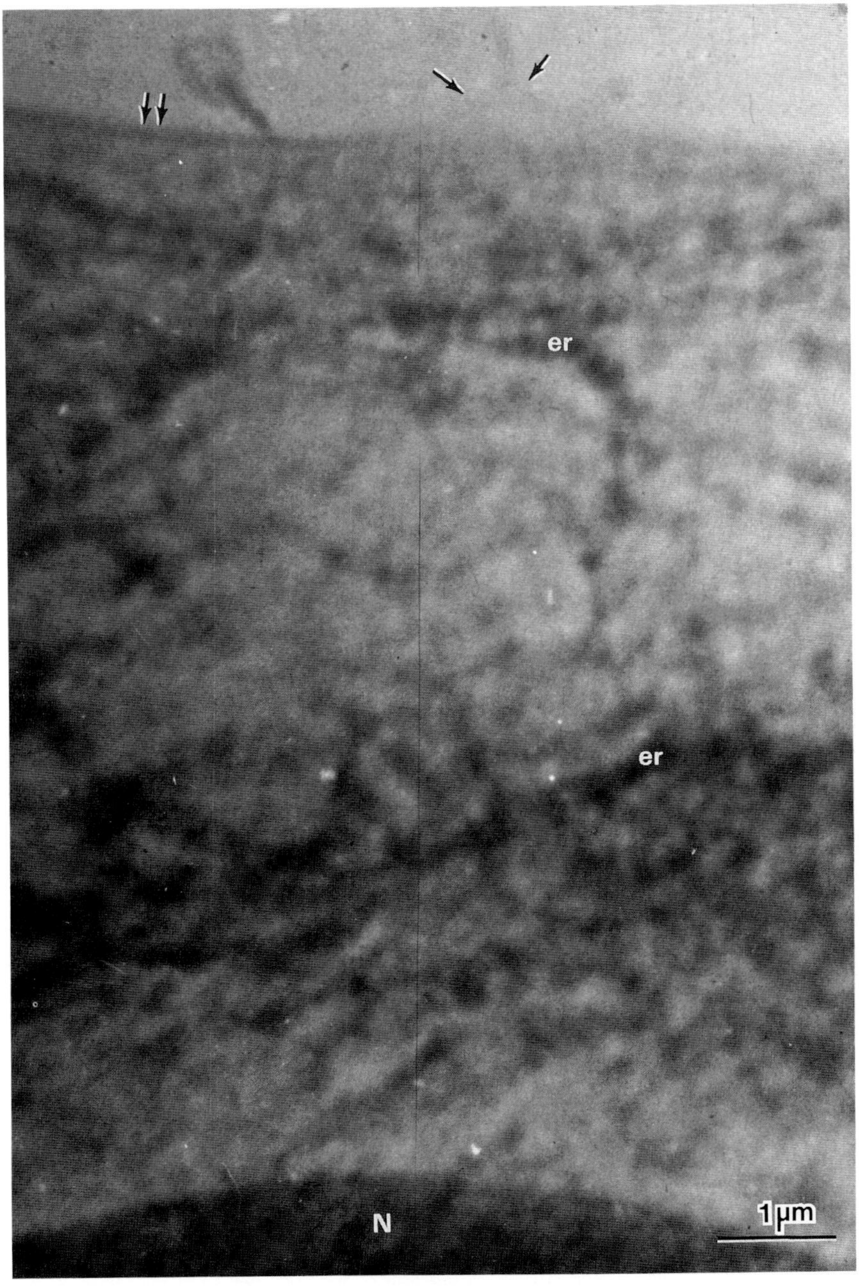

PLATE 14

X-ray image of a human fibroblast taken by the Göttingen x-ray microscope at 45Å The cell was cultured on a Formvar film, fixed in 1% glutaraldehyde, dehydrated in EtOh and critical-point dried. N: nucleus, No: nucleolus, er: endoplasmic reticulum, arrows: cell processes which attach to the culture substratum, double arrows: plasma membrane. (W. Meyer-Ilse, G. Nyakatura, P. Guttmann, B. Niemann, D. Rudolph, G. Schmahl and P. C. Cheng)

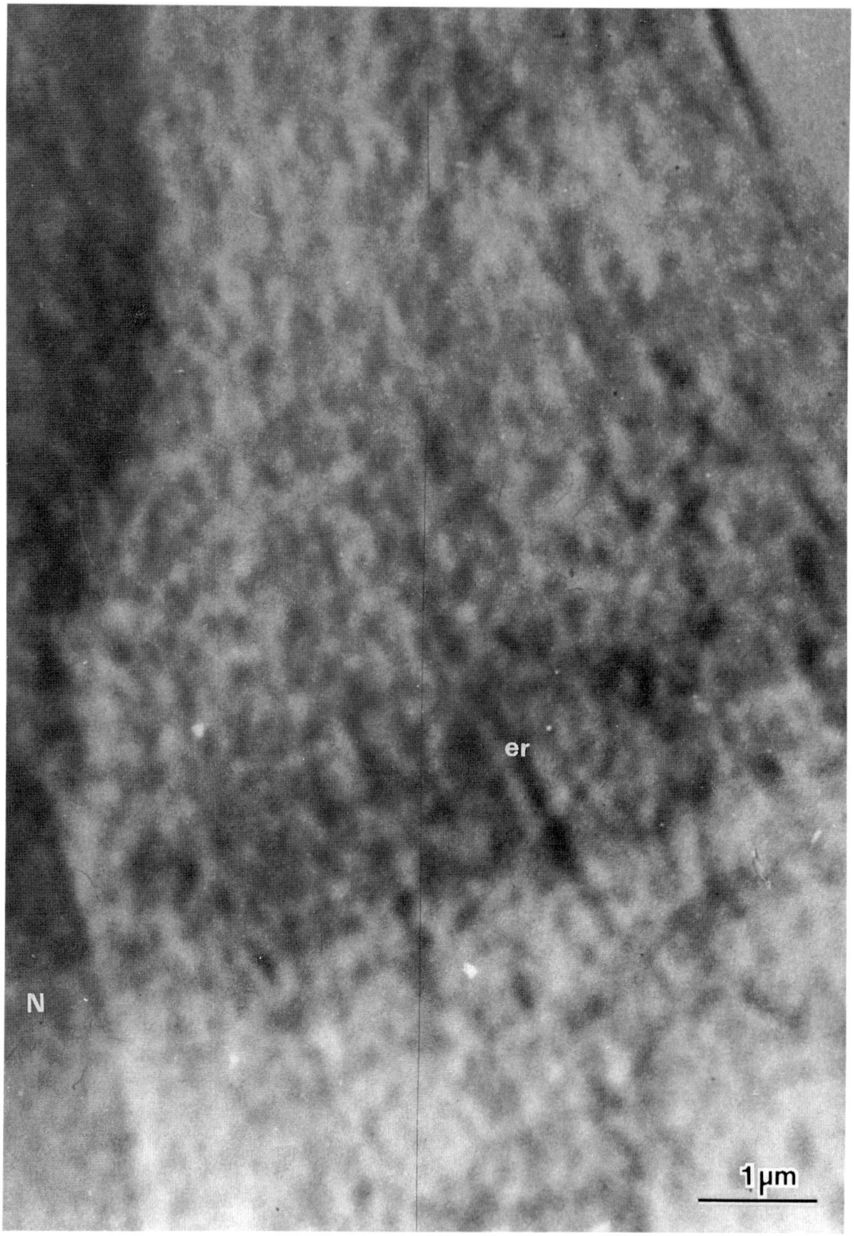

PLATE 15

Nuclear region of a human fibroblast imaged by the Göttingen x-ray microscope at 45Å The cell was cultured on a Formvar film, fixed in 1% glutaraldehyde, dehydrated in EtOh and critical-point dried. N: nucleus, er: endoplasmic reticulum. (W. Meyer-Ilse, G. Nyakatura, P. Guttmann, B. Niemann, D. Rudolph, G. Schmahl and P. C. Cheng)

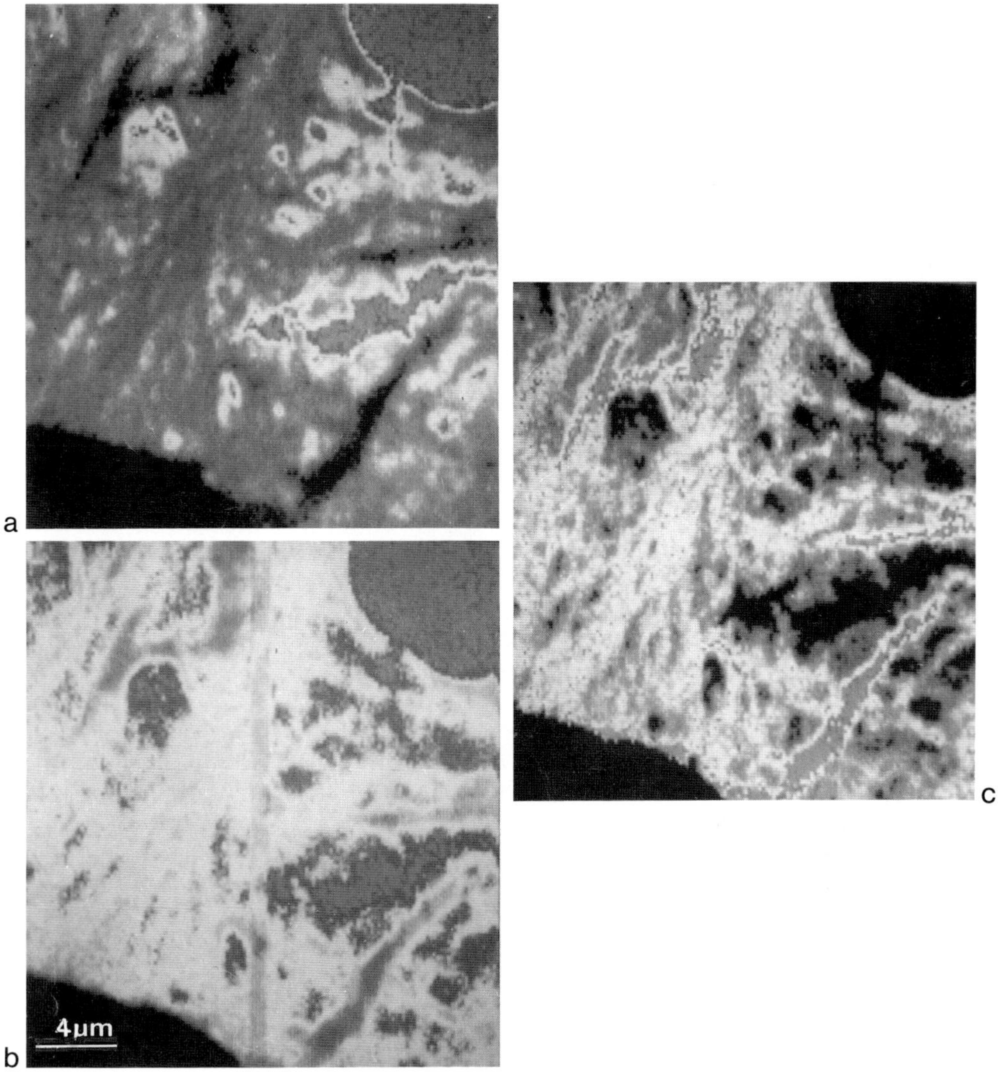

PLATE 16

X-ray micrographs of a 0.2μm thick epoxy section of human scull tissue. 140x140 pixels, 0.19μm each. (a): X-ray wavelength, 3.55nm, chosen for high absorption in calcium. (b): Similar to (a), except that x-ray wavelength, 3.58nm, has low absorption in calcium. (c): Calcium map constructed from data of (a) and (b). Higher calcium density is mapped into lighter shades. Specimen preparation is described in the article by Cinotti *et al.* in this volume. (J. Kirz)

Bibliography

This bibliography contains the cited references, and many uncited references relevant to x-ray microscopy. It is by no means complete, however, it represents one of the most extensive bibliography for this field. The editors acknowledge the help of S. P. Newberry for providing many of the early references.

Ade, H., Jacobsen, C., Kirz, J., McMulty, I., Rosser, R. J., Buckley, C., Burge, R. E., Browne, M. T., Cave, R., Charalamous, P., Duke, P. J., Kenney, J. M., Michette, A. G., Morison, G., Rarback, H., Shu, D., Feder, R., and Valdamirski, Y. (1986): Recent experiment in soft x-ray microscopy and zone plate replication. *Proc. ICXOMII.*

Al-Hadithi, Y., Eason, R. W., Michette, A. G., Hills, C., Ragoyski, A. M., O'Neill, F., Zammit, U., Turcu, I. C. E. (1986): Rutherford Appleton Laboratory Annual Report 1986: A4.16 - A4.l9.

Albery, W. J., and Knowles, J. R. (1976): Evolution of enzyme function and the development of catalytic efficiency. *Biochemistry*, 15: 5631-5640.

Alexander, A. E., and Sherwood, H. F. (1940): Radiography of cultured and natural pearls. *Photo Technique*, 3 (March): 50-52. Also *The Gemmologist*, 10 (Dec.): 45-52.

Allen, D. G., Blink, J. R., and Prendergast, F. G. (1977): Aequorin luminescence: Relation of light emission to calcium contration - A calcium-independent component. *Science*, 196: 996-998.

Amprino, R., and Engström, A. (1952): Studies on x-ray absorption and diffraction of bone tissue. *Acta anatomica*, 15: 1-22. Anderson, M. J., Cohen, M. W., and Zorychtat, E. (1977): Effects of innervation on the distribution of acetylcholine receptors on cultured muscle cells. *J. Physiol.* (London), 268: 731-756.

Anon. (1932): Flower radiography. *Radiography and Clinical Photography*, 8 (June): 12-13.

Anon. (1937): X-ray fingerprints. *Industrial and Engineering Chenistry (New Edition)*, 15 (Dec.): 541.

Anon. (1950): Radiography applied to microscopic specimens. *The Microscope*, 7 (Jan.): 295-298.

Anon. (1954): Microradiography and autoradiography. *Nature*, 173: 387-388.

Anon. (1954): A new compact device for microradiography in biology, medicine, mineralogy and metallurgy. *Norelco Reporter*, 1: 133.

Anon. (1947): Microradiography. *In*: Radiography in modern industry. Eastman Kodak Co., Rochester, NY. pp.45-46.

Aoki, S., and Kikuta, S. (1972): X-ray holographic microscopy, *Jpn. J. Appl. Phys.*, 11: 1857.

Aoki, S., and Kikuta, S. (1974): X-ray holographic microscopy. *Jpn. J. Appl. Phys.* 13:1385-1392

Aoki, S., and Kikuta, S. (1986): Soft x-ray intereferometry and holography. *In*: Short wavelength coherent radiation: generation and applications. eds Attwood, D. T., and Bokor, J. *Amer. Inst. of Physics*, New York, pp 49-56 (Optical science and engineering series, vol 7)

Applebaum, E. (1938): Grenz ray studies of the calcification of enamel. *J. Dental Research*, 17 (June):181-190.

Applebaum, E. (1938): Tissue changes in dental caries. *J. Amer. Dental Assoc.*, 25: 1223-1235.

Applebaum, E. (1940): The radiopaque surface layre of enamel and caries. *J. Dental Research*, 19 (Feb.): 41-46.

Applebaum, E. (1943): Grenz ray studies of enamel matrix formation and calcification. *J. Dental Research*, 22 (Feb.): 7-11.

Applebaum, E. (1948): Grenz-ray and histologic studies of tissue changes in dental caries. *Annals of Dentistry*, 7 (Sept.): 67-90.

Applebaum, E., and Adam, T. H. (1938): Decalcification versus mechanical injury in caries. *J. Dental Research*, 17 (April): 95-100.

Applebaum, E., Hollander, F., and Bödecker, C. F. (1933): Normal and pathographical variations in calcification of teeth as shown by the use of soft x-rays. *Dental Cosmos*, 75: 1097-1105. Also, *Transactions of the Dental Society of the State of New York*, pp 77-85.

Ardenne, M. von (1939): Zur Leistungsfähigkeit des Elektronen-Schattenmikroskopes und über ein Röntgenstrahlen- Schattenmikroskop. (The efficiency of the electron microscopes, An x-ray shadow microscope) *Naturwissenschaften*, 27: 485.

Ardran, G. M. (1953): Radiology in research. *In*, Modern trends in diagnostic radiology (second series). ed. McLaren, J. W., Paul B. Hoeber Inc., Medical Book Department of Harper & Brothers, NY. pp43-52.

Aritome, H., Nagata, K., and Namba, S. (1985): Fabrication of x-ray zone plates with a minimum zone width smaller than 100nm by electron beam lithography. *Microelectronic Engineering*, 3: 495-466.

Baez, A. V. (1952): A study in diffraction microscopy with special referenceto x-rays, *J. Opt. Sci. Amer.*, 42: 756-762.

Bailey, J., Ethinger, Y., Fisher, A., and Feder, R. (1982): Evaluation of the gas puff z-pinch as an x-ray lithography and microscopy source. *App. Phys. Lett.*, 40:33-35.

Baldini, M. G., Kim, B. K., Timmons, S., Feder, R., Sayre, D., and Banton, V. (1986): Soft x-ray microscopy of activated human platelets. *In*: NATO ASI Series, Improved methods for examing the submicron world, eds. Feder, R., McGowan, J. Wm., and D. M.Shinozaki, New York, Plenum Press.

Baldwin, F. and Diocee, M. S. (1985): Rapid specimen preparation using microwave energy. *Proc. 43rd Annual Meeting of the Electron Microscopy Society of America (ed. Bailey, G. W.),* 448-449.

Baldwin, K. G. H., and Rosser, R. J. (1987): Laser produced plasmas as soft x-ray sources. This proceedings.

Ball. L. W. (1945): X-ray micrography as a tool for foundary control. *Industrial Radiography*, 4 (Summer): 29-36.

Bancroft, G. M. and McGowan, J. Wm. (1980): Canadian synchrotron radiation facility. *Phys. in Can.*, 36: 53. Barnard, J. E. (1915): X-rays in relation to microscopy. *J. Roy. Microscop. Soc.*, Feb. 1915. 1-7.

Baranov, V. I., Zdanov, A. P., and Diesenrot-Mysovskaya, M. JU. (1944): Utilization of microradiographic methods for investigation of the nature of distribution of radioactive elements in the natural objects. (in Russian) *Bulletin de l'Academie des sciences de l'URSS*, 1: 20-28.

Barclay, A. E. (1947): Micro-arteriography. *British J. Radiography*, 20: 394-404.

Barclay, A. E. (1948): Micro-arteriography. *Amer. J. Roentgenology*, 60 (July): 1-12.

Barclay, A. E. (1949): Demonstration of iron in cells. *British J. Radiology* , 22: 268-269.

Barclay, A. E. (1951): Micro-arteriography. Charles C. Thomas, Springfield, IL; Blackwell Scientific Publcations Ltd., Oxford, UK, and The Ryerson Press, Toronto, Ont., Canada.

Barclay, A. E., and Bentley, F. H. (1949): The vascularisation of the human stomach. A preliminary note on the shunting effect of trauma. *British J. Radiology*, 22 (Feb.), 62-67. *Gastroenterology*, 12: 177.

Barclay, A. E., Daniel, P., Powell, H. M., and Prichard, M. M. L. (1946): Radio- micrography. A preliminary communication. *Proc. Physiol. Soc.*, (June 15) pp28P-29P. (Published with *J. Physiology*, 105).

Barclay, A. E., and Leatherdale, D. (1948): Microradiography applied to botany. *British J. Radiology*, 21: 544-551.

Barer, G. R. (1952): A radiographic study of tissue permeability in lymph drainage in inflammation. *British J. Experimental Pathology*, 33: 123-130.

Barnard, J. E. (1915): X-rays in relation to microscopy. *J. Royal Microscopical Society*, February, pp.1-7.

Barrett, C. S. (1945): A new microscopy and its potentialities. *Metals Technol.*, 12: Technical Publication No. 1865.

Barrett, C. S. (1945): New microscopy and its potentialities. *Transaction of the American Institute of Mining and Metallurgical Engineers*, 161: 15-64.

Barrett, C. S. (1949): X-ray microscopes. *Metal Progress*, 55:848, 890-892.

Bassous, E., Feder, R., Spiller, E., and Topalian, J. (1976): High Transmission X-ray Masks for Lithographic Applications, *Sol. State Technol.* 19: 55.

Baud, C. A., and Gossi, M. (1978): Quantitative microradiographic studies of the degree of mineralization of bone tissue; influence of age, sex and desease. *Amer. J. Roentgenol. Radium Ther. Nucl. Med.* 131:549 Abstract 4th Int. Conf. on Bone Measurement.

Baud, C. A., Pouezat, J. A., and Touchon-Danguy, H. J. (1976): Quantitative analysis of amorphous and crystalline bone tissue mineral in women with osteoporosis. *Calcif. Tiss. Res.* 21:suppl.452-456.

Bearden, J. A. (1967): X-ray wavelengths. *Rev. Mod. Phys.*, 39: 78-124.

Bearden, J. A., and Burr, A. F. (1967): Reevaluation of x-ray atomic energy levels. *Rev. Mod. Phys.*, 39: 125-142.

Beese, L., Feder, R., and Sayre, D. (1986): Contact x-ray microscopy: a new technique for imaging fine cellular structure. *J. Biophys.* 49:259-268

Bellman, S. (1952): Mikroangiografi. (Microangiography) *Svenska läkartidningen*, 49: 2753-2759.

Bellman, S. (1952): Mikroangiografi. (Microangiography) *Angéiologie et annales de la Sociéte fracaise d'angéiologie et d'histopathologie*, Supplement to *Gazette médicale de France*, 59: 35.

Bellman, S. (1953): Microangiography. *Acta radiologica Supplementum*, 102.

Bellman, S., Block, E., and Odeblad, E. (1953): Microangiographic study of the minute ovarian blood vessels in albino rats. *British J. Radiology*, 26: 584-588.

Bellman, S., and Engfledt, B. (1955): A microangiographic study on the effect of gonadotrophin upon the blood vessel system of the ovaries of rabbits. *Acta Radiologica*, 43: 459-468.

Bellman, S., and Engström, A. (1952): Microangiography. *Acta Radiologica*, 38: 98-110.

Beneke, G. (1966): Historadiography. *In*: Introduction to quantitative cytochemistry. ed Wied, G. L., Academic Press, pp. 107-151.

Bentley, F. H., and Barlow, T. E. (1951): Aetiology of chronic gastric ulcer. *Lancet*, 260: 470-471.

Berg, W. (1931): Über eine röntgenographische Methode zur Untersuchung von Gitterstörungen an Kristallen. (A radiographic method for investigation of the lattice deformation of crystals), *Naturwissenschaften*, 19: 391-396.

Berg, W. (1934): About the history of load of deformed crystals. *Zeitschrift für Kristallographie*, 89: 286-294.

Berghézan, A., Lacombe, P., and Chaudon, G. (1950): Étude microradiographique des contours de grains de l'aluminium pur et de la solution solide aluminium- zinc au cours de leur fusion commencante. (Microradiographic study of the grain boundaries of pure aluminum and aluminum-zinc solid solution during the beginning of fusion.) *Comptes rendus de l'Academie des sciences*, 231: 576-578.

Bergman, G., and Engfeldt, B. (1954): Studies on mineralized dental tissues. II. Microradiography as a method for studying dental tissue and its application to the study of caries. *Acta Odontologica Scandinavica*, 12: 99-132.

Bergman, G., and Engfeldt, B. (1954): Studies on mineralized dental tissue. III. Dental changes in a case of osteopetrosis. *Acta Odontologica Scandinavica*, 12: 133-144.

Bergman, G., and Engfeldt, B. (1954): Studies on mineralized dental tissues. IV. Biophysical studies on teeth and tooth-germs in osteogenesis imperfecta. *Acta Microbiologica Scandinavica*, 35: 537-548.

Bergman, G., and Engfeldt, B. (1955): Studies on mineralized dental tissues. V. Microradiographic investigation of experimentally produced caries and attrition in white rats. *Experimental Cell Research*, 12: 193-200.

Bergman, G., and Engfeldt, B. (1955): Studies on mineralized dental tissues. VI. The distribution of mineral salts in the dentine with special reference to the dental tubules. *Acta Odontologica Scandinavica*, 13: 1-8.

Berman, A. I., and Dutli, J. W. (1952): X-ray and electron radiography of thin specimens. *AECU-1885, LADC-1064.* (Published by the US Atomic Energy Commission)

Berman, A. I., and Dutli, J. W. (1955): Roentgen-ray and electron radiography of thin specimens. *Amer. J. Roentgenology*, 74: 518-525.

Berthold, R. (1941): Vergrösserte Röntgen-Raumbilder von Metalldunnschliffen. (Enlarged x-ray three-dimensional images of polished microsections of metals.) *Metallwirtschaft*, 20: 694-697.

Bertin, E. P., and Samber, R. L. (1967): Contact microradiography. *In*: Handbook of x-rays. ed. Kaelble, E. F., McGraw Hill, New York. pp. 45/3-45/27.

Bertwistle, A. P. (1932): Prehistoric bones. *British J. Radiology*, 5: 589-591.

Betteridge, W. (1950): Microradiography and metals. *Metal Industry*, 76 (Jan.): 5-8.

Betteridge, W., and Sharpe, R. S. (1948): The study of segregations and inclusions in steel by microradiography. *J. Iron Steel Institute*, 158: 185-191.

Betteridge, W., and Sharpe, R. S. (1948): Segrefations and inclusions; Their study in steel by microradiography. *Iron and Steel*, 21: 242-245.

Betteridge, W., Wilson, C., Haughton, M. A., and Morgan, W. (1949): The influence of over-ageing and annealing on the hardness and microstructure of an aluminum alloy to British Standard Specification L42. *J. Inst. Metals*, 75: 641-664.

Betz, H., Heinrich, K., Heuberger, A., Huber, H., and Oertel, H., (1986): Resolution limits in x-ray lithography calculated by means of x-ray lithography simulator-XMAS. *J. Vac. Sci. Technol.*, B4, 240-252.

Bohatirchuk, F. (1942): Die Fragen der Microrontgenographie. (The question of microradiography.) *Fortshritte auf dem Gebiete der Röntgenstrahlen*, 65: 253-261.

Bohatirchuk, F. (1943): Totale expérimentelle Makro-und Mikrovasographie, die Experimente mit kontrastierten Leber und Milz und der neue Röntgenographishe Test vom Zustand der activen physiologischen Gewebe. (Experimental macro- and microsvasography with contrast media in the liver and spleen and new radiographic test of the condition of the active physiological tissue.) *Fortschritte auf dem Gebiete der Röntgenstrahlen*, 68: 159-174.

Bohatirchuk, F. (1944): Über Ergebnisse der Mikroröntgenographie. (The results of microrentgenography.) *Acta radiologica*, 25: 351-365.

Bohatirchuk, F. (1944): Makro-und Mikrorontgenographie der Blutgefaessen bei experimentellen Geschwulsten. (Macro- and microradiography of the blood vessels in the case of experimental tumors.) *Virchow's Archiv fur Pathologische Anatomie und Physiologie und fur Klinische Medizin*, 313: 216-228.

Bohatirchuck, F. (1953): histotoentgenography (microradiography) of impregnated cancer tissue. *Amer. J. Roentgenology*, 70: 119-125.

Bohatirchuck, F. (1954): Some microradiographical data on bone ageing. *British J. Radiology*, 27: 177-182.

Bohatirchuck, F. (1955): The ageing vertebral column. (Macro- and Historadiographical study.) *British J. Radiology*, 28: 389-404.

Bohatirchuck, F. (1957): Applied microradiography in medical research: Its contemporary difficulties and prospective possibilities. *In*: X-ray microscopy and microradiography. eds. Cosslett, V. E., Engstrom, A., and Pattee, H. H., Academic Press, New York. pp. 259-266.

Bojarski, Z. (1951): Technika mikrorentgenografii absorpcyjnej i przyktady jej zastosowania do badan metalograficznych. (The technique of microradiography by absorption and some examples of application of this method of metallographic investigation.) *Prace Glownego Instytutu Metalurgii*, 3: 417-428.

Bond, A. F. (1935): Radiographs of prehistoric subjects. *X-ray Technician*, 7: 90-91.

Bosov, N. G. *et al.* (1966): Heating and disintegration of a plasma produced by a gigantic laser pulse focused on a solid target. *Zh. Eksper. Teor. Fiz.*, 51:989-1000.

Boss, G. H. (1948): macroscopic examination of spot welds in aluminum. *Metal Progress*, 53: 227-230.

Boss, G. H. (1948): Radiographic appraisal of spot welds in aluminum. *Metal Progress*, 53: 522-527.

Boss, G. H. (1948): A camera for microradiography. *Metal Progress*, 54: 689.

Botden, P. J. M., Combée, B., and Houtman, J. ((1952): An experimental x-ray apparatus with midget x-ray tube. *Philips Technical Review*, 14: 165-174.

Bourghardt, S. O., Brattgard, S. O., Hyden, H., Jiewertz, B., and Larsson, S. (1953): A microphotometer for microradiography. *J. Scientific Instruments*, 30: 464-466.

Bowen, D. K. (1984): *In*: Applications of Synchrotron Radiation to the Study of large Molecules of Chemical and biological Interest, ed. D.K. Bowen Daresbury Laboratory Report, DL/SCI/R13 (1982) P.37.

Boyes, E.D., (1984): High resolution at low voltage: The SEM philopher's stone. *Proc. 42nd Annual Meeting of the Electron Microscopy Society of America*, 446-449.

Brattgärd, S. O., and Hallén, O. (1952): X-ray microradiography as a quantitative method. *Boiochimica et Biophysica Acta*, 9: 488-495.

Brattgärd, S. O., Hallén, O., and Hydén, H. (1953): A critical evaluation of x-ray microradiography. *Biochimica et Biophysica Acta*, 10: 486-487.

Brattgärd, S. O., and Hallén, O., and Hydén, H. (1953): The elimination of errors in Roentgen microradiography. *Acta Radiologica*, 39: 494-498.

Brattgärd, S. O., and Hydén, H. (1952): Mass, lipids pentose nucleoproteins and proteins determined in nerve cells by x-ray microradiography. *Acta Radiologica, Supplementum*, 94.

Breton, C., and Papoular, R. (1973): VUV radiation of laser produced plasmas, *J. Opt. Soc. Amer.*, 63:1275.

Bronn, F. C., Bachrach, R., and Bianconi, A. (1978): Fine structure above the carbon K edge in methane and in the fluoromethanes. *Chem. Phys. Lett.*, 54: 425-429.

Buckley, C. J., Browne, M. T., and Charalambous, P. (1985): Contamination lithography for the fabrication of zone plate x-ray lenses. *Proc. Conf. E-Beam, X-ray & Ion Beam Tech*, 5: SPIE.

Buckley, C. J. (1987): The fabrication of gold zone plates and their application to x-ray microscopy, PhD thesis (to be published), KCL. Senate House, London University.

Buckley, C. J. (1986): Soft X-Ray Nano Lithography of Semitransparent Masks for the Generation of High-Resolution High-Contrast Zone Plates. Proc. Conf. Short wavelength coherent radiation: Generation & Applications. *Amer. Inst. Phys. Conf. Proc.*, No 147.

Burch, G. J. (1896): Plant structure revealed by Roentgen rays. *Nature*, 54: (June 4), p111.

Burge, R. E., Michette, A. G., Charalambous, P., Browne, M. T., Simpson, M. J., Buckley, C. J., and Duke, P. J. (1986): Development of a scanning x-ray microscope at the Daresbury Laboratory. *In*: Examining the submicron world (NATO ASI series). eds. Feder, R., McGowan, J. Wm., and Shinozaki, D. M., Plenum Press, New York. pp.265-276.

Burgess, D. D., Fawcett, B. C., and Peacock, N. J. (1967): Vacuum ultra-violet emission spectra from laser produced plasmas. *Proc. Phys. Soc.*, 92:805-816.

Burgess, M. D. J., Dragila, R., Luther-Davies, B., Nugent, K., and Tallents, G. (1984): The state of ultra-high intensity laser plasma interaction research in Australia. *In*: Laser Interaction and Related

Plasma Phenomena, Vol. 6. eds Hora, H. and Miley, G. H., Plenum Press, New York and London 461-477.

Burnett, H. C., Jr., and Vacher, H. C. (1950): A note on the formation and structure of iron dendrites in a magnesium alloy. *J. Research*, 44: 443-446.

Carroll, P. K., Kennedy, E. T., and Sullivan, G. O. (1980): Laser-produced continua for absorption spectroscopy in the VUV and XUV. *Appl. Optics*, 19: 1454-1462.

Cameron, J. R., Mazess, R. E., and Sorenson, J. A. (1968): Precision and accuracy of bone mineral determination by direct photon absorptometry. *Invest. Radiol.* 3:141-150.

Canals, M. (1926): Estéreo-radio-micrografiá (Stereoradiomicrographs.) *EI Progreso fotografico*, 7: 140-149.

Canals, M. (1927): Die Röntgenstereomikrographie. (X-ray stereo microradiography.) *Photographische Korrespondenz*, 1-6-12 (Jan.).

Carlson, T. A., and White, R. M. (1968): Measurement of the relative abundances and recoil energy spectra of fragment ions produced as the initial consequences of x-ray interaction with C_2H_5I, CH_3CD_2, and $Pb(CH_3)_4$. *J. Chem. Phys.*, 48: 5191-5194. Carlstrom, D., and Lundberg, B. (1958): The projection x-ray microscope for divergent beam diffraction. *J. Ultrastruct. Res.*, 2: 261-268.

Castel, P., Lamarque, P., and Turchini, J. (1936): Historadiographie et localisation histologique de substances médicamenteuses ou toxiques á poids atomique élevé. (Historadiography and histological localization of drugs or toxic substances of high atomic weight.) *Comptes rendus de la Société de biologie*, 123: 1051-1052.

Castellanos, I. (1927): Los Rayos X en la técnica policial. (Use of Roentgen rays by police officials.) *Revue de criminologie psiquiatrique*, 14: 84-96.

Castellanos, I. (1933): Radiography and dactyloscopy. *Sparks from the Anvil* (office publcation of the International Association for Identification), 1: 3-4.

Cauchois, Y. (1946): Formation of images by x-rays. Possibility of producing lenses and microscopes by the use of crystals. *Comptes rendus de l'Academie des sciences*, 223: 82-84.

Cauchois, Y. (1950): Formation of images with x-rays. *Revue d'optique*, 28: 151-163.

Chantraine, H. (1928): Ueber Weichstrahlaufnahmen mit der Gleichspannungsmaschine 'Trifas' der Elektrizitätsgesellschaft 'Sanitas.". (Soft-ray pictures with direct-current machine "Trifas.".) *Fortschritte auf dem Gebiete der Röntgenstrahlen*, 38: 534-541.

Cheavin, W. H. S. (1945): Photographing stamps by means of x-rays. (Letter Written to the president, The Royal Philatelic Society, Nov. 1, 1945.) *London Philatelist*, 54: 130-132.

Cheavin, W. H. S. (1947): Philately and x-rays. *Philately*, 1: 149-151.

Chen, C. P., and Lawrence J. M. (1986a): The ultrastructure of the plumula of the tooth of *Lytechinus vareigatus* ** (*Echinodermata: Echinoidea*). *Acta Zool.* 67(1): 33-41.

Chen, C. P., and Lawrence J. M. (1986b): Localization of carbonic anhydrase in the plumula of the tooth of *Lytechinus vareigatus* (*Echinodermata: Echinoidea*). *Acta Zool.*, 67(1): 27-32.

Chen, C. P., and Lawrence J. M. (1987): The role of carbonic anhydrase in the tooth of *Lytechinus vareigatus* (*Echinodermata: Echinoidea*). *Comp. Biochem. Physiol.* (in press).

Cheng, P. C., Walden, D. B., and Greyson, R. I. (1979): Improved plant microtechniques for TEM, SEM and LM specimen preparation. *Natl. Sci. Counc. Monthly, ROC* 7: 1000-1007.

Cheng, P. C., and Lin, M. I. (1980): Structure of Ubisch body and cuticle of *Caltha palustris* L. by LM, SEM, TEM and soft x-ray lithography. *Botany 80, BSA miscellaneous series* #158: p22.

Cheng, P. C., Findlay, W., McGowan, J. Wm., and Feder, R. (1980): Analysis of plant cells using soft x-ray lithography. *J. Cell Biol.* 87: 229a.

Cheng, P. C., and Lin, M. I. (1981): Technique and apparatus for serial ultra-thin sectioning. *Natl. Sci. Counc. Monthly, ROC*, 9: 15-22,

Cheng, P. C., Peng, H. B., Findlay, W., McGowan, J. Wm., and Feder, R. (1981): Soft x-ray lithography (microscopy) of cultured *Xenopus laevis* cell. 94th Annual Section of Amer. Assoc. of Anatomist. *The Anatomical Record* 199(3): 50a.

Cheng, P. C., Findlay, W., McGowan, J. Wm., and Feder, R. (1981): Soft x-ray images of plant cells- A new technique for the study of cells. *BSA miscellaneous series* #160: p7.

Cheng, P. C., Findlay, W., McGowan, J. Wm., and Feder, R. (1981): Soft x-ray microscopy- A new technique for the study of plant cell. 13th International Botanical Congress. Sydney. p39.

Cheng, P. C., and Peng, H. B. (1981): Letter to the editor: Soft x-ray contact microscopy. *Sci. Monthly (Taipei)* 12(9): 11-12.

Cheng, P. C., Peng, H. B., Feder, R. and McGowan, J. Wm. (1982): The use of transmission electron microscope as a viewing tool for high resolution soft x-ray contact microscopy. *Electron Microscopy* Vol. 1: 461-462.

Cheng, P. C., Tan, K. H., McGowan, J. Wm., Feder, R., Peng, H. B., and Shinozaki, D. M. (1984): Soft x-ray contact microscopy and microchemical analysis of biological specimens. In *Springer Series in Optical Sciences: X-ray microscopy*, eds. Schmahl, G., and Rudolph, D., Springer-Verlag, Berlin, 285-293.

Cheng, P. C. (1984): Applications and examples of biological soft x-ray microscopy. *Acta Crystallographica* A40.

Cheng, P. C. (1985): Soft x-ray contact microscopy and absorption properties of biological materials. PhD dissertation, University of Illinois at the Medical Center, Chicago, Illinois, USA.

Cheng, P. C., Feder, R., McGowan, J. Wm., Tan, K. H., and Shinozaki, D. M. (1985): 3D soft x-ray contact imaging and image interpretation. *Proceeding of the 43rd Ann. Meeting of EMSA*, 602-605.

Cheng, P. C., McGowan, J. Wm., Tan, K. H., Feder, R., and Shinosaki, D. M. (1986): Ultrasoft x-ray contact microscopy: A new tool for plant and animal cytology. *In*: Examining the submicron world (NATO ASI series). eds. Feder, R., McGowan, J. Wm., and Shinozaki, D. M., Plenum Press, New York. pp.299-350.

Cheng, P. C., Feder, R., Shinozaki, D. M., Tan, K. H., Eason, R. W., Michette, A., and Rosser, R. J. (1986): Soft x-ray contact microscopy. *Nuc. Inst. Methods in Phys. Res.*, A246: 668-674.

Cheng, L. M., and Michette, A. G. (1984): Three dimensional imaging in scanning soft x-ray microscopy. *J. Phys. (Paris) C2*, 45: 97-100.

Clark, G. L. (1939): X-ray photomicrography. *Photo Technique*, 1 (Dec.): 19-20.

Clark, G. L. (1944): Microradiography of colloidal materials. *In*: Colloid chemistry, ed. Alexander, F., Reinhold Publishing Co., NY. pp.146-151.

Clark, G. L. (1947): Medical, biological and industrial of monochromatic radiography and microradiography. *Radiology*, 49: 483-495.

Clark, G. L. (1955): Applied x-rays. 4th ed., McGraw-Hill, NY.

Clark, G. L., and Bicek, E. J. (1944): Microradiographic examination of biological materials. *In*: Medical physics. ed. Glasser, O., Year Book Publishers, Inc., Chicago, IL. pp.730-733.

Clark, G. L., and Bicek, E. J. (1950): Microradiography. *In*: Medical physics, 2nd ed., ed. Glasser, O., Year Book Publishers. inc., Chicago, IL, p.902.

Clark, G. L., and Eyler, R. W. (1943): A camera for stereoscopic microradiography. *Review Sic. Inst.*, 14: 277-278.

Clark, G. L., and Gross, S. T. (1942): Technique and applications of industrial microradiography. *Industrial Engineering Chemistry (Analytical Edition)*, 14: 676-683.

Clark, G. L., and Shafer, W. M. (1941): The technique of microradiography and its application to metals. *Transations of the American Society for Metals*, 29: 732-754.

Clemmons, J. J. (1955): Procedures and errors in quantitative historadiography. *Biochimica et biophysica acta*, 17: 297-321.

Clemmons, J. J., and Aprison, M. H. (1953): An improved historadiographic apparatus. *Review Sci. Instru.*, 24: 444-454.

Clemmons, J. J., Lalich, J. J., and Angevine, D. M. (1954): The technique of quantitative historadiography. *Laboratory Investigation*, 3: 19-32.

Clemmons, J. J., and Webster, T. C. (1953): An accurate reference system for historadiography. *Biochimica et biophysica acta*, 11: 464-470.

Cohen, E. (1950): Microradiography applied to the study of iron ores and sinters. *Metallurgia*, 41: 227-233.

Cohen, E., Hall, E., Leonard, L., and Ogilvie, R. (1955): Investigation of segregation in cast irons by radiographic techniques. *Nondestructive Testing*, 13: 33-34.

Collett, C. T., Hughes, J. E., and Morey, F. C. (1950): Measurement of the internal diameters of metallic capillary tubes. *J. Res. Natl. Bureau Standards*, 45: 283-288.

Collette, J. M. (1953): La microradiographie; bases et principes de la technique; une nouvelle application: la microlymphographie. (Microradiography; Foundations and principles of the technique, A new application: Microloymphography.) *J. belge de radiologie*, 36: 293-312.

Combée, B. (1955): An x-ray tube for microradiography. *Philips Technical Review*, 17: 45-46.

Combée, B., and Botden, P. J. M. (1951): Special x-ray tubes. *Philips Technical Review*, 13: 71-80.

Combée, B., and Engström, A. (1954): A new device for microradiography and a simplified technique for the determination of the mass of cytological structures. *Biochimica et biophysica acta*, 14: 432-434.

Combée, B., and Recourt, A. (1957/58): A simple apparatus for contact microradiography between 1.5 and 5kV. *Philip Tech. Rev.*, 19: 221-233.

Compton, A. H., and Allison, S. K. (1935): X-rays in theory and experiment. D. van Nostrand, New York.

Cosslett, V. E., and Nixon, W. C. (1951): X-ray shadow microscope. *Nature*, 168: 24-25.

Cosslett, V. E., and Nixon, W. C. (1952): X-ray shadow microscopy. *Nature*, 170: 436-438.

Cosslett, V. E., and Nixon, W. C. (1952): An experimental x-ray shadow microscope. *Proc. Royal Society*, 140B: 422-431.

Cosslett, V. E. (1953): High resolution x-radiography. *British J. Radiology* , 26 (May): 258-260.

Cosslett, V. E., and Nixon, W. C. (1953): The x-ray shadow microscope. *J. Appl. Phys.*, 24: 616-623.

Cosslett, V. E. (1954): X-ray microscopy. *Medical Biological Illustration* , 4: 95-103.

Cosslett, V. E. (1954): Microscopy with x-rays. *J. Photographic Science*, 2 (July-August): 125-130.

Cosslett, V. E., and Nixon, W. C. (1954): A point x-ray focus for shadow microscopy. *Electron Physics*, (National Bureau of Standards Circular) 527 (March 17): pp 257-264.

Cosslett, V. E., and Pearson, H. E. (1954): An improved x-ray shadow projection microscope. *J. Scientific Instruments*, 31 (July): 255-257.

Cosslett, V. E., Engström, A., and Pattee, H. A. (1957): X-ray microscopy and microradiography. Academic Press, New York.

Cosslett, V. E., and Nixon, W. C. (1960): X-ray microscopy. University Press, Cambridge

Cowden, R. R. (1985): Ultrasoft x-ray historadiography and new developments in ultrasoft x-ray imaging. *In*: Advances in microscopy. eds. Cowden, R. R., and Harrison, F. W. Harrison, Alan R. Liss, Inc., pp289-298.

Dahl, A. D., Rowley, J. R., Stein, D. L., and Wegstedt, L. (1957): The intracellular distribution of mass during ontogeny of pollen in *Tradescantia* L.. *Exptl. Cell Research*, 13: 31-46. Dainty, J. C. (1984): Laser speckle and related phenomena, *Topics in Applied Physics*, 2nd edition, Springer-Verlag, Berlin.

Devaney, A. J. (1985): Diffraction tomography. *In*: Inverse methods in electromagnetic imaging. eds. Boerner, W. M., and Reidel, D., Dordrecht, pp 1107-1135

Davey, W. P. (1919): Some interesting applications of the coolidge x-ray tube. *In*: X-ray studies (G. E. Publication), Schenectady, NY, p231-237.

Dauvillier, A. (1927): Sur un tube á rayons X de longueur d'onde effective égale á 8 unités Angstrom. (An x-ray tube that produces x-rays with an effective wavelength of 8Å.) *Comptes rendus de l'Academie des sciences*, 185: 1460-1462. Dauvillier, A. (1930): Réalisation de la microradiographie intégrale. (Integral microradiography.) *Comptes rendus de l'Académie des sciences*, 190: 1287-1289.

Davey, W. P. (1919): Some interesting applications of the Coolidge x-ray tube. *In*: X-ray studies. Published by research Laboratories, General Electric Co., Schenectady, NY. pp. 231-237.

Davies, H. G., Engström, A., and Lindström, B. (1953): A comparison between the x-ray absorption and optical interference methods for the mass determination of biological structures. *Nature*, 172: 1041.

Dershem, E. (1938): Radiography of thin microtome sections by means of fluorescence x-radiation. *Physiological Reviews*, 53: 202.

Dershem, E. (1939): Photomicrographs of thin bone sections by the use of fluorescence x-radiation. *Proc. Natl. Academy Sci.*, 25: 6-10.

Dershem, E. (1939): X-ray photograph by means of fluorescence x-rays. *J. Optical Soc. Amer.*, 29: 41-42.

Dershem, E. (1939): X-ray photographs of biological sections. *British J. Photography*, 86: 508.

Djurle, E., and Hallén, O. (1953): A double slit interferometer for determination of the thickness of the reference system used in x-ray microradiography. *Experimental Cell Res.*, 5: 301-310.

Dole, M. (1973): Polymers that degrad on irradiation. *In*: The Radiation Chemistry of Macromolecules, ed. M. Dole, Vol. 2, Academic Press. pp.97-117.

Doniach, S., Eisenberger, P., and Hodgson, K. O. (1980): X-ray absorption spectroscopy of biological molecules. *In*L: Synchrotron radiation research, eds. Winick, H., and Doniach, S., New York, Plenum Press, pp. 425-458.

Doran, F. S. A. (1951): Aetiology of chronic gastric ulcer. Observation on the blood supply of the human gastric mucosa with a note on the arteriovenous shunt. *Lancet*, 260: 199-202.

Doran, F. S. A. (1951): Aetiology of chronic gastric ulcer. *Lancet*, 260: 635-636.

Douzou, P. (1977): Cryobiochemistry: An introduction. Academic Press, London.

Duke, P. J. (1984): X-ray microscopy at the Daresbury Laboratory. *In*: Springer Series in Optical Sciences, X-ray Microscopy, eds. Schmahl, G. and Rudolph, D., Berlin, Springer-Verlag, pp. 232-241.

Dubochet, J., Adrian, M., Lepault, J. and McDowall, A. W. (1985): Cryo-electron microscopy fo vatrified biological specimens. *TIBS*, 4: 143-146.

Dyson, J. (1952): optics of the focusing x-ray microscope. *Proc. Phys. Soc.*, 65: 580-589.

Eason, R. W., Cheng, P. C., Feder, R., Michette. A. G., Rosser, R. J., 0'Neill, F., Owadano, Y., Rumsby, P. T., Shaw, M. J., and Turcu, I. C. E. (1986): Laser x-ray microscopy. *Optica Acta* 33: 501-516.

Eason, R. W., Cheng, P. C., Feder, R., Michette, A. G., Rosser, R. J., O'Neill, F., Owadano, Y., and Turcu, I. C. E. (1986a): Source characteristics on Sprite and Vulcan. *In*: Rutherford Appleton Laboratory Annual Report 1986: A4.5-A4.7.

Eason, R. W., Michette, A. G., and Rosser, R. J. (1986): Scaling of x-ray microscopy source brightness with laser energy on Sprite and Vulcan. *Rutherford Appleton Laboratory Annual Report 1986*: A4.14-A4.16.

Eder, J. M., and Valenta, E. (1896): Versuche uber Photographie mittels der Röntgenschen Strahlen. (Experiment in photography with the aid of x-rays.) Published by R. Lechner, Vienna, Austria, and Wilhelm Knapp, Halle a.S., Germany.

Egerton, R. F., and Cheng, S. C. (1985): Thickness measurement by EELS. *Proc. 43rd Annual Meeting of the Electron Microscopy of America*, 398-399.

Ehrenberg, W. (1949): X-ray optics: The production of converging beams by total reflection. *J. Optical Soc. Amer.*, 39: 741-746.

Ehrenberg, W. (1949): X-ray optics: Imperfections of optical flats and their effect on the reflection of x-rays. *J. Optical Soc. Amer.*, 39: 746-751.

Ehrenberg, W., and Spear, W. E. (1951): X-ray microradiography of biological specimens. *Nature*, 168: 513-514.

Ehrenberg, W., and Spear, W. E. (1951): An electrostatic focusing system and its application to a fine focus x-ray tube. *Proc. Phys. Soc. London*, 64 (Jan.): 67-75.

Ehrenberg, W., and White, M. (1957): Some physical aspects of contact microradiography. *In*: X-ray microscopy and microradiography. eds. Cosslett, V. E., Engström, A., and Pattee, H. H., Academic Press, New York. pp 213-217.

Eidlin, L. M. (1933): Röntgenographischer Nachweis des Metallringes am Eincschuss. (Radiographic detection of metal rings around a bullet hole.) *Deutsche Zeitschrift fur die gesamte gerichtliche Medizin*, 22: 204-219.

Elliott, W. J. (1943): X-rays in scientific examination of paintings. *Amer. J. Roentgenology*, 50: 779-790.

Elliott, J. C., Bowen, D. K., and Dover, S. D. (1986): Biological Trace Element Research (to be published).

Ely, R. (1952): Grenz Radiography. *Medical and Biological Illustration*, 2: 158-159.

Ely, R. V. (1954): Instrumentation in microradiography. *J. Photographic Sci.*, 2: 119-124.

Ely, R. V. (1957): Fine focus x-ray tube with turret window and rotatable target. *In*: X-ray microscopy and microradiography. eds. Cosslett, V. E., Engström, A., and Pattee, H. H., Academic Press, New York. pp 59-63.

Endicott, K. M., and Yagoda, H. (1947): Microscopic historadiographic technic for locating and quantitaing radioactive elements in tissues. *Proc. Soc. Experimental Biol. Medicine*, 64: 170-172.

Engfeldt, B., Bergman, G., and Hammarlund-Essler, E. (1954): Studies on mineralized dental tissue. I. A microradiographic and autoradiographic investigation of teeth and tooth germs of normal dogs. *Experimental Cell Res.*, 7: 381-392.

Engfeldt, B., Engström, A., Helander, C. G., Wilton, Å, and Zétterstrom, R. (1952): Biophysical studies on bone tissue. I. Paget's disease. *Acta Pathologica Microbiologica Scandinavica*, 31: 256-261.

Engström, A. (1946): Quantitative micro- and histochemical elementary analysis by Roentgen absorption spectrography. *Acta Radiol. Scand., Suppl.*, 63, 1-106.

Engström, A. (1946): Quantitative microchemical and histochemical analysis of elements by x-rays. *Nature*, 158: 664-665.

Engström, A. (1947): Qualitative microchemical analysis by microradiography with fluorescenct screen. *Experientia*, 3: 208-209.

Engström, A. (1947): A new differential x-ray absorption method for elementary chemical analysis. *Review Sci. Instru.*, 18: 681.

Engström, A., and Lindström, B. (1947): Histochemical analysis by x-rays of long wavelengths. *Experientia*, 3: 191-193.

Engström, A. (1949): Microradiography. *Acta Radiologica*, 31: 503-521.

Engström, A., Hamberger, A., and Welin, S. (1949): A Röntgen-ray microspectrographic investigation of the inflammatory destruction of the mastoid bone. *British J. Radiology*, 22: 309-324.

Engström, A., and Lindström, B. (1949): A new method for determining the weight of cellular structures. *Nature*, 163: 563-564.

Engström, A., and Lüthy, H. (1949): Die Massenverteilung in der markaltigen Nervenfaser, bestimmt durch Röntgenabsorptionsmessung. (The mass distribution in the medullated nerve fiber determined by x-ray absorption measurement.) *Experientia*, 5: 244-245.

Engström, A. (1950): Some comparison between x-ray and ultra-violet absorption in different tissue and cell structures. *Discussion of the Faraday Society*, 9: 427-431.

Engström, A. (1950): Submicroscopic structure of striated muscle. *Scandinavian J. Clinical and Laboratory Investigation*, 2: 252-256.

Engström, A. (1950): Use of soft x-rays in the assay of biological material. *In*: Progress in biophysics and biophysical chemistry. eds. Butler, J. A. V., and Randall, J. T., Academic Press, NY, pp.164-196.

Engström, A., and Glick, D. (1950): The mass of gastric mucosa cells measured by x-ray absorption. *Science*, 111: 379-380.

Engström, A., and Lindström, B. (1950): A method for the determination of the mass of extremely small biological objects. *Biochimica et Biophysica Acta*, 4: 351-373.

Engström, A., and Lüthy, H. (1950): The distribution of mass and lipids in the single nerve fiber. *Experimental Cell Res.*, 1: 81-96.

Engström, A. (1951): Stereomicroradiography. *Acta Radiologica*, 36: 305-310.

Engström, A. (1951): X-ray absorption methods in quantitative histochemistry. *Conf. on Microspectrophotometry of cells*, Massachusetts Institute of Technology, Cambridge, MA, pp.44-45.

Engström, A., and Lindström, B. (1951): The properties of fine-grained photographic emulsions for microradiography. *Acta radiologica*, 35: 33-44.

Engström, A., and Ruch, F. (1951): Distribution of mass in salivary gland chromosomes. *Proc. Natl. Academy Sci.*, 37: 459-461.

Engström, A., and Wegstedt, L. (1951): Equipment for microradiography with soft Roentgen rays. *Acta Radiologia*, 35: 345-355.

Engström, A. (1952): X-ray absorption methods in histochemistry. *Laboratory Investigation*, 1: 278-285.

Engström, A. (1952): Microradiography and micro- x-ray diffraction. *Transaction of Instruments and Measurements Conference*, Stockholm, Published by Svensk Teknologforeningen, Stockhlom, Sweden, pp.88-92.

Engström, A., and Malmström, B. (1952): X-ray absorption studies on dog gastric mucosa cells. *Acta Physiologica Scandinavica*, 27: 91-96.

Engström, A., and Engfeldt, B. (1953): Lamellar structure of osteons demonstrated by microradiography. *Experientia*, 9: 19.

Engström, A., and Josephson, B. (1953): Historadiographic demonstration of diodrast in the rabbit kidney. *Amer. J. Physiology*, 174: 61-N.

Engström, A., and Lundberg, B. (1957): A simple midget x-ray tube for high resolution microradiography. *Exper. Cell Res.*, 12: 198-200.

Engström, A. (1957): Contact microradiography: A general survey. *In*: X-ray microscopy and microradiography. eds. Cosslett, V. E., Engström, A., and Pattee, H. H., Academic Press, New York. pp. 24-33.

Engström, A., Greulih, R. C., Henke, B. L., and Lundberg, B. (1957): High resolution contact microradiography with ultrasoft polychromatic x-rays. *In*: X-ray microscopy and microradiography. eds. Cosslett, V. E., Engström, A., and Pattee, H. H., Academic Press, New York. pp. 218-233.

Engström, A. (1962): X-ray Microanalysis in Biology and Medicine. Amsterdam, Elsevier Publishing Co..

Engström, A. (1966): X-ray microscopy and x-ray absorption analysis. *In*: Physical Techniques in Biological Research, Vol. III, Part A, ed. Pollister, A. W. (2nd edition). Academic Press, New York, pp. 87-171.

Engström, A. (1967): Contact microradiography with ultrasoft x-rays. *In*: Handbook of x-rays. ed. Kaelble, E. F., McGraw Hill, New York. pp. 46/1-46/10.

Engström, A. (1980): X-ray microscopy, past, present, and future. *Ann. N. Y. Acad. Sci.*, Vol. 342: 392-400.

Epstein, H. M. (1985): Ultrasoft x-ray historadiography and new developments in ultrasoft x-ray imaging. B: New approaches to ultrasoft x-ray imaging. *In*: Advances in microscopy. eds. Cowden R. R., and Harrison, F. W., Alan R. Liss, Inc., pp 299-311.

Erasmus, S. J. (1987): Damage to resist structures during scanning electron microscope inspection. *J. Vac. Sci. Technol.*, B5(1), 409-413.

Farrant, J. L. (1950): Wavelengths suitable for x-ray microscopy of biological specimens. *J. Appl. Phys.*, 21: 63-64.

Feder, R., Spiller, E., and Topalian, J. (1975): Replication of 0.1 μm Geometries with X-ray Lithography, *J. Vac. Sci. Technol.*, 12: 1322.

Feder, R., Sayre, D., Spiller, E., Topalian, J., and Kirz, J. (1976): Specimen Replication for Electron Microscopy Using X-rays and X-ray Resist, ,*J. Appl. Phys.*, 47: 1192.

Feder, R., Spiller, E., Topalian, J., Broers, A. N., Gudat, W., Panessa, B., Zadunaisky, J. A., and Sedat, J. (1977): High Resolution Soft X-ray Microscopy, *Science* 197: 259.

Feder, R., and Sayre, D. (1980): Recent developments in x-ray contact microscopy. *Ann. N. Y. Acad. Sci.*, 342: 213-234.

Feder, R., Costal, J. L., Chaudhari, P., and Sayre, D. (1981): Improved detail in biological soft x-ray microscopy: Study of blood platelets. *Science* 212: 1398-1400.

Feder, R., Mayne-Banton, V., Sayre, D., Costa, J., Kim, B. K., Baldini, M. G., and Cheng, P. C. (1984): Recent developments in x-ray contact microscopy. *In*, Springer Series in Optical Sciences: X-ray microscopy, eds. Schmahl, G. and Rudolph, D., Berlin, Springer-Verlag, pp.279-284.

Feder, R., Pearlman, J. S,, Riordan, J. C., and Costa, J. L. (1984): Flash x-ray microscopy with a gas jet plasma source. *J. Microscopy*, 135:347-351.

Feder, R., Banton, V., Sayre, D., Costa, J., Baldini, M. and Kim, B. (1985): Direct imaging of live human platelets by flash x-ray microscopy. *Science*, 227: 63-64.

Feder, R. and Mayne-Banton, V. (1985): X-ray contact microscopy. *In*: Examining the submicron world. eds. Feder, R., McGowan, J. Wm., and Shinozaki, D. M., New York, Plenum Press, pp.277-298.

Feder, R., Banton, V., Sayre, D., Costa, J. L., Baldini, M. G., and Kim, B. K. (1985): Direct imaging of live human platelets by flash x-ray microscopy. *Science* 227:63-64

Fenton, F. A., and Waite, W. W. (1932): Detecting pink bollworms in cotton seeds by the x-ray. *J. Agri. Res.*, 45: 347-348.

Fernandez-Moran, H. (1965): Formes of water in biologic system-discussion. *Ann. N. Y. Acad. Sci.* Vol. 125 Part 2: 739-753.

Fisher, R. C., and Tasker, H. S. (1940): The detection of wood-boring insects by means of x-rays. *Annals Applied Biology*, 27: 92-100.

Fisher, S. B. (1970): On the temperature rises on electron irradiated foils. *Radiation Effects*, 5, 239-243.

Fitzgerald, P. J., and Engström, A. (1950): Microradiographic and ultraviolet absorption studies of human thyroid gland. *6th Int. Congr. Radiology, London*, pp.167.

Fitzgerald, P. J., and Engström, A. (1952): The use of ultraviolet-microscopy , Roentgen-ray-absorption, and radioautographic techniques in the study of neoplasmic disease. A discussion of these cytophysical techniques. *Cancer*, 5: 643-677.

Fitzgerald, P. J. (1957): A continuously evacuated, water cooled, copper target, x-ray tube producing soft x-rays for x-ray absorption studies. *In*: X-ray microscopy and microradiography. eds. Cosslett, V. E., Engström, A., and Pattee, H. H., Academic Press, New York, pp. 49-58.

Flanders, D. C. (1980): Replication of 170Å lines and spaces in PMMA using x-ray lithography. *Appl.Phys.Lett.* 36(1), 93-96.

Flanders, D. C., Shaver, D. C., Hawryluk, A. M., and Smith, H. I. (1980): Progress in the fabrication of artifical microstructures for x-ray imaging and spectroscopy. *Ann. N. Y. Acad. Sci.* 342: 203-212.

Fournier, F. (1938): Technique et possibilités de la microradiographie. (Technique and possible applications of microradiography.) *Revue de métallurgie*, 35: 349-355.

Fournier, F. (1949): Microradiographie. Application á l'étude des alliages légers. (Microradiography. Application to the study of light alloys.) *Revue de métallurgie*, 46: 360-362.

Fournier, F., and Perrot, M. (1944-1945): Considérations techniques sur l'historadiographie. (Technical considerations in historadiography.) *J. de Radiologie et d'électrologie*, 26: 12-14.

Fox, R. B. (1967): Photodegradation of high polymers. *In*: Progressive Polymer Science, New York, Plenum Press.

Franke, H. W., and Fries, P. (1954): Rontgenformen. (Pictorial Radiographs.) *Photo Magazin*, 6: 41 (11)- 45(11).

Fricke, H. (1932): The radiography of insects. *Radiography and Clinical Photography*, 8: 12-13.

Gabor, D. (1961): Light and information, *Progr. in Optics Vol.1*, North Holland, New York, pp.109.

Gerritsen, H. C., van Brug, H., Bijkerk, F., and van der Wiel, M. J. (1986): Laser generated plasma as soft x-ray source. *J. Appl. Phys.*, 59:2337-2344.

Ginter M. L. (1984): Laser produced plasma VUV and soft x-ray light sources. *In*: Springer Series in Optical Sciences: X-ray Microscopy, eds. Schmahl, G. and Rudolph, D., Berlin, Springer-Verlag, pp.25-29.

Ginter, M.L., and McIlrath, T. J. (1986): Laser produced plasma light sources for high resolution XUV and VUV spectroscopy. *Nuc. Inst. and Methods in Phys. Res.*, A246: 779-781.

Glaeser, R. M. (1971): Limitations to significant information in biological electron microscopy as a result of radiation damage. *J. Ultrastruct. Res.* 36: 466-482.

Glick, D., Engström, A., and Malström, B. G. (1951): A critical evaluation of quantitative histo- and cytochemical microscopic techniques. *Science*, 114: 253-258.

Goby, P. (1913a): Une application nouvelle des rayons x, Ia microradiographie. (A new application of x-rays; Microradiography.) *Compt. Rend. Acad. Sci. Paris*, 156: 686-688.

Goby, P. (1913b): A new application of Roentgen rays, microradiography. *J. Roy. Mic. Soc.*, August, p373-375.

Goby, P. (1913): la microradiographie et ses applications á l'anatomie végétale. (Microradiography and its application to vagetable anatomy.) *Bulletin Société francaise de photographie*, 4 (Oct.): 310-312.

Goby, P. (1913): Microradiography. *Archives of the Roentgen Ray*, 18 (Dec.): 247-250.

Goby, P. (1925): La microradiographie stéréoscopique en relief et en pseudo-relief. La stéréomicroradiographie. (Stereomicroradiographs in relief and in pseudo relief.) *Comptes rendus de l'Académie des sciences*, 180: 735-737.

Goby, P. (1925): Applications nouvelles et prefectionnements de la microradiographie. (New applications and improvements of microradiography.) *In*: VIe Congrés International de Photographie, 1925. Published by Société francaise de photographie, 6 (Sept. 1): 250-252. and 6 (Sept.):260-264.

Goby, P. (1941): Applications Nouvelles de la microradiographie á l'entomologie par l'utilisation des rayons X. (New applications of microradiography to entomology.) *Bulletin Société francaise de photographie*, 5 (June): 196-197.

Godlewski, M. (1953): Essai radiographique sur la peau de l'homme et de quelques animaux. (Radiographic tests of human and animal skin.) *J. de Radiologie et d'électrologie*, 34: 239-241.

Gogoberidze, D. B. (1953): Micro Roentgenography and x-ray microscopy (in Russian), *Uspekhi Fizicheskikh Nauk*, 50: 577-599.

Goldspiel, S., and Bernstein, F. (1953): Some industrial applications for microradiography. *Non Destructive Testing*, 11: 15-19.

Goldstein, J.I., and Yakowitz, H. (eds.), (1976): Practical Scanning Electron Microscopy, Plenum Press, N.Y..

Goodhead, D. T., and Thacker, J. (1977): Inactivation and mutation of cultured mammalian cells by aluminum characteristic x-rays. *Int. J. Radiat. Biol.* 31: 561-576.

Goodhead, D. T. (1982): An assessment of the role of microdosimetry in radiobiology. *Radiation Research*, 91: 45-76.

Goodman, J. W. (1968): Introduction to Fourier optics. McGraw-Hill, New York

Goodman, J. W. (1985): *Statistical Optics*, J. Wiley, New York.

Gouy, M. G. (1923): Improving the microscope by the use of x-rays. *Comptes rendus de l'Academie des sciences*, 176: 807-808.

Graham, J. G. (1955): A technique of microradiography. *Radiography*, 21: 120-127.

Grechishkin, S. V. (1937): Rentgenologischeskoe issledovanie povrezhdennykn semian. (The Roentgen Examination of seeds.) *Vestnik rentgenologii i radiologii*, 18: 41-48.

Grechishkin, S. V. (1938): mikrorentgenografiia. (Microroentgenography.) *Vestnik rentgenologii i radiologii*, 20: 397-408.

Greenwood, H. W. (1939): Micro-methods in radiography and x-ray analysis *Light Metals*, 2 (Jan.): 8-9.

Gross, S. T., and Clark, G. L. (1942): Technique and practice of microradiography. *Industrial Radiography*, 1 (Fall): 21-25.

Gross, S. T., and Clark, G. L. (1943): Microradiography of light alloy spot welds. *Iron Age*, 152 (July): 44-48.

Grubb, D.T. (1970): D.Phil. Thesis, University of Oxford, U.K..

Gudat, W. (1982): Soft x-ray contact microradiography. *In*: Uses of synchrotron radiation in biology, ed. Stuhrmann, H. B., New York, Academic Press, pp23-50.

Guettel, C. L. (1945): Miscellaneous techniques in spectrographic analysis, X-ray diffraction and microradiography. Rock Island Arsenal Laboratory, Rock Island, IL. (Publication Board Report 4404, April 26, 1945).

Guttmann, P (1983): Construction of a micro zone plate and evaluation of imaging properties. *In*: X-Ray Microscopy. eds. Schmahl, G., and Rudolph, D., Springer-Verlag, Berlin, pp.75-90 (Series in Optical Sciences vol 43)

Guttmann, P. (1987): Examination of detector systems for high photon rates in the ultrasoft x-ray region, this volume. Guttmann, P. (1986): Untersuchungen an Detektorsystemen zur Verarbeitung hoher Photonenraten im Bereich Ultraweicher Röntgenstrahlung", Thesis, Göttingen.

Haelbich, R. -P., Segmüller, A., and Spiller, E. (1979): Smooth multilayer films suitable for x-ray mirrors, *Appl. Phys. Lett.*, 34: 184.

Haelbich, R. P., Staehr, W., and Kunz, C. (1980): A scanning ultrasoft x-ray microscope with large aperture reflection optics for use with synchrotron radiation. *Ann. N. Y. Acad. Sci.*, 342: 148-157.

Hall, T. A., Röckert, H. O. E., and Saunders, R. L. (1972): Morphological microradiography. *In*: X-ray microscopy in clinical and experimental medicine. ed. Kugelmas, I. N., Charles Thomas Publish, IL, pp.3-95.

Hall-Edwards, J. (1914): The radiography of flowers. *Archives of the Roentgen Ray*, 19 (June): 31.

Hallén, O. (1953): A simplified methods for the preparation of the reference system used in x-ray microradiography. *Experimental Cell Res.*, 4 (May): 494-495.

Hallén, O., and Hyden, H. (1957): Quantification and automation in contact microradiography. *In*: X-ray microscopy and microradiography. eds. Cosslett, V. E., Engström, A., and Pattee, H. H., Academic Press, New York, pp. 249-258.

Hallén, O., and Ingelstam, E. (1952): Determination by interferometric methods of the thickness of the reference system used in microradiography. *Experimental Cell Research*, 3 (Feb.): 248.

Haller, I., Feder, R., Hatzakis, M., and Spiller, E. (1979): Copolymers of methyl methacrylate and methacrylic acid and their heavy metal salts as radiation sensitive resists, *J. Electrochem. Soc.*, 126: 154.

Halpern, A. (1982): Damage to biomolecules and cells by low-energy x-rays and vacuum ultraviolet light. *In*: Uses of Synchrotron radiation in biology, ed. Stuhrmann, H. B., New York, Academic Press, pp. 255-283.

Hamos, L. v. (1937): The x-ray microscope. *Nature*, 140, 30. Also, *J. Sci. Instru.*, 15:87-94 (1938).

Hamos, L. v. (1938): X-ray image method of crystal analysis. *Amer. Mineralogist*, 23: 215-226.

Harrison, J. E., Williams, W. C., Watts, J., and McNeill, K. G. (1975): A bone calcium index based on partial body calcium measurements by *in vivo* activation analysis. *J. Nucl. Med.*, 16:116-122.

Harrison, R. G., Astbury, W. T., and Rudall, K. M. (1940): An attempt at x-ray analysis of embryonic processes. *J. Experimenatl Zool.*, 85 (Dec.): 339-363.

Harrison, R. G., and Barclay, A. E. (1948): The distribution of the testicular artery (internal spermatic artery) to the human testis. *British J. Urology*, 20 (June): 57-66.

Hedberg, E. (1953): The chemical composition of the human ovarian oocyte. *Acta Endocrinologica, Supp.*, 15. (Accompanies Vol. 14), pp.1-89.

Hedenberg, I., Engfeldt, B., and Engström, A. (1953): X-ray absorption and diffraction studies on experimental vesical calculi. *British J. Urology*, 25 (March): 33-37.

Heinrich, K. (1981): Electron beam x-ray microanalysis. Van Nostrand Reinhold Co., New York.

Henke, B. L., White, R., and Lundberg, B. (1957): Semiempirical determination of mass absorption coefficients for the 5 to 50 Angstrom x-ray region. *J. Appl. Phys.*, 28: 98-105.

Henke, B. L. (1957): Monochromatic sources of ultrasoft x-radiation for quantitative microradiographic analysis. *In*: X-ray microscopy and microradiography. eds. Cosslett, V. E., Engström, A., and Pattee, H. H., Academic Press, New York, pp. 72-88.

Henke, B. L., Lundberg, B., and Engström, A. (1957): Conditions for optimum visual and photometric contrast in microradiograms. *In*: X-ray microscopy and microradiography. eds. Cosslett, V. E., Engström, A., and Pattee, H. H., Academic Press, New York, pp. 240-248.

Henke, B. L., and Elgin, R. L. (1970): X-ray absorption tables for the 2 to 200Å region. *Adv. X-ray Analysis*, 13: 639-665.

Henke, B., and Ebisu E. S.: Low energy x-ray and electron absorption in solids (100-1500 eV region). *In*: Advances in x-ray Analysis, Vol. 17, eds. C. L. Grant, C. S. Barrett, J. B. Newkirk and C. O. Ruud, New York, Plenum Press, 1974. Henke, B. L., Lee, P., Tanaka, T.J., Shimabukuro, R. L., Fujikawa, B. K. (1981): The atomic scattering factor, $f_1 + if_2$, for 94 elements and for the 100 to 2000 eV photon energy region. *In*: Low Energy X-Ray Diagnostics - 1981. eds. Attwood, D. T., and Henke, B. L., *AIP Conf. Proc.* No 75:340-388.

Henke, B. L., Lee, P., Tanaka, T. J., Shimabukuro, R. L., and Fujikawa, B. K. (1982): Low-energy x-ray interaction coefficients: photoabsorption, scattering, and reflection. E=100-2000 eV. Z = 1 - 94. *Atomic Data nad Nuclear Data Tables*, 27: 1-144.

Herres, N., and Lang, A. R. (1983): X-ray topography of natural beryl using synchrotron and conventional sources. *J. Appl. Cryst.* 16: 47.

Hounsfield, G. N. (1973): *Br. J. Radiobiol.* 46: 1016.

Hexter, R. M. (1980): X-ray microscopy of thin sections of striated muscle and of polystyrene microspheres. *Ann. N. Y. Acad. Sci.*, 342: 330- 349.

Heycock, C. T., and Neville, F. H. (1898): Röntgen ray photography applied to alloys. *Phil. Trans. Roy. Soc. London*, 73, p714-723.

Heycock, C. T., and Neville, F. H. (1900): gold-aluminium alloys. *Philosophical Transactions of the Royal Society of London*, 194: 201-232.

Hicks, B. (1896): Another application of the x-rays. *British J. Photography*, 43 (Oct.): 691.

Hilkenbach, R., and Thieme, J. (1986): Phase zone plates as condensers for the Göttingen scanning x-ray microscope. *SPIE Proc.* 733: in press.

Hiraoka, H., (1977): Radiation chemistry of poly(methacrylate), *IBM J. Res. Develop.*, (March), 121-130.

Hoh, F. C., and Lindstrom, B. (1959): On the theory of quantitative microradiography in biology. *J. Ultrastruct. Res.*, 2: 512-544.

Hollander, F., Bödecker, C. F., Applebaum, E., and Saper, E. (1935): A study of the bands of Schreger by histological and Grenz-ray methods. *Dental Cosmso*, 77 (Jan.): 12-20.

Hollander, F., and Saper, E. (1935): The apparent radiopaque surface layer of the enamel. *Transaction of the dental Society of the State of new York* , pp. 124-134.

Holweck, M. (1923): Propriétés optiques des rayons longueur d'onde. (Optical Properties of Long-wavelength X-rays.) *Comptes rendus de l'Acadämie des sciences*, 176: 570-573.

Homes, G. A., and Gouzou, J. (1951): les ressources de la microradiographie dans les problémes métallurgiques. (Suitability of microradiographic methods for metallurgical problems.) *Revue de métallurgie*, 48 (April): 251-261.

Honeycombe, R. W. K. (1951): A simple method of x-ray microscopy and its application to the study of deformed metals. *J. Inst. Metals*, 80 (sept.) :39-44.

Honeycombe, R. W. K. (1951): Inhomogeneities in the plastic deformation of metal crystals; II, X-ray and optical micrography of aluminum. *J. Inst. Metals*, 80 (Oct.): 49-56.

Hoppe, W., and Trurnit, H. J. (1947): Geometric und Kristallphysik einer Röntgenlinse für monochromatische Strahlung. (Geometric and crystal physics of x-ray lenses for monochromatic radiation). *Zeitschrift für Naturforschung*, 2a, (Nov.-Dec.) 608-615.

Horn, H., and Waltinger, H. (1973): Röntgenmikrographie und Röntgenabsorptionsanalyse im Rasterlektronenmikroskope. *Beitr. Elektronenmikroskop. Direktabb. Oberfl.*, 6: 163-168.

Horn, H., and Waltinger, H. (1978): How to obtain and use x-ray projection microscopy in the SEM. *Scanning*, 1: 100-108.

Horowitz, P., and Howell, J. A. (1972): A scanning x-ray microscope using synchrotron radiation. *Science*, 178: 608-611.

Howe, R. J. (1983): Absorption properties of polymers using synchrotron radiation, Master thesis, University of Western Ontario, London, Ontario, Canada.

Howells, M. R. (1984): Possibilities for x-ray holography using synchrotron radiation, *In*: X-ray Microscopy, eds. Schmahl, G., and Rudolph, D., Springer-Verlag, Berlin, p.318.

Howells, M. R., Kirz, J., Sayre, D., and Schmahl, G. (1985): Soft-x-ray microscopes. *Physics Today* 38(8):22-32

Howells, M. R., Iarocci, M. A., and Kirz, J. (1986): Experiments in x-ray holographic microscopy using synchrotron radiation. *J. Opt. Soc. Amer.* B3: in press

Hren, J. J., Goldstein, J. I., and Joy, D. C., (1979): Introduction to Analytical Electron Microscopy, Plenum Press, N.Y.

Hudson, B. and Makin, M. J. (1970): The optimum tilt angle for electron stereo-microscopy. *J. Phy. E: Sci. Instru.*, 3: 311.

Hurd, J. M., Lucht, C. M., and Smoluchowski, R. (1945): Composition and diffraction effects in x-ray microradiographs. *Physical Review*, 60 (August): 100.

Isaacson, M., and Murray, A. (1981): In situ vaporization of very low molecular weight resists using 1/2nm diameter electron beams. *J. Vac. Sci. Technol.*, 19: 1117-1120.

Ivanov, K. A. (1949): Microradiography of metals and alloys. (in Russian) *J. Technical Physics of the USSR*, 19 (Nov.): 1217-1230.

Jackson, D. F. and Hawks, D. J. (1981): X-ray attenuation coefficients of elements and mixtures. *Physics Reports*, 70: 171-233.

Jacobsen, C., Kenney, J. M., Kirz, J., McNulty, I., Rosser, R. J., Cinotti, F., Rarback, H., and Shu, D. (1986): Microanalysis with a soft x-ray scanning microprobe. Annals N. Y. Acad. Sci., 483: 463-469.

Jacobsen, C., and Rarback, H. (1986): *Proc. Int. Conf. on Insertion Devices for the production of Synchrotron Radiation, SPIE* Vol. 58.

Jaubert de Beaujeu, A. (1939): Utilisation du rayonnement de fluorescence X. (Use of x-ray fluorescence (in historadiography).) *Bulletins et mémoires de la Société d'électro- radiologie médicale de France*, 27 (June): 427-429.

Jowsey, J. (1955): Use of the milling machine for preparing bone sections for microradiography and microautoradiography. *J. Sci. Instru.*, 32 (May): 159-163.

Joy, D.C. (1984): Resolution in the low voltage SEM. *Proc. 42nd Annual Meeting of the Electron Microscopy Society of America*, 444-445.

Kagayama, M., Takahashi, N., Kindaichi, K., and Akita, H. (1981): Three dimensional image of Golgi-stained neurons by x-ray microscopy. *Acta Anat. Nippon.*, 56: 322.

Kallen, E., Dijkstra, J. H., Bartlett, R., Kitchens, T., Tatchyn, R, Hecht, M., and Lindau, I. (1981): Analysis of the throughput of a grazing incidence monochromator using transmission gratings. Brookhaven Meeting, #315-329.

Karlsson, K., Engström, A., Engström, H. (1954): microradiographic studies of the auditory ossicles (Malleus and Incus) and of the ossesus labyrinth. *Acta Radiologica*, 42: 381-390.

Katz, R., Lee, M., and Milner, M. (1950): X-ray inspection of wheat. *Non Destructive Testing*, 9: 16-18.

Keller, F., and Bossert R. A. (1950): Revealing the microstructure of 24S alloy. *Metal Progress*, 41: 63-72.

Kenney, J. M. (1985): Elemental analysis using a scanning transmission x-ray microscope - mapping calcium concentration in human bone. PhD Thesis, SUNY, Stony Brook, New York, USA.

Kenney, J. M., Jacobsen, C., Kirz, J., Rarback, H., Cinotti, F., Thomlinson, W., Rosser, R., and Schidlowsky, G. (1985): Absorption microanalysis with a scanning soft x-ray microscope: mapping the distribution of calcium in bone. *J. Microsc.*, 138:321-328.

Kenney, J. M., Browne, M. T., Buckley, C. J., Burge, R. E., Cave, R. C., Charalambous, P., Duke, P. J., Michette, A. G., Morrison, G. R., Ogawa, K., and Rogoyski, R. (1987): The beginning of scanning x-ray microscopy at Daresbury. This volume.

Key, J. A. (1950): Blood vessels of a gastric ulcer. *British Medical Journal* , 2: 1464-1465.

Kihara, H. (1985): Physiological time and diffusion. *Protein Nucleic Acid Enzyme Suppl.*, 28: 37-44.

Kirkpatrick, P. (1949): The x-ray microscope. *Sci, Amer.*, 180: 44-47.

Kirkpatrick, P. (1949): X-ray images by refractive focusing. *J. Opt. Soc. Amer.*, 39: 796.

Kirkpatrick, P. (1950): Approach to x-ray microscopy. *Nature*, 166: 251-253.

Kirkpatrick, P., and Baez, A. V. (1948): Formation of optical images by x-rays. *J. Opt. Soc. Amer.*, 38: 766-774.

Kirkpatrick, P., and Pattee, H. H. (1953): Approaches to x-ray microscopy. *In*: Advances in Biological and Medical Physics. Vol. 3. Academic Press, NY. pp 247-283.

Kirz, J. (1980): Specimen damage considerations in biological microprobe analysis. *Scanning Electr. Microsc.* 1980/II:239-250.

Kirz, J. (1980): Mapping the distribution of particular atomic species. *Ann. N. Y. Acad. Sci.*, 342: 273-287.

Kirz, J., and Sayre, D. (1980): Soft x-ray microscopy of biological specimens. *In*: Synchrotron Radiation Research. eds. Doniach, S., and Winick, H., Plenum Press, New York, pp. 277-318.

Kirz, J.,and Sayre, D. (1980): Synchrotron Radiation Research, eds. Winick, and Doniach, S., Plenum Press, N.Y., 277-322.

Kirz, J., and Sayre, D. (1984): Prospects and problems in x-ray microscopy. *In*: Springer Series in Optical Sciences: X-ray microscopy, eds. Schmahl, G. and Rudolph, D., Springer-Verlag, Berlin, pp. 262-267.

Kirz, J., and Rarback, H. (1985): Soft x-ray microscopes. *Rev. Sci. Instrum.*, 56: 1-13.

Kirz, J. (1986): Scanning soft x-ray microscopy at the Brookhaven Light Source. *In*: Examining the submicron world (NATO ASI series). eds. Feder, R., McGowan, J. Wm., and Shinozaki, D. M., Plenum Press, New York. pp. 251-264.

Koch, E. E., Kunz, C., and Sonntag, B. (1977): Electronic states in solids investigated by means of synchrotron radiation. *Physics Report*, 27: 153-231.

Kremser, K. (1952): Röntgen-Diagnostik in der Entomologie. (X-ray diagnostics in entomology.) *Fortschritte auf dem Gebiete der Rontegenstrahlen*, 76: 393-396.

Krinsky, S., Perlman, M. L., and Watson, R. E. (1983): Characteristics of synchrotron radiation and its sources. *In*: Handbook on Synchrotron Radiation, ed. Koch, E. E., North Holland, Amsterdam, Ch.2, pp.65-171.

Kura, J. G., Eastwood, L. W., and Doig, J. R. (1951): Microradiography applied to leaded copper-based alloys. *Foundary*, 70 (Sept.): 90-91.

Ladd, W. A., Hess, W. M., and Ladd, M. W. (1956): High resolution microradiography. *Science*, 123: 370-371.

Ladd, W. A., and Ladd, M. W. (1957): High resolution x-ray microscopy. *In*: X-ray microscopy and microradiography. eds. Cosslett, V. E., Engström, A., and Pattee, H. H., Academic Press, New York, pp. 383-389.

Lahodny, A., and Nonveiller, F. (1951): Quelques examples d'application de la semi-microradiographie á l'examen d'alliages d'aluminium coules. (Some examples of the application of semi-microradiography in the examination of aluminum alloy castings.) *Revue de métallurgie*, 48 (Oct.): 778-786.

Lamarque, P. (1936): Technique de l'historadiographie. (The technique of historadiography.) *Comptes rendus de l'Association des anatomistes*, 31: 197-206.

Lamarque, P. (1936): Historadiographie. (Historadiography.) *Comptes rendus de l'académie des sciences*, 202 (Feb.): 684-685.

Lamarque, P. (1936): L'historadiographie. Nouvelle technique d'examen des coupes microscopiques. (Historadiography. New technique for examination of microscopic sections.) *La presse médicale*, 44 (March): 478-481.

Lamarque, P. (1936): L'historadiographie. (Historadiography.) *Archieve de la Société des sciences medicales et biologiques de Montpellier et du Languedoc Méditerranéen*, 17 (May): 180-181. Also *Bulletins et memoires de la Société de radiologie médicale de France*, 24 (May): 515.

Lamarque, P. (1936): La technique de l'historadiographie. (The technique of historadiography.) *J. de radiologie et d'electrologie*, 20 (June): 325-330.

Lamarque, P. (1936): Historadiography. New application of x-rays. *Radiology*, 27 (Nov.): 563-568.

Lamarque, P. (1936): Radiosensibilité cellulaire et historadiographie. (Radiation sensitivity of cells and historadiography.) *La presse médicale*, 44 (Dec.): 1959.

Lamarque, P. (1936): Historadiographie et localisation histologique de substances médicamenteuses ou toxiques á poids atomique élevé. (Historadiography and histological localization of drugs or toxic substances of high atomic weight.) *La presse médicale*, 44 (Dec.): 2069.

Lamarque, P., and Sirc, R. (1936): Quelques réflexions á propos de la technique de l'historadiographie. (Several considerations concerning the technique of historadiography.) *Bulletins et mémoires de la Société medicale de France*, 24 (June): 597-598.

Lamarque, P., and Turchini, J. (1936): Premiers résultats généraux obtenus par la technique de l'historadiographie. Dégre de perméabilité des constituants tissulaires aux rayons X mous de λ = 25Å. (First results obtained by the technique of historadiography; Degree of transmission of tissue componts for soft x-rays of wavelength of λ=25Å) 1Comptes rendus de la Société de biologie, 122: 294-295.

Lamarque, P., and Turchini, J. (1936): Historadiographie et historadiogrammes. (Historadiography and historadiographs.) *Comptes rendus de l'Association des anatomistes*, 31: 341.

Lamarque, P. (1937): L'historadiographie- principe technique. (Technical principle of historadiography.) *Bulletin d'histologie appliquee a la physiologie et a la pathologie*, 14 (Jan.): 5-16.

Lamarque, P. (1937): Une nonvelle méthode d'utilisation des rayons X, I'historadiographie. (Historadiography, new application of Roentgen rays.) *Montpellier médical*, 11 (Feb.): 47-68.

Lamarque, P. (1937): Radiosensibilité cellulaire et histo-radiographie. (Cellular radio-sensitivity and historadiography.) *J. de Radiologie et d'électrologie*, 21 (April): 157-158.

Lamarque, P. (1937): Possibilitiés des méthodes d'analyse fine par l'historadiographie. (Possible extension of the methods of fine analysis by historadiography.) *Bulletins et mémoires de la Société de radiologie médicale de France*, 25 (June): 51-53.

Lamarque, P., Turchini, J., and Castel, P. (1937): Historadiographie et localisations médicamenteuses; détection des sels d'or. (Historadiography and medical localization: Detection of gold salts.) *Archives de la Societe des sciences medicales et biologiques de Montpelliet et du Languedoc Mediterraneen*, 18 (Jan.): 27-29.

Lamarque, P. (1938): Historadiography. *British J. Radiology*, 11 (July): 425-435.

Lamarque, P. (1939): L'historadiographie. (Historadiography.) *Acta de l'Union internationale contre le cancer*, 4: 205-220.

Lindstrom, B (1955): Roentgen absorption spectrophotometry in quantitative cytochemistry. *Acta Radiol. Suppl.*, 125.

Lambot, H. J. (1946): Une application de la microradiographie: la détermination des charges dans le papier. (An application of microradiography: determination of loading in paper.) *Bulletin de la Société royale des sciences de Liége*, 21 (Nov.): pp473-476.

Lambot, H. J. (1947): Determination of the "loading" in paper by microradiography. *Nature*, 159: 676.

Lambot, H. J. (1947): Sur la microradiographie des papiers. (Microradiography of paper.) *L'industrie chimique belge*, 5 (Oct.): 150-154.

Laue, M. v. (1914): Die Freiheitsgrade von Lichtbündeln, *Ann. Physik*, 44: 1197.

Lazzari, E. P. (1976): Dental Biochemistry. Lea and Febiger, Philadelphia.

Lefkowitz, W. (1940): Histological evidence of the harmful effect of cement under orthodontic bands. *J. Dental Res.*, 19 (Feb.): 47-55.

Lepault, J. (1985): Cryo-electron microscopy of helical particles TMV anf T4 polyheads. *J. Microsc.*, 140: 73-80.

Lepault, J. and Leonard, K. (1985): Three-dimensional structure of unstained, frozen-hydrated extended tails of bacteriophage T4. *J. Mol. Biol.*, 182: 431-441.

Levy, D. M. (1930): Bucky's soft rays (Infra-Roentgen or Granz rays). Biology, physics and therapeutic application. *Nederlandsch tijdscrift voor geneeskunde*, 74 (Jan.): 381-392.

Lim, W. K. (1930): Radiography in zoology. *Radiography and Clinical Photography*, 6 (Oct.): 8-9.

Lindsley, C. H., Fisher, E. K., and Brant, J. H. (1949): X-ray microradiography of fibers and fabrics. *Textile Research Journal*, 19 (Nov.): 686-698.

Lindström, B. (1952): The accuracy of cytologic mass determination by Roentgen absorption. *Acta radiologica*, 38: 355-358.

Lindström, B. (1955): Roentgen absorption spectrophotometry in quantitative cytochemistry. *Acta Radio. Scand. Suppl.*, 125: 1-206.

Long, J. V. P., and Cosslett, V. E. (1957): Some methods of x-ray microchemical analysis. *In*: X-ray microscopy and microradiography. eds. Cosslett, V. E., Engström, A., and Pattee, H. H., Academic Press, New York, pp. 435-442.

Lotz, W. (1970): Electron binding energies in free atoms. *J. Opt. Soc. Amer.*, 60: 206-210.

Lovas, I., Santy, W., Spiller, E., Tibbetts, R., and Wilczynski, J. (1981): Design and assembly of a high resolution Schwarzschild microscope for soft x-rays, *Proc. SPIE*, 316: 90.

Löwe-Beer, S. (1938): Untersuchungen mit Grenzstrahlen. (Investigations with Grenz rays.), *Fortschritte auf dem Gebiete der Röntgenstrahlen*, 87 (April): 394.

Lucht, C. M., Mann, M., and Smoluchowski, R. (1946): X-ray microradiographic study of strains in crystals. *Physical Review*, 69 (March): 256.

Lucht, C. M., and Harker, D. (1951): An x-ray microscope using mirrors of adjustable curvature. *Rev. Sci. Insturm.*, 22: 392-395.

Lurie, W. B. (1949): A new advance in historadiography. Microradiography portrayal of human tissue. *Cathode Press*, 6 (Summer): 26-28.

Mackie, S., and Beaumont, S. P.(1985): Materials and processes for nanometer lithography. *Sol. St. Technol.*, (Aug), 117-122.

MacLeod, G. F., and Sherwood, H. F. (1937): Grenz radiographs of sulfur dispersion on foliage. *J. Economic Entomology*, 30 (June): 395-398.

McMaster, R. C., and Begovich, H. A. (1945): Spotwelding of heavier gages of aluminum alloy: Part II. New portable instruments for rapid measurement of current and tip force in spot-welding machines. *Welding Journal*, 24 (Dec.): 644S-652S.

McMaster, R. C., and Grover, H. J. (1947): Radiography and the fatique strength of spot welds in aluminum alloys. *Welding Journal*, 26 (March): 223-232.

Maddigan, S. E. (1944): The technique of microradiography. *J. Appl. Phys.*, 15 (Jan.): 43-54.

Maddigan, S. E. (1946): Microradiography. *Industrial Radiography*, 4 (Spring): 28-30.

Maezawa, H., Suzuki, Y., Kitamura, H., and Sasaki, T. (1986): Characterization of undulator radiation at the Photon Factory. *KEK Preprint*, 85-73, January A/M.

Mallet, L. (1945): Les acquisitions nouvelles en historadiographie. (New achievements in histradiography.) *J. de radiogie et d'électrologie*, 26: 14-16.

Marinelli, L., Ferlazzo, G., Filzgerald, P. J., and Foote, F. W. (1949): Historadiography; A study of chemical elements in histologic sections by means of x-ray absorption. (Abstract of paper read at 46th Annual Meeting of the American Association of Pathologists and Bacteriologists, Boston, MA, April 15-16, 1949). *Amer. J. Pathology*, 25 (July): 811-812.

Martienssen, W. (1966): Holographie, *In*: Plenarvorträge der Physikertagung. München, 92-114, B.G. Teubner, Stuttgart.

Martienssen, W., and Spiller, E. (1967): Holographic reconstruction without granulation, *Phys. Lett. A*, 24: 126.

Manuelidis, L., Sedat, J., and Feder, R. (1980): Soft x-ray lithographic studies of interphase chromosomes. *Ann. N. Y. Acad. Sci.*, 342: 304-325.

Morton, L. (1954): X-ray shadow microscope. *Electron Physics* (National Bureau of Standards Circular 527), (March): pp 265-270.

Matthews, D. L. *et al.* (1983): Characterization of laser produced plasma x-ray sources for use in x-radiography. *J. Appl. Phys.*, 54:4260-4268.

Matthews, D. L. *et al.* (1985): Demonstration of soft x-ray amplifier. *Phys. Rev. Lett.*, 54:110-113.

McConnell, D. (1973): Apatite: its crystal chemistry, mineralogy, utilization and geologic and biologic occurrences. Springer-Verlag, Wien and New York.

McCorkle, R. A. (1980): High intensity pulsed electron beam plasma source. *Ann. N. Y. Acad. Sci.*, 342: 53-64.

McGowan, J. Wm., Borwein, B., Medeiros, J. A., Beveridge, T., Brown, J. D., Spiller, E., Feder, R., Topalian, J., and Gudat, W. (1979): High resolution microchemical analysis using soft x-ray lithographic techniques. *J. Cell Biol.* 80: 732-735.

McGowan, J. Wm., and Malachowski, M. J. (1980): Soft x-ray replication of biological material- X-ray microscopy and microchemical analysis of cells. *Ann. N. Y. Acad. Sci.*, 342: 288-303.

Mead, W. C. *et al.* (1981): Laser-plasma interactions at 0.53μm for disk targets of varying Z. *Phys. Rev. Lett.*, 47:1289-1292.

Meisels, E. L., and Heller, J. (1927): Über die röntgenologische Beobachtung der Metamorphose bei Insekten. (The radiological observations of the metamorphoses of insects.) *Fortschritte auf dem Gebiete der Röntgenstrahlen*, 36: 104-109.

Meunier, P. J., Coupron, P., Eduard, C., Alexandre, C., Bressot, C., Lips, P., and Boyce, B. F. (1979): Bone histomorphometry in osteoporotic states. *In*: Osteoporosis. ed. Barzel, U. S., Grune and Stratt, New York, pp 27-47.

Meunier, P., Courpron, P., Edouard, C., Benard, J., Bringuer, J. P., and Vignon, G. (1973): Physiological senile involution and pathological rarefaction of bone - quantitative and comparative histological data. *Clin-Endocrinol-Metab.*, 2:239-256.

Meyer-Ilse, W., Nyakatura, G., Guttmann, P., Niemann, B., Rudolph, D., Schmahl, G., and Cheng, P. C. (1987): Status of X-Ray Microscopy Experiments at the BESSY Laboratory. This volume.

Michette, A. G., Cheng, P. C., Eason, R. W., O'Neill, F., Owadano, Y., Rosser, R., Rumsby, P., and Shaw, M. J. (1986): Soft x-ray contact microscopy using laser plasma sources. *J. Phys. D.*, 19: 363-371.

Miller, J. (1954): The microradiographic appearance of dentine. *British Dental Journal*, 97 (July): 7-9.

Miller, R. R. (1953): X-rays aid in classifying animals. *General Electric X-ray News*, 25 (Dec.): 6.

Milner, M., Lee, M. R., and Katz, R. (1950): Application of x-ray technic to the detection of internal insect infestion of grain. *J. Economic Entomology*, 43: 933-935.

Milner, M., Lee, M. R., and Katz, R. (1952): Radiography applied to grain and seeds. *Food Technology*, 6 (Feb.): 44-45.

Mitchell, G. A. G. (1950): Microradiographic demonstration of tissues treated by metalic impregnation. *Nature*, 165: 429-430.

Mitchell, G. A. G. (1951): Microradiography in biology. *British J. Radiology*, 24 (Feb.):110-117.

Mitchell, G. A. G. (1954): An evalution of microradiography in biology. *J. Photographic Science*, 2 (July): 113-118.

Modine, N., and O'Connor, D. T. (1953): Radiographic-micrographic examination of aluminum resistance weld. *In*: Papers on non-destructive testing. American Society for Testing Materials, Philadelphia, PA, pp.28-29.

Moore, H. B. (1955): X-rays reveal the inner beauty of shells. *National Geographic*, 107 (March): 427-434.

Montel, M. (1954): Systéme de miroirs sphériques permettant d'obtenir des images agrandies en lumiére X. (System of mirrors for obtaining enlarged x-ray images (X-ray microscope)). *Comptes rendus de l'Académie des sciences*, 239 (July): 39-41.

Morton, E. R. (1913): Radiograph of flowers. *Archives of Röntgen Ray*, 18, (Oct. 1913), p182.

Müller, E. A. W. (1950): Röntgen-mikrographie. (Microradiography.) *Archiv Fur Technisches Messen, V9114-12*, T66-T68.

Müller, E. A. W. (1950): Milrographie mit Röntgenstrahlen. (Microradiography.) *Umschau*, 50 (March):184-186.

Nagel, D. J., *et al.* (1974): X-ray emission from laser produced plasmss. *Phys. Rev. Lett.*, 33:743-746.

Nagel, D. J. (1980): Comparison of x-ray sources for exposure of photoresists. *Ann. N. Y. Acad. Sci.*, 342: 235-251.

Nagel, D. J., (1984): Characteristics and uses of x-radiation from laser heated plasmas. *Proc. SPIE Int. Soc. Opt. Eng.*, 448:17-24.

Nagel, D. J. *et al.*, (1984): Repetilively pulsed plasma soft x-ray source. *Appl. Optics*, 23:1428-1433.

Nath, A, Klein, M. P., Kundig, W. and Lichtenstein, D (1970): Mossbauer studies of after-effects of auger ionization following electron capture in cobalt complexes. *Radiation Effects*, 2: 211-234.

Newberry, S. P., and Summers, S. E. (1954): The General Electric shadow x-ray microscope. *Proc. Int. Conf. Electron Microscopy*, 305-307.

Newberry, S. P. (1985): System for producing 3-D views in contact microradiography. *Proc. 43rd Annual Meeting, Electron Microscopy Soc. Amer.*, 614-615.

Nicholson, J. F. (1954): Some applications of radiography to the examination of foods and drugs. *J. of the Association of Official Agricultural Chemists*, 37 (Feb.): 148-153.

Nicholson, J. F., Milner, M., Munday, W. H., Kurtz, O. L., and Harris, K. L. (1953): An evalution of five procedures for the determination of internal insect infestation of wheat. V. The use of x-rays. *J. of the Association of Official Agricultural Chemists*, 36 (Feb.): 150-155.

Niemann, B., Rudolph, D., and Schmahl, G. (1976): X-ray microscopy with synchrotron radiation. *Appl. Optics*, 15: 1883.

Niemann, B., Schmahl, G., and Rudolph, D. (1981): Status of the Scanning X-Ray Microscope. *Proc. SPIE*, 316: 106-108.

Niemann, B. (1984): The Göttingen Scanning X-ray Microscope. *In*: X-Ray Microscopy, eds. Schmahl, G., and Rudolph, D., Springer Series in Optical Sciences, Vol 43, Springer-Verlag, Heidelberg, pp. 217-225.

Niemann, B., Sarafis, V., Rudolph, D., Schmahl, G., and Meyer-Ilse W, Guttmann P. (1986): X-ray microscopy with synchrotron radiation at the electron storage ring BESSY in Berlin. *Nucl. Instrum. Method* A246: 675-680.

Niemann, B. (1987): Current Status of the Göttingen scanning x-ray microscope experiments at the Bessy Storage Ring. (this Volume)

Nixon, W. C. (1955): Improved resolution with x-ray projection microscope. *Nature*, 175: 1078-1079.

Nurnberger, J., Engström, A., and Lindström, B. (1952): A study of the ventral horn cells of the adult cat by two independent cytochemical microabsorption techniques. *J. Cellular & Comparative Physiology*, 39 (April): 215-251.

O'Connor, D. T., and Eagleson, E. W. (1950): Plastics evalution by soft x-rays. *Non Destructure Testing*, 9 (Summer): 15-22.

O'Hara, C. E., and Osterburg, J. W. (1949): An introduction to criminalistics. The application of the physical sciences to the detection of crime. MacMillan Co., NY, NY.

Omnell, K. -A., Lindström, B., Hoh, F. C., and Hammarlund-Essler, E. (1960): Method for non-destructive determination of inorganic and organic material in mineralized tissue. *Acta Radiol.*, 54: 209-219.

Ong, P. S. (1967): X-ray projection microscopy. *In*: Handbook of x-rays. ed. Kaelble, E. F., McGraw Hill, New York. pp. 47/1-47/24.

Oosawa, F. (1985): Biological function and its time scale: loose coupling mechanism of molecular machine. *Protein Nucleic Acid Enzyme Suppl.*, 28: 2-13.

Oosawa, F., and Masai, J. (1982): Mechanism of flagellar motor rotation in bacteria. *J. Phy. Soc. Jpn.*, 51: 631-641.

Osipov, K. A., and Fedotov, S. G. (1951): microradiographic study of the distribution of alloy components in solid nickel solution. (in Russian) *Doklady Akademiia Nauk SSSR*, 78: 51-53.

Osswald, E. (1949): Röntgenmikroskopische Untersuchung an Leichmetall- Legierungen. (microradiographic investigation of light metal alloys.) *Zeitschrift für Metalkund*, 40 (Jan.): 12-15.

Paic, M. (1941): Semi-microradiographie, moyen d'investigation en métallographie. (Semi-microradiography. Means of investigation in metallography.) *Comptes rendus de l'Académie des sciences*, 213 (oct.): 572-574.

Paic, M. (1941): Microanalyse topologique á l'aide des rayons X. (Microanalysis by means of x-rays.) *J. de physique et le radium*, 12 (Oct.): 41A-52A.

Paic, M. (1944): Application de la radioanalyse á la métallurgie. (Application of x-ray analysis to metallurgy.) *Revue de métallurgie*, 41 (June): 169-179.

Panessa-Warren, B. J., and Warren, J. B. (1980): Determining biological fine structure by differential absorption of soft x-rays. *Ann. N. Y. Acad. Sci.*, 342: 350-367.

Panessa, B. J., Warren, J. B., Hoffman, P., and Feder, R. (1980a): Imaging unstained proteoglycan aggregates by soft x-ray contact microscopy. *Ultramicroscopy*, 5: 267-274.

Panessa, B. J., Warren, J. B., Feder, R., Sayre, D., and Hoffman, P. (1980b): Ultrastructural and elemental imaging of biological specimens by soft x-ray contact microscopy. *Scanning Electron Microscopy*, II: 107-116.

Panessa, B. J., McCorkle, R. A., Hoffman, P., Warren, J. B., and Coleman, G. (1981): Ultrastructure of hydrated proteoglycans using a pulsed plasma source. *Ultramicroscopy* 6:139-148

Panessa-Warren B. J. (1984): Biological applications of x-ray contact microscopy, *In*: Springer Series in Optical Sciences: X-ray microscopy, eds. Schmahl, G., and Rudolph, D., Springer- Verlag, Berlin.

Parsons, D. F., Matricardi, V. R., Moretz, R. C., and Turner, J. N. (1974): Electron microscopy and diffraction of wet unstained and unfixed biological objects. *In*: Advances in biological and medical physics, eds. Lawrance, J. H., Gofman, J. W., and Hayes, T. L., Academic Press, New York.

Parsons, D. F. (ed) (1980): Ultrasoft x-ray microscopy. *NY Acad. Sci.*, New York (Annals, vol 342)

Rarback, H. (1983): The development of a scanning x-ray microscope. PhD Thesis, SUNY, Stony Brook, New York, USA.

Rarback, H., Kenney, J. M., Kirz, J., Howells, M. R., Chang, P., Coane, P. J., Feder, R., Houzego, P. J., Kern, D. P., and Sayre, D. (1984): Recent results from the Stony Brook scanning microscope. *In:* X-ray Microscopy. eds. Schmahl, G., and Rudolph, D., Springer - Verlag, Berlin, pp 206-216.

Pattee, H. H. (1953): The scanning x-ray microscope. *J. Opt. Soc. Amer.*, 43: 61-62.

Pattee, H. H. (1960): High resolution radiosensitive materials for microradiography at long wavelengths. *In*: X-ray Microscopy and Microanalysis, eds. Engström, A., Cosslett, V., and Pattee, H., Elsevier Publishing Co., Amsterdam, pp. 61-65.

Pearlman, J. S. and Riordan, J. C. (1981): X-ray lithography using a pulse plasma source. *J. Vac. Sci. Technol.*, 19: 1190-1193.

Pelgroms, J. D. (1952): Study of the structure of paper by means of soft x-rays. *Paper Trade Journal*, 134 (Jan.): 25-32.

Peyer, B. (1934): Über die Röntgenuntersuchung von Fossilien, hauptsächlich von Vertebraten. (X-ray examination of fossils, particularly the vertebrates.) *Acta radiologica*, 15: 364-379.

Pfannkuch, F., Holder, D. and Baumgartel H. (1984): Possible applications of x-ray microscopy in pathology. *In*: Springer Series in Optical Sciences: X-ray microscopy, eds. Schmahl, G. and Rudolph, D., Springer-Verlag, Berlin.

Piergrossi, L. (1920): La Microradiografia. (Microradiography.) *Il Progresso Fotografico*, 27: 10-12.

Pollack, H. C., and Bridgman, C. F. (1952): X-ray philately, Part I. *S. P. A. Journal*, (Published by the Society of Philatelic Americans, ed. J. Elmer Zinsmeister), 14 (August): 601-604.

Pollack, H. C., and Bridgman, C. F. (1954): X-ray in philately. *Radiology*, 62 (Feb.): 259-261.

Pollack, H. C., Bridgman, C. F., and Splettstrosser, H. R. (1955): The x-ray investigation of postage stamps. *Medical Radiography and Photography*, 31: 74-78.

Polack, P., and Lowenthal S.: Photoelectron x-ray microscopy: Recent developments. *In*: Springer Series in Optical Sciences: X-ray Microscopy, Berlin, Springer-Verlag, 1984.

Poole, M. W., Munro, I. H., Taylor, D. G., Walker, R. P., and Marr, G. V. (1983): Development of an undulator for the SRS. *Nucl. Instr. Meth.*, 208: 143.

Prives, M. G. (1938): Krovosnabzhenie dlinnykh trubchatykh Kostei. (The intra-osseous blood supply of long bones.) *Vesnik rentgenologii i radiologii*, 20: 175-198.

Prives, M. G., and Grechishkin, S. V. (1935): Miagkie rentgenovy luchi (Bukki) v anatomii i embriologii. (Soft Roentgen rays (Bucky) as applied in anatomy and embryology.) *Vestnik rentgenologii i radiologii*, 14: 201-206.

Ranwez, F. (1896): Application de la photographie par les rayons Röntgen aux recherches analytiques des matiéres végétales. (Application of photography by Roentgen rays to the analytical research of vegetable matters.) *Comptes rendus de l'Academie des sciences*, 122 (April): 841-842.

Rarback, H., Kenney, J. M., Kirz, J., Howell, M. R., Chang, P., Coane, P. J., Feder, R., Houzego, P. J., Kern, D. P. and Sayre, D. (1984): Recent result from the Stony Brook scanning microscope. *In*: Springer Series in Optical Sciences: X-ray microscopy, eds. Schmahl, G., and Rudolph, D., Springer-Verlag, Berlin.

Raup, D. (1966): The endoskeleton. *In*: Physiology of Echinodermata, ed. Boolootian, R. A., Interscience, New York., pp 379-394.

Recourt, A. (1957): Contact microradiography below 1kV. *In*: X-ray microscopy and microradiography. eds. Cosslett, V. E., Engström, A., and Pattee, H. H., Academic Press, New York, pp. 234-239.

Recourt, A. (1957): Osmium tetroxide fixation observed by means of microradiography. *In*: X-ray microscopy and microradiography. eds. Cosslett, V. E., Engström, A., and Pattee, H. H., Academic Press, New York, pp. 475-483.

Rees-Jones, S., and Ritchie, P. D. (1937): Radiographic detection of lyctus larvae in situ. *Technical Studies in the Field of Fine Arts*, 5 (Jan.): 179-181.

Reimer, L., (1984), Transmission Electron Microscopy, Springer-Verlag, Berlin.

Rishton, S. A., Beumont, S. P., and Wilkinson, C. D. (1982): Exposure range of low energy electrons in PMMA, *Proc. 10th Symp. Electron & Ion Beam Sci. Tech.*, Electrochem. Soc. p211-218.

Robinson, A. L. (1982): High resolution imaging with soft x-rays. *Science*, 215: 150-152.

Robinson, A. L. (1985): A rescue for Wisconsin's synchrotron source. *Science*, 227: 154-155.

Robertson, J. D. (1936): The function of the calciferous glands of earthworms. *J. Experimental Biology*, 13 (July): 279-297.

Robinson, A. (1985): Momentum Builds for New Synchrotron Source, *Science*, 227:, 396.

Röntgen, W. C. (1895): *Sitzungsber. der Würzburger Physik-Medic. Gesellsch. Jahrg.*. Reprinted in *Ann. der Phys.*, 64, I (1898). Translation by A. Stanton in *Science*, 3: 227 and 726 (1896).

Rodman, G. H. (1920): X-rays approached from a popular standpoint. *The Photographic Journal*, 60 (Feb.): 57-60.

Rogers, T. H., and Lurie, W. B. (1950): Histochemical and microchemical analysis by x-rays. *Cathode Press*, 7 (Fall): 6-9, 30-33.

Röhrl, W. (1952): Radiographische Untersuchunge am Gerassystem des Kaninchenohres. (Radiographic investigations on the vascular system of rabbit's ears.) *Strahlentherapie*, 88: 276.

Rogers, T. H. (1945): A high-intensity source of long wavelength x-rays. *Industrial Radiography*, 4 (Winter): 35-39.

Rogers, T. H. (1952): Production of monochromatic x-radiation for microradiography by excitation of fluorescent characteristic radiation. *J. Applied Physics*, 23 (August): 881-887.

Rollier, M. A. (1947): Microradiografia. (Microradiography.) *Metallurgia Italiana*, 39 (Sept.): 219-220.

Rose, A., (1948): Television tubes and the problem of vision. *Adv. in Electronics*, 1, 131-166.

Rosen, M. D., *et al.* (1985): Exploding-foil technique for achieving a soft x-ray laser. *Phys. Rev. Lett.*, 54: 106-109.

Rosengren, B. H. O. (1959): Determination of cell mass by direct x-ray absorption. *Acta Radiol. Scand. Suppl.,* 178.

Prince, E. (1950): The resolving power of an x-ray microscope. *J. Appl. Phys.*, 21: 698.

Ramachandran, G. N. (1952): New method of obtaining focused images with x-rays. *Phy. Rev.*, 85: 378-379.

Rosser, R. J. (1984): Soft x-ray microscopy at Imperial College. *In*: X-ray Microscopy. eds. Schmahl, G., and Rudolf, D., Springer-Verlag, Berlin, pp242-250.

Rosser, R. J., and Baldwin, K. G. H. (1984): Soft x-ray microscopy. *Rutherford Appleton Laboratory Annual Report 1984*, A4.33-4.34.

Rosser, R. J., Baldwin, K. G. H., Feder, R., Bassett, D., Coles, A., and Eason, R. W. (1985a): Soft x-ray contact microscopy with nanosecond exposure times. *J. Microscopy*, 138:311-319.

Rosser, R. J., Feder, R., Ng, A., and Celliers, P. (1985b): Single shot soft x-ray microscopy using plasmas generated by a laboratory sized Nd-glsss laser. *J Microscopy*, 140:RPl-RP2.

Reuter, B., and Mahr, H. (1976): Experiments with Fourier transform holograms using 4.48 nm x-rays, *J. Phys. E*, 9: 746.

Rosser, R. J., Baldwin, K. G. H., Feder, R., Bassett, D., Coles, A., and Eason, R. W. (1985a): Soft x-ray contact microscopy with nanosecond exposure times. *J. Microscopy*, 138: 311-319.

Rudolph, D. and Schmahl, G. (1980): High power zone plates for a soft x-ray microscope. *Ann N. Y. Acad. Sci.*, 342: 94-104.

Rudolph, D., Niemann, B., and Schmahl, G. (1981): Status of the sputtered sliced zone plates for x-ray microscopy. *High Resolution X-ray Optics, Proc. SPIE*, 316: 103-105.

Rudolph, D., Niemann, B., Schmahl, G., and Christ, O. (1984): The Göttingen x-ray microscope and x-ray microscopy experiments at the BESSY Storage Ring. *In*: X-Ray Microscopy. eds. Schmahl, G., and Rudolph, D., Springer-Verlag, Berlin, 192-202 (Series in Optical Sciences vol 43).

Rudolph, D., Niemann, B., Schmahl, G., and Christ, O. (1984): The Göttingen x-ray microscope and x-ray microscopy experiments at the BESSY storage ring. In: X-ray microscopy. eds. Schmahl, G., and Rudolph, D., Springer-Verlag, Berlin Heidelberg New York, pp 192-202 (Springer series in optical sciences, vol 43)

Rudolph, D., Schmahl, G., Niemann, B., Meyer-Ilse, W., and Thieme, J. (1986): X-ray microscopy - State of the art and expected developments. Annals N. Y. Acad. Sci., 483: 457-462.

Rumsby, P. T. (1985): Laser produced plasmas as intense x-ray sources for microscopy at the Central Laser Facility. *J. Microscopy*, 138: 245-265.

Rumsby, P. T. (1985): Laser produced plasmas as intense x-ray sources for microscopy at the Central Lsser Facility. *J. Microscopy*, 138:245-265.

Sainte-Marie, L. G., Charon, S. A., Edouard, L., Chapuy, M. C., and Meunier, P. J. (1984): Iliac bone histomorphometry in adults and csildren with osteogenesis imperfecta. *J. Clin. Pathol.*, 37:1081-1089.

Salkovitz, E. I. (1946): Microradiography with routine x-ray equipmqnt. *Metal Progress*, 11, 1091-1096.

Salkovotz, E. I., Schaum, J. H., and Von Batchelder, F. W. (1947): Microradiography of gray cast iron. *Transcations of the American Foundrymen's Association*, 55: 82-88.

Salkovitz, E. I.: Micro-radiography manual. Publication Board Report 109531. Published by Office of Technical Services, Washington D. C.

Sarafis V. (1984): X-ray microscopy as a possible tool for the investigation of plant cells. *In*: Springer Series in Optical Sciences: X-ray microscopy, eds. Schmahl, G., and Rudolph, D., Springer-Verlag, Berlin.

Sasaki, T. (1980): Status of the SOR-RING and the experimental program for 1979/80. *Nucl. Instr. and Meth.*, 177: 19-25.

Sayre, D., Kirz, J., Feder, R., Kim, D. M., and Spiller, E. (1977): Potential operating region for ultrasoft x-ray microscopy of biological objects. *Science* 196: 1339.

Sayre, D., Kirz, J., Feder, R., Kim, D. M., and Spiller, E. (1977): Assessment of the potential of ultrasoft x-ray microscopy. *Annals N. Y. Acad. Sci.* 306: 286.

Sayre, D., Kirz, J., Feder, R., Kim, D. M. and Spiller, E. (1977): Potential operating region for ultrasoft x-ray microscopy of biological material. *Science*, 196: 1339.

Sayre, D., Kirz, J., Feder, R., Kim, D. M., and Spiller, E. (1977): Transmission microscopy of unmodified biological materials: Comparative radiation dosages with electrons and ultrasoft x-ray photons. *Ultramicroscopy* 2: 337-349.

Sayre, D., Kirz, J., Feder, R., Kim, D. M., Spiller, E. (1978): Assessment of the potential of ultrasoft x-ray microscopy. *Ann N. Y. Acad. Sci.*, 306, 286-290.

Sayre, D., and Feder, R. (1979): Exposure and development of x-ray resist in microscopy. *IBM Research Report RC-7498*, Yorktown Heights, New York.

Sayre, D. (1980): Review of image formation methods with the soft x-ray photon. *In*: Ultrasoft x-ray microscopy: Its application to biological and physical sciences. ed. Parson, D. F., *Ann. N. Y. Acad. Sci.*, 342, pp. 387-391.

Sayre, D., and Feder, R. (1980): Additional technical aspects: Wet specimens, thick specimens, and factors affecting convenience. *Ann. N. Y. Acad. Sci.*, 342: 268-272.

Sayre, D., Haelbich, R.-P., Kirz, J. and Yun, W. B. (1984): On the possibility of imaging microstructures by soft x-ray diffraction pattern analysis. *In*: Springer Series in Optical Sciences: X-ray microscopy, eds. Schmahl, G., and Rudolph, D., Springer- Verlag, Berlin.

Sayre, D. (1986): A phasing method for near-field diffraction data. In: abstracts of American Crystallographic Association meeting, 22-27 June 1986, McMaster University. Available from Polycrystal Book Service, Western Springs, Illinois

Scherzer, O. (1978): Limits of resolving power of electron microscopes. *Electron microscopy* (Proceedings of 12th Internation Congress on Electron Microscopy), Vol. 3, p123.

Schmahl, G., Rudolph, D., and Niemann, B. (1980): Zone-plate x-ray microscopy. *Quart. Rev. Biophys.*, 13: 297-315.

Schmahl, G., Rudolph, D., Niemann, B., and Christ, O. (1980): X-ray microscopy of biological specimens with a zone plate microscope. *In*: Ultrasoft X-Ray Microscopy: Its Application to Biological and Physical Sciences. ed. Parsons, D. F., *Annals N. Y. Acad. Sci.*, 342: 368-386.

Schmahl, G., Rudolph, D., Guttmann, P., and Christ, O. (1983): Zone plates for x-ray microscopy. *In*: X-Ray Microscopy. eds. Schmahl, G., and Rudolph, D., Springer-Verlag, Berlin, pp 63-74 (Series in Optical Sciences vol. 43).

Schmahl, G. (1983): X-ray microscopy. *Nucl. Instr. and Meth.*, 208: 361-364.

Schmahl, G., and Rudolph, D. (eds) (1984): X-ray microscopy. Springer-Verlag, Berlin Heidelberg New York (Springer series in optical sciences, vol 43)

Schmahl, G., Rudolph, D., Guttmann, P., and Christ, O. (1984): Zone plates for x-ray microscopy. *In*: X-ray microscopy, eds. Schmahl, G., and Rudolph, D., Springer series in optical sciences, Vol 43, Springer-Verlag, Heidelberg, pp. 63-74.

Schmahl, G., and Rudolph, D. (eds)(1984): X-ray Microscopy. Springer Series in Optical Sciences, Vol. 43, Springer Verlag, Berlin.

Schmahl, G., Rudolph, D., Niemann, B., Guttmann, P., and Thieme, J. (1986): Lenses for soft x-rays and x-ray microscopy experiments. *In*: Examining the submicron world (NATO ASI series). eds. Feder, R., McGowan, J. Wm., and Shinozaki, D. M., Plenum Press, New York. pp. 231-250.

Schmahl, G., Rudolph, D., Niemann, B., and Meyer-Ilse, W. (1986): X-ray microscopy with synchrotron radiation. *J. Microsc. Spectrosc Electron*, 11: 389-396.

Schmahl, G., and Rudolph, D. (1987): Proposal for a phase contrast x-ray microscope. (this volume)

Schmidt, R. A. M. (1952): Microradiography of microfossils with x-ray diffraction equipment. *Science*, 115: 94-95.

Schupp, O. E., Jr., and Boller, E. R. (1938): Sodium silicate adhesive bonds in corrugated fiberboard. *Industrial and Engineering Chemistry*, 30 (June): 603-611.

Scott, G. W., Jr., Sutton, L. G., and Widmyer, J. H. (1944): Radiography for development and control of aluminum spot welding. *Welding Journal*, 23 (Nov.): 560S-570S.

Seemann, H. E. (1943): Miscellaneous applications of radiography and fluoroscopy. *In*: Symposium on radiography. American Society for Testing Materials, Philadelphia, PA.

Seki, K., Nakagawa, H., Fukui, K., Ishiguro, E., Kato, R., Mori, T., Sakai, K., and Watanabe, M. (1986): A plane-grating monochromator for 2eV $\leq h\nu \leq$ 150 eV. *Nucl. Inst. Meth.*, A246: 264-266.

Sharpe, R. S. (1979): Projection microradiography. *J. Microscop.*, 117: 143.

Sherwood, H. F. (1934): The radiography of small biological specimens. *Radiography and Clinical Photography*, 10: 10-13.

Sherwood, H. F. (1936): The radiography of cloth. *Rayon and Meliand Textile Monthly*, 17 (May): 51-53.

Sherwood, H. F. (1937): The radiography of small biological specimens. *J. Biological Photographic Association*, 5 (March): 138-145.

Sherwood, H. F. (1937): Some radiographic applications of soft x-rays. *J. Photographic Society of American*, 3 (June): 1-5.

Sherwood, H. F. (1937): Soft x-ray motion pictures of small biological specimens. *J. Society of Motion Picture Engineers*, 28 (June): 614-618.

Sherwood, H. F. (1937): Stereoscopic radiography of cloth. *Rayon and Melliand Textile Monthly*, 18 (August): 53-54.

Sherwood, H. F. (1937): A binocular microscope for viewing long wavelength stereoradiographs. *J. Biological Photographic Association*, 6 (Dec.): 78-84.

Sherwood, H. F. (1937): Soft x-ray radiography. *Electronics*, 10 (Dec.): 78-84.

Sherwood, H. F. (1938): Stereoscopic soft x-ray examination of parchment antiphonaries. *Technical Studies in the Field of Fine Arts*, 6 (April): 276-280.

Sherwood, H. F. (1947): Vacuum exposure holder for microradiography. *Review of Scientific Instruments*, 18: 80-83.

Sherwood, H. F. (1953): Methods for eliminating the image on one side of double-coated x-ray film. *Photographic Science and Technique*, 4 (Nov.): 151-152.

Shevchenko, M. I. (1937): Revealing seed pests by means of x-rays. (in Russion, English summary). *Zashchita rastenii ot vreditelei (Plant Protection)* , 14: 14-25.

Shinohara, K., Aoki, S., Yanagihara, M., Yagishita, A., Iguchi, Y., and Tanaka, A. (1986): A new approach to the observation of the resist in x-ray contact microscopy. *Photochemistry and Photobiology*,

Shinozaki, D. M., McGowan, J. Wm., Cheng, P. C., and Tan, K. H.: Properties of polymer resists. In *NATO ASI series: Improved methods for examining the submicron world*, eds. McGowan, J. Wm., Feder, R., and Schinozaki, D. M., pp. 191-214. New York, Plenum Press, 1985.

Shinozaki, D. M., Cheng, P. C., Tan, K. H., and McGowan, J. Wm.: Experimental absorption coefficients of resists and biological materials. *Proceeding of the 43rd Ann. Meeting of EMSA*, 606-609.

Shinozaki, D. M., Cheng, P. C., and Feder, R.: Soft x-ray induced surface roughness in PMMA. *Proceedings of the XI International Congress on Electron Microscopy*. 1986.

Sievert, R. M., and Forssberg, A. (1936): The time factor in the biological action of Roentgen rays. *Acta radiologica*, 17: 290-298.

Sievert, R. M. (1936): Two methods of Roentgen micro-photography. *Acta radiologica*, 17: 299-309.

Silk, J. K. (1980): A grazing incidence microscope for x-ray imaging applications. *Ann. N. Y. Acad. Sci.*, 342: 116-129.

Simpson, M. (1984): Design considerations of zone plate optics for a scanning x-ray microscope. PhD thesis. QEC. Senate House, London University.

Sissons, H. A. (1949): Intermittent periosteal activity. *Nature*, 163: 1001-1002.

Sissons, H. A. (1950): Microradiography of bone. *British J. Radiology*, 23 (Jan.): 2-7.

Smith, C. B., and Smith, D. A. (1976): An x-ray diffraction investigation of age-related changes in the crystal structure of bone apatite. *Calcif. Tiss. Res.*, 22:219-226.

Smith, D. W., and Keller, F. (1943): Correlation of metallographic and radiographic examination of spot welds in aluminum alloys. *Welding Journal*, 21 (Dec.): 573S-583S. Also *Industrial Radiography*, 1 (April): 31-44.

Smoluchowski, R., Lucht, C. M., and Hurd, J. M. (1946): Composition and diffraction effects in x-ray microradiographs. *J. Applied Physics*, 17 (Nov.): 864-870.

Smoluchowski, R., Lucht, C. M., and Mann, M. (1946): X-ray microradiography and distorted crystals. *Phys. Rev.*, 70: 318-322.

Sokolov, A. A., and Ternov, I. M. (1968): Synchrotron radiation. Pergamon Press, Oxford.

Solem, J. C., and Baldwin, G. C. (1982): Microholography of living organisms. *Science*, 218:229-235.

Solem, J. C., and Chapline, F. F. (1984): X-ray biomicroholography, *Optical Engineering*, 23: 193.

Solem, J. C. (1984): High intensity x-ray imaging of biological specimens, *Ultramicroscopy*.

Sparks, Jr. C. J. (1980): X-ray fluorescence microprobe for chemical analysis. In: Synchrotron radiation research, eds: Winick, H., and Doniach, S., Plenum Press, New York, p459.

Spear, W. E. (1951): Investigation of electron optical properties of an electrostatic focusing system. *Proc. Phys. Soc. London*, 64: 233-243.

Spiller, E., Feder, R., Topalian, J., Eastman, D., Gudat, W., and Sayre, D. (1976): X-ray microscopy of biological objects with carbon Kα and with synchrotron radiation, Science *191*, 1172-1174.

Spiller, E. (1976): Reflecting multilayer coatings for the far UV, *Appl. Opt.*, 15: 2333.

Spiller, E., Feder, R., and Topalian, J. (1977): X-ray lithography and x-ray microscopy. *Phys. Bl.*, 32: 564, and *Phys. Technol.*, 8: 22.

Spiller, E., and Feder, R. (1977): X-Ray Optics, *Topics in App. Phys.* Vol 22. Springer-Verlag, Berlin, p 35-92.

Spiller, E. and Feder, R. (1977): X-ray lithography. *IBM Research Report, RC 6568 (#28324).*

Spiller, E., and Feder, R. (1978): The optics of long-wavelength x-rays. *Scientific American* 239-5: 70-78.

Spiller, E. (1978): A zone plate monochromator for synchrotron radiation, Workshop on x-ray instrumentation for synchrotron radiation research, ed., by Winick, H., and Brown, G., Stanford, p. 6, 44. SSRL Report No. 78/04.

Spiller, E., Feder, R., and Topalian, J. (1978): Soft x-rays for biological and industrial pattern replications, *J. de Physique, collq.*, 39: C4-205.

Spiller, E., Feder, R., and Sayre, D. (1979): X-ray lithography and microscopy, *In*: Synchrotron radiation applied to biophysical, biochemical and biomedical research. eds. Castellani, A., and Quercia, I. F., Plenum Press, New York, pp.377-387.

Spiller, E. (1979): X-ray microscopy, present art and future possibilities, *Proceedings of the workshop on x-ray microscopy and lithography, Stanford, Feb. 21-22, 1979* ed. by Daunemiller, C. R., SSRL Report No. 79102, p. 41.

Spiller, E. (1980): Recent developments towards high resolution x-ray imaging. *Nuclear Instr. Methods*, 177: 187.

Spiller, E. (1980): The scanning x-ray microscope - Potential realizations and applications, *In*: Scanned image microscopy, ed. Ash, E., Academic Press, pp. 365-391.

Spiller, E., Segmuller, A., and Haelbich, R-P.: The fabrication of multilayer x-ray mirrows. *Ann. N. Y. Acad. Sci.*, 342: 188-198.

Spiller, E. (1983): Soft x-ray optics and microscopy, Chapter 12, *In*: Handbook on synchrotron radiation, Vol.1, ed. Koch, E.E., North Holland, Amsterdam 1983, pp. 1093-1130.

Spiller, E. (1983): The design of multilayer coatings for soft x-rays and their application for imaging and spectroscopy *In* New techniques in x-ray and XUV optics, eds. Kent, B. Y., and Patchett, B. E., Rutherford Appleton Lab., Chilton, U.K.

Spiller, E. (1984): A scanning soft x-ray microscope using normal incidence optics. *In* X-ray microscopy, eds. Schmahl, G., and Rudolph, D., Springer-Verlag, Berlin, 226-231.

Spiller, E. (1985): X-ray microholgraphy - Exciting possibility or impossible dream? *Sympos. Applications of X-Ray Lasers*, Lawrence Livermore Lab., Feb. 1985, (IBM RC 11099).

Splettstosser, H. R., and Seemann, H. E. (1952): Application of fluorescence x-rays to metallurgical microradiography. *J. Applied Physics*, 23: 1217-1222.

Spurr, A. R. (1969): A low-viscosity epoxy resin embedding medium for electron microscopy. *Ultrastruct. Res.* 26: 31-43.

Stewart, M. and Lepault, J. (1985): Cryo-elcetron microscopy of topomyosin magnesium paracrystals. *J. Microsc.,* 138: 53-60.

Suckewer *et al.*, (1985): Amplification of stimulated soft x-ray emission in a confined plasma column. *Phys. Rev. Lett.*, 55:1753-1756.

Sylvia, R. K., Rush, A. D., Clarke, D. T., and Myring, W. J. (1986): Soft x-ray contact microscopy using synchrotron radiation of thin sectioned, lead-contaminated chloragogenous tissue of the earthworm *Dendrobaena rubido. J. Microsc.*, 142(1): 1.

Takahashi, N., Takahashi, S., Kagayama, M., and Yada, K. (1983): Three-dimensional visualization of Golgi-stained neurons by a projection x-ray microscope converted from a scanning electron microscope. *Tohoku J. Exp. Med.*, 141: 249-256.

Tan, K. H., Bancroft, L. L., Coatsworth, L. L., and Yates, B. W. (1982): Mark IV "Grasshopper" grazing incidence monochromator for the Canadian Synchrotron Radiation Facility (CSRF). *Can. J. Phys.*, 60: 131-136.

Tan, K. H., Cheng, P. C., Bancroft, G. M., and McGowan, J. Wm.: The Canadian Synchrotron Radiation Facility: Soft x-ray contact microscopy of biological material. *Can. J. Spectroscopy* 29(5), 134-137, 1984.

Tasker, D. L. (1942): X-ray goes pictorial. *Popular Photography*, 10 (March): 33, 78-79.

Taylor, A. (1943): Radiography of aircraft spot-welds. *Aero Digest*, 43 (July): 263, 265-266.

Taylor, A. (1951): Intense gallium x-rays for microradiography and diffraction investigations. *Nature*, 168 (Sept.): 471-472.

Tevis, M. (1919): X-ray photographs of plants and their parts. (A report of work done by Rieder, H. in Umschau Germany in 1916). *Sci. Amer.*, 121 (Oct. 1919), p354.

Thevenard, P. (1955): Direct kineradiography on 35mm film and one application in biological research. *Medical and Biological Illustration*, 5 (April): 66-70.

Thewlis, J. (1934): X-ray examination of human teeth. *British Dental Journal*, 57 (Nov.): 457-466.

Thewlis, J. (1937): The calcification of enamel and dentine. *British Dental Journal*, 62 (March): 303-313.

Thieme, J. (1984): Construction of Condenser Zone Plates for a Scanning X-Ray Microscope. *In*: X-Ray Microscopy, eds. Schmahl, G., and Rudolph, D., Springer Series in Optical Sciences, Vol 43, Springer-Verlag, Heidelberg, pp. 91-96.

Thieme, J. (1984): Theoretische Untersuchungen über Kondensorzonenplatten als abbildende Systeme für weiche Röntgenstrahlung", Diplomarbeit, Göttingen.

Thomas, G., and Goringe, M.J., (1979): Transmission Electron Microscopy of Materials, Wiley, N.Y..

Thompson, B. J., Ward, J. H., and Zinky, W. (1967): Application of hologram techniques for particle size analysis, *Appl. Optics*, 6: 519.

Thompson, L. F., Wilson, C. G., and Bowden, M. J. (1983): Introduction to microlithography, *A.C.S Symposium Series*, 219. p98-107. A.C.S.

Tirman, W. S., and Banker, H. W. (1951): A simple method of microradiography using oridnary diagnostic x-ray equipment. *Science*, 113: 530-531.

Tirman, W. S., Caylor, C. E., Banker, H. W., and Caylor, T. E. (1951): Microradiography: Its application to the study of the vascular anatomy of certain organs of the rabbit. *Radiology*, 57 (July): 70-80.

Trillat, J. J. (1940): Microradiométallographie. (Microradiometallography.) *Revue scientifique*, 78 (April): 212-221.

Trillat, J. J. (1942): La microradiométallographie. (Microradiometallography.) *Science et industries photographiques*, 13 (Jan.): 1-11.

Trillat, J. J. (1942): Radiographie et microradiographie par réflexion. (Radiography and microradiography by reflection.) *Revue scientifique*, 80 (Sept.): 364-370.

Trillat, J. J. (1943): Radiographie et microradiographie par transmission et par réflexion. (Radiography and microradiography by transmission and reflection.) *Bulletin d la Société francaise des électriciens (Series 6)*, 3 (May): 145-164.

Trillat, J. J. (1945): Sur l'étude des mélanges soufre-caouthouc par microradiographie. (Study of sulfur impregnated rubber by microradiography.) *Comptes rendus de l'Académie des sciences*, 220 (March): 456-458.

Trillat, J. J. (1945): L'étude par la microradiographie de la dispersion et du broyage des pigments dans les peintures. (Microradiographic study of the dispersion and the grinding of the pigments in paints.) *Peintures, Pigments, vernis*, 21 (June): 178-180.

Trillat, J. J. (1945): Étude radiographique des mélanges soufre- caoutchouc. (Radiographic study of sulfur impregnated rubber.) *Revue generale du caoutchouc*, 22: 63-66.

Trillat, J. J. (1947): La microradiographie. (Microradiography.) *La nature*, 75 (March): 82-86.

Trillat, J. J., and Urbain, P. (1943): Sur l'étude de minerais par microradiographie. (The study of minerals by microradiography.) *Comptes rendus de l'Académie des sciences*, 216: 534-535.

Trout, E. D., Gager, R. M., and Pace, A. L. (1948): Possible industrial applications of soft x-radiation - 15 to 100 kilovolts. *Non Destructive Testing*, 7 (Winter): 20-24.

Trueta, J., Barclay, A. E., Franklin, K. J., Daniel, P. M., and Prichard, M. L. (1947): Studies of the renal circulation. Charles C. Thomas Co., Springfield, IL, and Blackwell Scientific Publications, Oxford, UK.

Trurnit, H. J., and Hoppe, W. (1946): Über Linsen für Röntgenstrahlen. (On x-ray lenses). *Nachrichten der Akademie der Wissenschaften in Göttingen, marthematisch-physicalishe Klasse*, pp 29-32. No. 1.

Turchini, J. (1936): L'historadiographie - principales applications histologiques. (Historadiography - Principal histological applications.) *Comptes rendus de l'Association des anatomistes*, 31: 314-322.

Turchini, J. (1937): L'historadiographie - applications. (Applications of historadiography.) *Bulletin d'histologie appliquée la physiologie et a pathologie*, 14 (Jan.): 17-28.

Turchini, J. (1937): Historadiographie et fixations histologiques. (Historadiography and histologic fixations.) *Archives de la Société des sciences médicales et biologiques de Montpellier et du Languedoc Méditerranéen*, 18 (Feb.): 113-115.

Turchini, J. (1942): Un nonveau moyen d'investigation physique appliqué á l'histologie; la radiographie histologique. (A new physical method of investigation applied to histology; Historadiography.) *Schweizerische Zeitschrift f&ür Pathologie und Bakteriologie*, 5: 137-149.

Uchida, M. (1951): Observations on producing pictures by soft x-rays. *J. Nippon Hashasen Lijutsn Gakkai*, 1: 4-12.

Underwood, J. H., and Attwood, D. T. (1984): The Renaissance of X-Ray Optics, *Phys. Today*, 37: 44.

Vallebona, A. (1931): Radiography with great enlargment (microradiography) and a technical method for radiographic dissociation of shadow. *Radiology*, 17: 233-243.

Vallebona, A. (1933): Radiographie du poumon par la technique de la microradiographie. (Radiography of the lung by the technique of microradiography). *Bulletins et mémoirs de la Sociéte de radiologie médicale de France*, 21 (Feb.): 242-243.

Van Den Broek, S. L. (1957): A simple contact microradiography unit with sealed off x-ray tube. *In*: X-ray microscopy and microradiography. eds. Cosslett, V. E., Engström, A., and Pattee, H. H., Academic Press, New York, pp. 64-71.

Van Huysen, G., Hodge, H. C., Warren, S. L., and Bishop, F. W. (1933): Quantitative Roentgen-ray study of certain pathological changes in dentin. *Dental Cosmos*, 75 (August): 729-738.

Veis, D. J., Albinger, T. M., Clohisy, J., Rahima, M., Sabsay, B., and Veis, A. (1986): Matrix proteins of the teeth of the sea urchin *Lytechinus variegatus*. *** *J. Exp. Zool.* 240: 35-46.

Vladimirski, Y., Källne, E., and Spiller, E. (1984): The fabrication of free-standing zone plates and transmission gratings, *SPIE*, 448: 25-37.

Victoreen, J. A. (1943): Probable x-ray mass absorption coefficients for wavelengths shorter than the K critical absorption wavelength. *J. Appl. Phys.*, 14: 95-102.

Victoreen, J. A. (1948): The absorption of incident quanta by atoms as defined by the mass photoelectric absorption coefficient and the mass scattering coefficient. *J. Appl. Phys.*, 19: 855-860.

Victoreen, J. A. (1949): The calculation of x-ray mass absorption coefficients. *J. Appl. Phys.*, 20: 1141-1147.

Voigt, O. (1933): Die Funktion der Regenwurm - Kalkdrüsen. (The function of earthworms - chalk glands.) *Zoologische Jahrbücher - Abteilung für Allgemeine Zoologie*, 52 (July): 677-708.

Von Batchelder, F. W., and Schaum, J. H. (1940): Microradiography. *Iron Age*, 160 (Dec.): 94-97.

Walker, J. T. (1940): Bullet holes and chemical residues in shooting cases. *Amer. J. Police Sci.*, 31 (Nov.): 497-521.

Wall, J. C., Chatterji, S., and Jeffery, J. W. (1979): Age related changes in density and strength of femoral cortical bone. *Calcif. Tiss. Int.*, 27:105-108.

Warren, S. L., Bishop, F. W., Hodge, H. C., and Van Huysen, G. (1934): A quantitative method for studying the Roentgen-ray absorption of tooth slabs. *Amer. J. Roentgenology*, 31 (May): 663-672.

Watanabe, M. (1986): The UVSOR facility: status in 1985. *Nucl. Inst. Meth.*, A246: 15-17.

Whatuore, R. W., Goddard, P. A., Tanner, B. K., and Clark, G. F. (1982): Direct imaging of travelling Rayleigh waves by stroboscopic x-ray topography. *Nature*, 299: 44.

Williams, W. M., and Smith, C. S. (1952): A study of grain shape in an aluminum alloy and other applications of stereoscopic microradiography. *J. Metals*, 4 (July): 755-765.

Wolfe, K. J. B., and Robinson, I. R. (1949): Effect of metallic segregation on the cutting power of high-speed steel tools. *Iron and Coal Trades Review*, 158 (Feb.): 337-345.

Wolfe, K. J. B., and Robinson, I. R. (1949): Metallic segregation in ferritic steels; An application of microradiography. *Metal Treatment and Drop Forging*, 16 (Winter): 209-218.

Wollon, E. O., and Copton, A. H. (1934): *J. Opt. Soc. Amer.*, 24: p229.

Wolosewick, J. J. (1980): The application of polyethylene glycol (PEG) to electron microscopy. *J. Cell Biol.*, 86: 675-681.

Wolter, V. H. (1952): Spiegelsysteme streifenden Einfalls als abbildende Optiken fur Rontgenstrahlen. *Ann. Phys.*, 10: 94-114.

Wolter, V. H. (1952): General Swarzschild mirror system of low angle reflection as optics for x-rays. *Annalen der Physik*, 10: 286-295.

Wood, R. C., Barrett, J. C., and Dietz, T. W. (1943): Radiography of spot welds. I. *Metals and Alloys*, 16 (Sept.): 442-447. Also *Industrial Radiography*, 1 (Jan.): 42-47. II. *Metals and Alloys*, 16 (Dec.): 1080-1083. Also *Industrial Radiography*, 1 (April): 17-20.

Woods, R. C., and Cetrone, V. C. (1943): Radiography of light alloys. *Iron Age*, 151 (March): 52-58.

Woods, R. C., and Cetrone, V. C. (1943): Microradiography of alloys, A new industrial tool. *Metals and Alloys*, 18: 1320-1325.

Woods, R. C., and Rich, S. L. (1942): Some radiographic aspects of aluminum spot welds. *Welding Journal*, 21 (Feb.): 101-104.

Wooster, N., and Wooster, W. A. (1945): X-ray topographs. *Nature* , 155: 786-787.

Wrighton, W. J., and Jarrett, T. C. (1937): Radiography of composite gold plate. *Metals and Alloys*, 8 (march): 78-79.

Yaakobi, B. *et al.* (1981): High x-ray conversion efficiency with target irradiated by frequency tripled Nd glass laser. *Opt. Comm.*, 38: 196-200.

Yaakobi, B. *et al.* (1984): Submicron x-ray lithography using laser-produced plasma as a source. *Appl. Phys. Lett.*, 43:686-688.

Yada, K., and Ishikawa, H. (1980): Transmission type x-ray shadow microscope converted from scanning electron microscope. *Bull. Res. Inst. Sci. Meas. Tohoku Univ.*, 29: 25-42. (in Japaness with English abstract)

Yashida, U., and Tanaka, H. (1934): A method of taking stereoscopic microradiographs. *Memoirs of the College of Sciences, Kyoto Imperial University*, 17: 401-404.

Yuasa, H. (1926): On the advantage of the x-ray examination of certain classes of materials and insects subject to the plant quarantine regulations. *Proc. 3rd Pan-Pacific Science Congress, Tokyo*, 1: 1141.

Yun, W. B., Kirz, J., and Sayre, D. (1987): Observation of the soft x-ray pattern of a single diatom. *Acta Cryst.* A43: in press

Zunick, M. J., and Pace, A. (1954): X-rays show more than insects. *Food Engineering*, 26 (May): 64-65, 139-140.

Subject Index